home plumbing handbook

by charles mc connell

THEODORE AUDEL & CO.
a division of
HOWARD W. SAMS & CO., INC.
4300 West 62nd Street
Indianapolis, Indiana 46268

FIRST EDITION

FIRST PRINTING—1976

International Standard Book Number: 0-672-23239-1
Library of Congress Catalog Card Number: 76-1133

CONTENTS

PREFACE

This is the age of "do-it-yourself". Here, for the first time, is a complete and comprehensive book on plumbing repair and installation. Detailed instructions and illustrations take you step-by-step through various projects. Advice on tool buying, plumbing terms, parts, and parts buying are included to help you take an intelligent approach to the job at hand. You will learn what to repair, what to replace, and how to do it.

Because of the increasing popularity of garbage disposers, and dishwasher, instructions on installing these appliances are included. The satisfaction of knowing that the job is well done and that considerable money has been saved in the process, will make your work worthwhile. YOU CAN DO-IT-YOURSELF!

Some of the subjects covered in the book require disconnection and reconnection of electrical wiring, or the installation of new wiring. Before starting on any project where electrical wiring is involved, *remove* the fuse or trip the circuit breaker in the electrical circuit which is involved. The installations and repairs covered in this book must conform to local codes and local laws.

About the Author

Charles McConnell is a licensed master plumber actively engaged in the plumbing trade, and holds an award for over 30 years continuous membership in the United Association of Journeymen Plumbers and Steamfitters. He has designed, installed and supervised plumbing and heating installations in residental, commercial, industrial and public buildings. He has also trained many apprentices to the plumbing trade in the State of Indiana.

The author wishes to thank the following companies for their assistance in furnishing information and drawings on their products:

AMERICAN STANDARD
 P.O. Box 2003 New Brunswick, New Jersey 08903

DELTA FAUCET CO.
 Greensburg, Indiana 47420

FLINT & WALLING CO.
 Kendallville, Indiana 46755

FLUIDMASTER, INC.
 P.O. Box 4204 Anaheim, California 92803

J. H. INDUSTRIES
 1712 Newport Circle Santa Ana, CA 92705

JET AERATION CO.
 750 Alpha Drive Cleveland, Ohio 44143

KOHLER CO.
 Kohler, Wisconsin 53044

NIBCO
 Elkhart, Indiana 46514

RADIATOR SPECIALTY CO.
 P.O. Box 10628 Charlotte, N.C. 28237

RIDGE TOOL COMPANY
 400 Clark Street Elyria, Ohio 44032

The author also wishes to thank his wife, Joyce, for her help in preparing the manuscript and proofreading the text.

The parts catalogs from which the reproductions of *American Standard, Delta,* and *Kohler* products were made are not ordinarily available to the general public. The reproductions were made through the courtesy of, and with the permission of, the manufacturers of these products. The reproductions shown in the book are not the complete line of merchandise manufactured by these companies; they are, however, representative of some types of products offered by these companies.

There are, of course, many other brand name products of equally excellent quality. The exclusion of other brand name goods does not reflect on the quality or use of these products. Other parts catalogs were not available for reproduction or inclusion in this book.

Chapter 1

General Plumbing Information

The plumbing systems which we, in the United States, have in our homes are the most modern in the world. They are so trouble-free that we tend to take our plumbing for granted. Take the time now to learn some basic facts about the plumbing system in your home. When repairs become necessary or if a remodeling project requires some plumbing changes, you will be able to cope with the problems.

Technically, the plumbing system of a building consists of the water supply to the fixtures in a building, and the drainage piping from the fixtures in the building to the building sewer. Gas or oil piping, while often used, is not actually part of the plumbing system. It is a part of the piping system. The plumbing and piping systems are not hard to understand or follow, they are all installed in a logical sequence.

Water Piping

The water service pipe is the cold-water line from a water main in the street or a well or other source, such as a spring, a lake, etc., into a building. A valve should be installed on the cold-water pipe at the point where it enters the building. This valve will control all of the water in the building. Learn where this valve is, mark it or tag it so that it can be easily located in an emergency. The cold-water pipe will continue from the point where it enters the building to the fixtures requiring cold-water connections. At the location where the water heater (and softener, if used) is located, a tee will be installed to provide a cold-water supply to the water heater.

The valve on the inlet or cold-water supply to a water heater will control or shut off the cold-water supply to the water heater. Thus, it also controls *all* of the hot water to the building. In the event of a leak in the hot-water piping or a leak in the hot-water heater, the valve on the *cold-water* supply to the heater can be shut off,

shutting off the *hot water* but leaving the building supplied with *cold water*.

Drainage Piping

The drainage system consists of the building drain and the waste and vent stacks with branches. The building drain is that part of the lowest horizontal piping of the sanitary drainage system inside the walls of the building which receives the discharge from soil or waste stacks or branches and conveys this discharge to a point outside the building where the building drain connects to the building sewer. This point may be three, five, or ten feet outside the building, depending on local codes.

In (A) of Fig. 1, that part of the drainage system inside the heavy wavy line is the building drain. In this case, it is under the basement floor. As shown at (B) of Fig. 1 the soil and waste stack receives the discharge from a water closet, bathtub, and lavatory and conveys this discharge to the building drain. A sanitary tee with 1½'' or 2'' tappings is used to connect the bath drain to the waste stack, the use of this fitting ensures proper venting of the bath drain.

The branch waste line shown at (C) of Fig. 1 receives the discharge from the automatic washer in the basement and the kitchen sink on the first floor. The piping extending above the drain connections at the sink (D) and at the washer (E) is the vent piping. The water closet and the bathtub are vented by a *wet vent*. A wet vent is a vent pipe from a water closet or bathtub, or both, which also receives the discharge of a lavatory

Vent Piping

Proper venting of soil and waste piping is extremely important. A vent pipe is a continuation of a soil or waste pipe and prevents trap syphonage and back pressure in the soil and waste piping. The main soil and waste stack

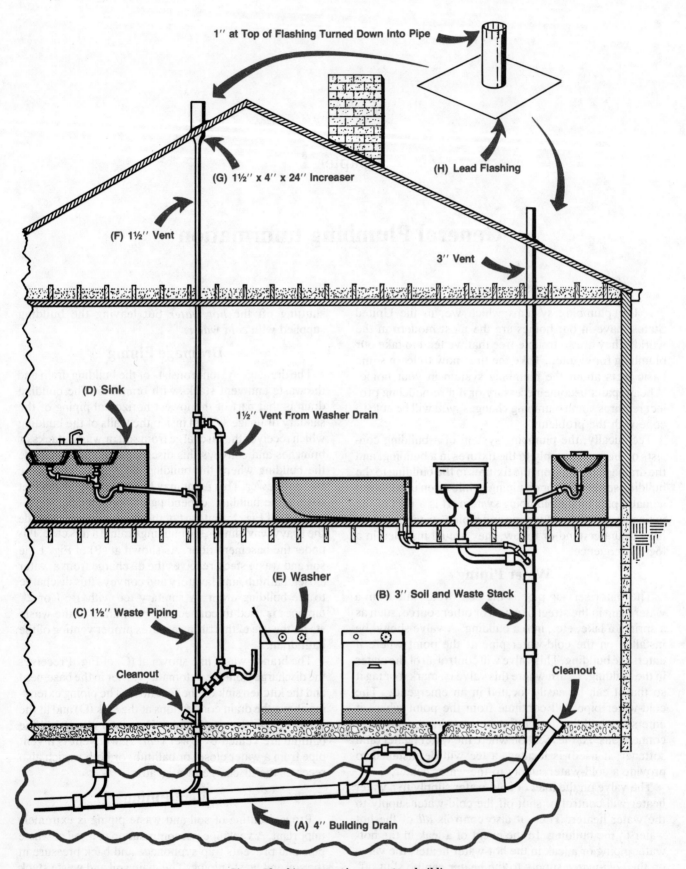

1" at Top of Flashing Turned Down Into Pipe

(H) Lead Flashing

(G) 1½" x 4" x 24" Increaser

(F) 1½" Vent

3" Vent

(D) Sink

1½" Vent From Washer Drain

(C) 1½" Waste Piping

(E) Washer

(B) 3" Soil and Waste Stack

Cleanout

Cleanout

(A) 4" Building Drain

Fig. 1. Plumbing system in a one story building.

becomes a vent stack above the waste connection to the highest fixture on the stack. The main vent stack must be continued full size through the roof. The branch vent shown at (F) in Fig. 1 may be extended through the roof or may be connected to the main vent stack. If the branch vent is extended through the roof in areas where there is danger of frost building up on the inside of the pipe and closing the pipe, an increaser should be used to extend the vent pipe through the roof. The increaser should be 24 inches long so that the change in pipe size is made 12 inches below the roof line. The increaser may be 1½'' x 3'', 1½'' x 4'', 2'' x 3'', or 2'' x 4'', depending on local codes. A typical increaser is shown at (G) in Fig. 1.

Fig. 1 shows a typical single story and basement home with the soil and waste piping and vent piping. At (H) the method of installing a lead roof flashing is shown. The base of the flashing is under the roof shingles; the top of the flashing is turned down into the vent stack. The piping sizes shown on the drawings are intended to serve as a guide and in general represent minimum acceptable sizes. In many areas, homeowners are permitted to do repair work or new work on their own homes but they are subject to the same requirements of permits, inspections, and tests, as the licensed plumber. Local ordinances and codes must be followed.

Tools

When you are repairing or installing plumbing, half of the battle is won if you have the proper tools to work with. If you plan to do the work yourself, you may, at one time or another, use most or all of the common plumbing tools shown in Figs. 2 through 25. You probably already have blade and Phillips type screwdrivers, pliers, and a hammer or two.

The following list of tools are those which will be needed at some time when doing repair work:

Fig.   4 Adjustable wrenches
Fig.   5 A 6 ft. rule
Fig.   6 Pipe wrenches
Fig.   7 Closet auger
Fig.   8 Propane tank and burner
Fig. 11 Hacksaw
Fig. 12 Tubing cutters
Fig. 16 Flaring tool
Fig. 17 Offset hex wrench
Fig. 18 Basin wrench
Fig. 24 Plungers
Fig. 25 Seat wrench

A 10'' pipe wrench and a 14'' pipe wrench should be included in a basic tool set for use in the home.

Tools are not cheap but neither is a plumber's labor. If these tools are purchased as you need them and are taken care of, they will pay for themselves in labor costs saved — and you will still have the tools for the next repair job.

For opening clogged or sluggish drains, where the stoppage is not over 25 feet from the nearest cleanout or fixture opening in the drain line, the spinner type cable is a good investment. The larger tools shown, such as pipe dies, vises, pipe cutters, soil-pipe tools, etc., can be rented at tool rental shops when and if they are needed, for a fraction of their initial cost.

Slip-joint plier (Fig. 2) is almost indispensable for the home-owner-repairman as well as the journeyman plumber. The tool is lightweight and the long handles give extra leverage when gripping pipe, fittings, etc. The teeth are hardened for toughness and long life.

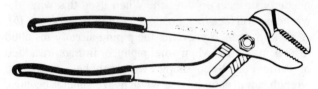

Courtesy Ridge Tool Co.

Fig. 2. Slip-joint pliers.

The ball-pein hammer (Fig. 3) is the all-purpose hammer for plumbing work. The 12 oz. and 16 oz. sizes in these hammers are the most useful.

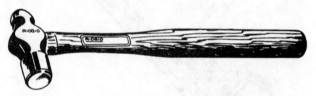

Courtesy Ridge Tool Co.

Fig. 3. Ball-pein hammer.

The adjustable smooth jaw wrench (Fig. 4) is used to grip and turn squared surfaces of valves, nuts and fittings.

Seven Sizes 4'' Through 18''

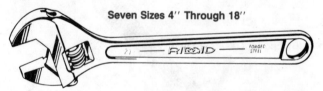

Courtesy Ridge Tool Co.

Fig. 4. Adjustable wrench.

A 6 ft. inside reading rule (Fig. 5) is very useful when measuring small repair jobs.

Pipe wrenches (Fig. 6) are used to grip pipe and fittings when installing and removing piping. Pipe wrenches have two sets of jaws; each jaw has teeth designed to grip the pipe as pressure is applied to the wrench handle. The upper or hook jaw is adjustable, the lower or heel jaw is fixed. When installing or removing piping there are many times when two wrenches are needed, one wrench to tighten or loosen a pipe or fitting, the other wrench is

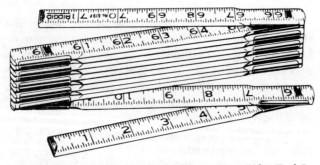

Fig. 5. A 6-foot folding rule.

End type (or offset) pipe wrenches (Fig. 6B) are made for use in tight places: pipes against a wall, or pipes that cannot be turned with the straight pipe wrench.The closet auger (Fig. 7) is used to remove or force out a foreign object from a water closet bowl. See chapter on

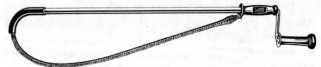

Fig. 7. A closet auger.

needed as a backup wrench. When used this way, the wrenches will face in opposite directions (Fig. 6C and D). A backup wrench prevents the piping already installed from being turned when other piping or fittings are added or removed. When using two wrenches, a backup wrench can usually be one wrench size smaller because the piping being held with the backup wrench is already tightened.

What To Do About Stopped-Up Drains.

Propane torches (Fig. 8) using replaceable fuel cylinders are useful for many purposes. A pencil type burner is used to solder copper tubing and fittings.

A faucet reseating tool (Fig. 9) is available with a large range of cutters. This is a special tool designed to do one job, and do it well. See chapter on **Faucet Construction and Repairs.**

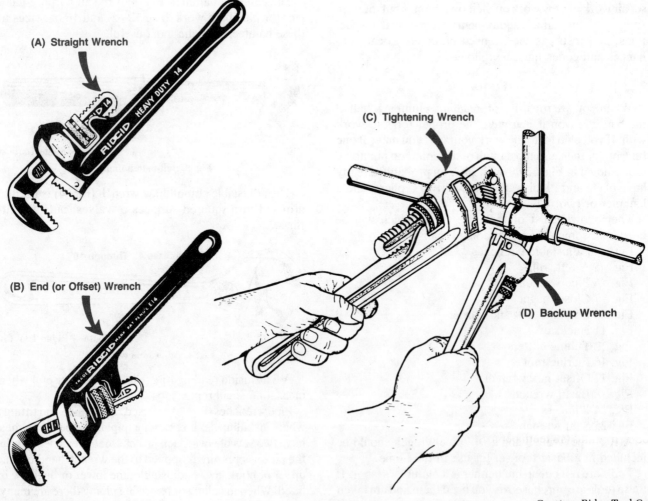

(A) Straight Wrench

(B) End (or Offset) Wrench

(C) Tightening Wrench

(D) Backup Wrench

Fig. 6. Pipe wrenches.

Fig. 8. Propane torch.

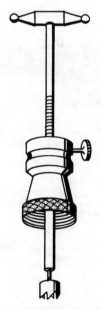

Fig. 9. Faucet reseating tool.

Pipe Taps (Fig. 10) are used to cut female threads, for cleaning rust or corrosion from female threads and straightening out damaged threads.

The hacksaw (Fig. 11) is a metal-cutting saw. Coarse-toothed blades (18 teeth per inch) are used for steel bars, pipe, or heavy metals. Medium blades (24 teeth per inch) are for pipe or metal up to ⅛″ thickness. Fine blades (32 teeth per inch) are used for thin metals, thin tubing, etc.

A tubing cutter (Fig. 12) is used to cut copper, brass or aluminum tubing and thin wall conduit. Two sizes will

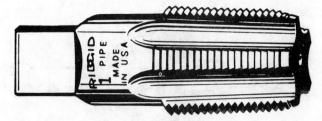

Courtesy Ridge Tool Co.

Fig. 10. Pipe taps.

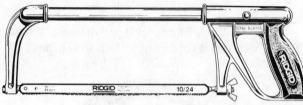

Courtesy Ridge Tool Co.

Fig. 11. Hacksaw.

cover normal household repair requirements, No. 10, ¼″ through 1″; No. 20, ⅝″ through 2⅛″. All sizes are O.D. (outside diameter of tubing).

There are many times when it is necessary to thread piping in place. Ratchet-type dies (Fig. 13) can be used for this purpose, as well as threading pipe held in a vise or power-drive machine. Ratchet dies are available in sizes for ⅛″ IPS to, and including, 2″ IPS.

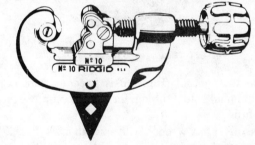

Courtesy Ridge Tool Co.

Fig. 12. A tubing cutter.

Courtesy Ridge Tool Co.

Fig. 13. Ratchet-type pipe dies.

Three-way threaders (Fig. 14) are used to thread steel pipe held in a vise or power-drive machine. They are available in ⅜″, ½″, ¾″, iron pipe sizes, and ½″, ¾″, and 1″ iron pipe sizes.

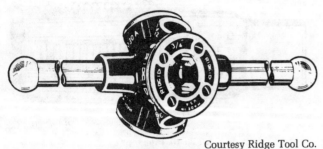

Courtesy Ridge Tool Co.

Fig. 14. A three-way threader.

The 1″ to 2″ (IPS) ratchet-type adjustable threader (Fig. 15) uses one set of dies for all four sizes of pipe. (1″ - 1¼″ - 1½″ - 2″).

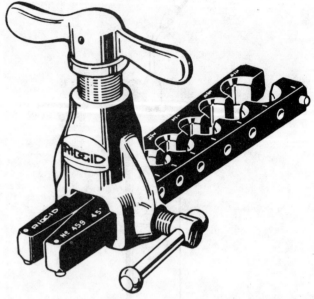

Courtesy Ridge Tool Co.

Fig. 16. Flaring tool.

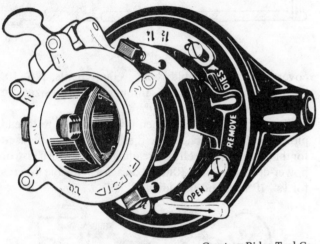

Courtesy Ridge Tool Co.

Fig. 15. A ratchet-type threader.

Handles Up Through 1½″ Drain Nuts

Courtesy Ridge Tool Co.

Fig. 17. The offset hex wrench.

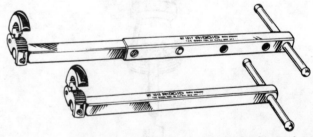

Courtesy Ridge Tool Co.

Fig. 18. A basin wrench.

The flaring tool (Fig. 16) is used to flare copper, aluminum and steel tubing, for use in compression flare type fittings.

The offset hex wrench with smooth jaws (Fig. 17) can be used on chrome-plated nuts, fittings, etc. The smooth jaws will not mar the chrome finish.

The basin wrench (Fig. 18) is used primarily for removing and installing sink faucets and supplies. The jaws are hinged to turn 180 degrees for either tightening or removing nuts, the hook jaw is spring loaded to close on a nut or fitting. The shank telescopes to desired lengths from 10 inches to 17 inches.

The hand spinner cable (Fig. 19) is used to open plugged or sluggish drain piping.

When pipe or tubing is cut off with a wheel-type cutter, the inside edge of the pipe is pressed inward, leaving a sharp burr when the cut is made. The burr, if left in the pipe, would restrict the flow of water or waste. The reamer illustrated in Fig. 20 is a ratchet-type reamer and will remove the burred edge from ⅛ inch pipe to and including, 2 inch pipe.

Pipe cutter shown in Fig. 21 is used for cutting pipe either by hand or is used with a power vise. It will cut pipe from ⅛ inch through 2 inches. It can be converted to a 3-wheel cutter for cutting pipe in place by removing the rollers and inserting cutter wheels.

The vise stand (Fig. 22) is used to hold pipe, for cutting or threading by hand, or to hold pipe while tightening or removing fittings. Vise stands use chain or yoke type vises.

Column vises (shown in Figs. 22 and 23) clamp onto a column and need no vise stand. They are very useful for small as well as large repair jobs and they are available either in chain type or yoke type.

Plungers (or force cups) are used to open clogged drains. Plungers (Fig. 24) are made in several sizes, the large bell-shaped plunger is used to open a clogged water closet bowl.

A seat wrench, shown in Fig. 25, is used to remove and replace bibb washer seats.

Courtesy Ridge Tool Co.

Fig. 19. A hand spinner.

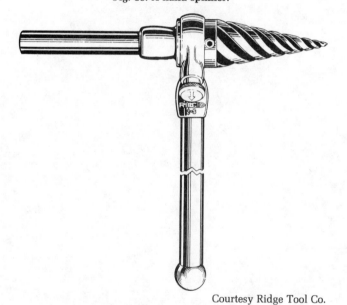

Courtesy Ridge Tool Co.

Fig. 20. A pipe or tubing reamer.

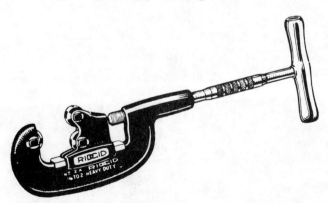

Courtesy Ridge Tool Co.

Fig. 21. Pipe cutter.

Courtesy Ridge Tool Co.

Fig. 22. A vise stand.

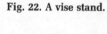

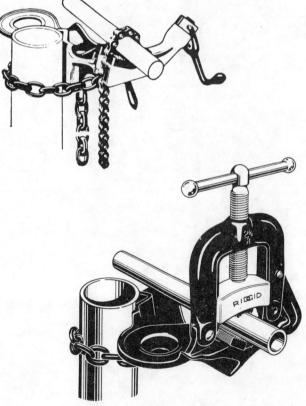

Courtesy Ridge Tool Co.

Fig. 23. Column vice clamp.

Fig. 24. A rubber plunger and force cup.

Fig. 25. A seat wrench.

Chapter **2**

Plumbing Valves and Their Uses

Each type of valve is designed to be used for a specific purpose (Fig. 1). Basically valves serve four purposes:
1. A flow is turned on,
2. The flow is throttled or regulated,
3. The flow is turned off,
4. Check valves permit flow only in one direction.

Gate valves should be used only as turn-on or stop valves. When the gate is raised and the valve is opened, a straight through full-on flow is permitted—there is virtually no restriction to full flow. Gate valves are not made for throttling or regulating use and are not intended for frequent operation.

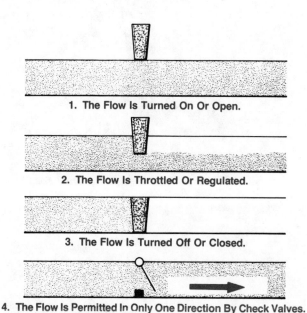

1. The Flow Is Turned On Or Open.

2. The Flow Is Throttled Or Regulated.

3. The Flow Is Turned Off Or Closed.

4. The Flow Is Permitted In Only One Direction By Check Valves.

Fig. 1. Four functions of specific valves.

Globe valves are very good to use as control valves. They have good throttling characteristics. A globe valve operates with fewer turns of the hand wheel and shorter disc travel than a gate valve. Variations of globe valves, called compression stops, are most commonly used around the home. Full flow of water is *not* usually necessary in the home and the compression stop is much less expensive to manufacture when compared to the gate valve.

The homeowner should be familiar with the locations and functions of the valves in the home. There are three or four valves of special importance. One of them is not a plumbing valve but deserves mention.

The Main Water Valve

This valve is usually located at the point where the cold water pipe enters the home. If the water meter is located in the house, the valve should be on the inlet (or street) side of the meter. If the meter is outside, there should be a main shut-off valve at the point where the piping enters the home, or the nearest accessible location. This valve will control all of the water in the home, both hot and cold, its location should be known and tagged if necessary, for use in an emergency.

Hot Water Valve

This valve controls all of the hot water to the home. In the event of a leak in the water heater, a broken or leaking hot water pipe, or a hot water faucet which leaks badly, shutting of this valve will shut off the hot water to the home but will still leave the cold water on.

Main Gas and/or Oil Valve

Fire or a leaking pipe could require the immediate shut off of either or both of these valves. If the gas supply is city gas, the shut-off valve should be at the meter location. If it is bottled gas, the shut-off should be on the piping close to the tank. If heating oil is used, the shut-off valve should be on the oil line at the tank, or at the point where the oil piping enters the home.

Main Electrical Disconnect Switch

This is a plumbing book, true, but the main electrical disconnect switch is a valve and in an emergency, such as trouble with an electric water heater or severe electrical problems, everyone should know how to locate and turn off the main electric disconnect switch, at the main electric panel.

Most city and state plumbing codes require the installation of shut-off valves on the piping to every plumbing fixture. These valves should be exposed underneath each lavatory, toilet and sink. The valves on the hot and cold water piping to bathtubs should be behind an access panel built into a wall at the low (or drain) end of the tub. The valves on a built-in shower stall can be under the floor (in a home with a basement) or can be built into the shower valve. The installation of these valves is important, in the event of trouble with any one plumbing fixture, it should be possible to shut off the water to that particular fixture without affecting the rest of the fixtures.

Stop and Waste Valves

Stop and waste valves (Fig. 2) are globe valves with a drain feature built into them. Automatic stop and waste valves are made to automatically drain on the downstream (or no pressure side) when the water is turned off. Button type stop and waste valves will drain when the valve is turned off and the button or cap on the side is opened. Stop and waste valves are normally lo-

cated inside a heated or protected area and are used to drain piping which could freeze, such as a regular sillcock.

Sillcocks

A common sillcock (Fig. 3) is a compression stop valve with a hose thread. Water is normally present at the valve and a common sillcock is subject to freezing unless a stop valve is located in the piping to the sillcock. The stop valve must be shut off and the sillcock opened and drained in cold weather to prevent freezing. Antifreeze type sillcocks are made with a long body, the shut-off point is inside the home, or in a protected area. (Fig. 4). Since there is no water in the sillcock when it is shut off there is no danger of the sillcock freezing and bursting.

There is an important point to remember with either type of sillcock. DO NOT LEAVE A HOSE CONNECTED TO A SILLCOCK IN FREEZING WEATHER! A garden hose can hold water up in either type of sillcock and the water in the sillcock or the pipe to the sillcock can freeze and burst the pipe or the sillcock.

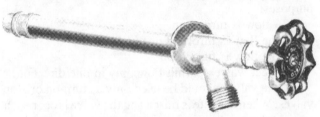

Courtesy Nibco, Inc.

Fig. 4. Frost-proof sillcock.

Courtesy Nibco, Inc.

Fig. 3. Angle sillcock.

Courtesy Nibco, Inc.

Fig. 2. Stop and waste valves.

Chapter **3**

Faucet Construction and Repair

Basic construction of a faucet is shown in Fig. 1 to enable you to disassemble and repair or replace it if needed. Specific instructions are shown in chapters on Sink, Lavatory and Bathtub faucets. In order to repair almost any faucet, the handle must first be removed from the stem.

The outside, visible parts of most faucets — the handles, escutcheons, flanges, skirts and stems, are chromium-plated. The makers of good quality faucets use brass for the base metals of these parts, and apply chromium plating to these brass parts for a quality finished product. The makers of cheaper, competitive faucets, use "pot" metal as a base metal for handles, escutcheons, flanges and skirts. Pot metal, which is an alloy of zinc and other metals, is cheaper than brass. Pot metal will corrode when in contact with dissimilar metals. Thus a pot metal handle, mounted on a brass stem, or a pot metal flange screwed on to a faucet body, will often be very hard to remove. The corrosion between these parts is occasionally so severe that an old faucet cannot be taken apart by normal methods — it must literally be cut apart using a chisel or a hacksaw. A faucet in this condition is not repairable and must be replaced with a new unit. When a pot metal handle is corroded on a stem, it may be necessary to insert a screwdriver between the bottom of the handle and the flange to pry the handle up in order to get at the stem. This usually causes the bottom of the handle to crumble and break.

The basic steps for repairing any faucet are:
1. Remove the handle — Fig. 2 illustrates different methods by which handles are mounted on stems.
2. Remove the stem nut or packing nut — turn the stem nut, or the packing nut counterclockwise to loosen and remove it.

3. Remove the stem — turn the stem in the direction of opening the faucet (turning on the water) to loosen and remove the stem.
4. If the faucet leaks around the stem — replace or add to, the graphited packing, or replace the O-rings.
5. If the faucet drips — inspect the bibb washer retainer, replace it if it is damaged.

Turn the bibb washer screw counterclockwise to unscrew and remove it. Install a new bibb washer and replace the bibb screw. Use a soft rubber bibb washer and make certain it is the correct size. The correct size bibb washer will fit easily, without forcing, into the bibb washer retainer.

When the O-ring on a faucet stem becomes worn, water will leak around the stem. To repair this type leak, the O-ring must be replaced. The replacement O-ring must be the same exact size as the original O-ring. If your faucet has a brand name on it, consult the yellow pages of your phone book under "Plumbing Fixtures and Supplies" for a dealer handling this brand merchandise, for O-rings, or any other repair parts for brand name products. When reassembling a faucet stem using O-rings, apply a thin coating of waterproof grease to the O-rings. To aid in selecting the correct O-ring, full-size illustrations are shown in the Appendix portion at the rear of this book.

Index caps are so called because they indicate the temperature of the water, either hot or cold. On older types of china faucet handles they were screwed into the top of the handle, and tightened with a right hand (clockwise) thread. Chrome plated index caps now in use are pressed into the handle and can be removed by inserting a small thin screwdriver blade under the edge of the cap and prying the cap out. After removing the screw

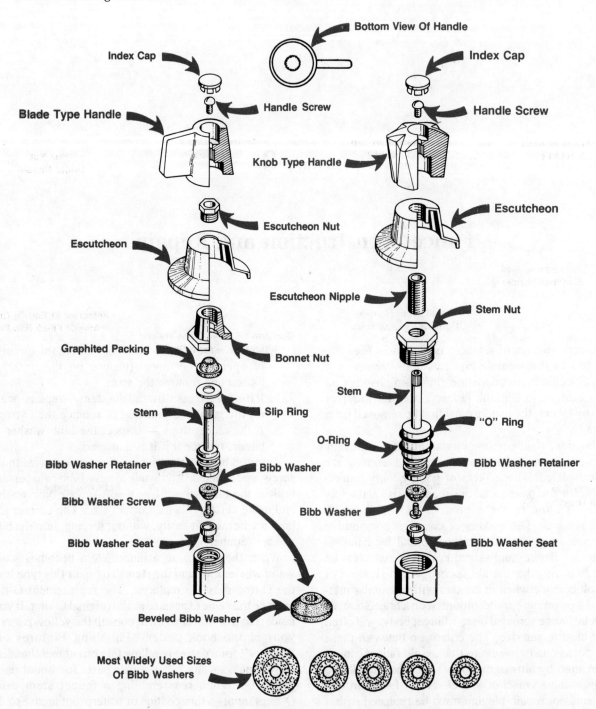

Fig. 1. Basic construction of a faucet.

which holds the handle, the handle should lift off; if it does not, tap the underside of the handle until the handle loosens. See this chapter (toward the end) explaining repairs to *Delta* faucets for instruction on disassembly.

If a faucet handle becomes worn and turns on the stem, a fit-all handle can be used as a replacement. Fit-all handles can be bought in the plumbing department of most hardware stores.

Wall-Mounted Sink Faucets
(Faucets bearing no brand name Fig. 3)

Turn off the hot and cold water supply to the faucet, either under the sink or at the main shut-off valve, before starting repair work. A steady dripping from the spout indicates a replacement of the hot or cold (or both) bibb washers is needed. With the hot and cold stems re-

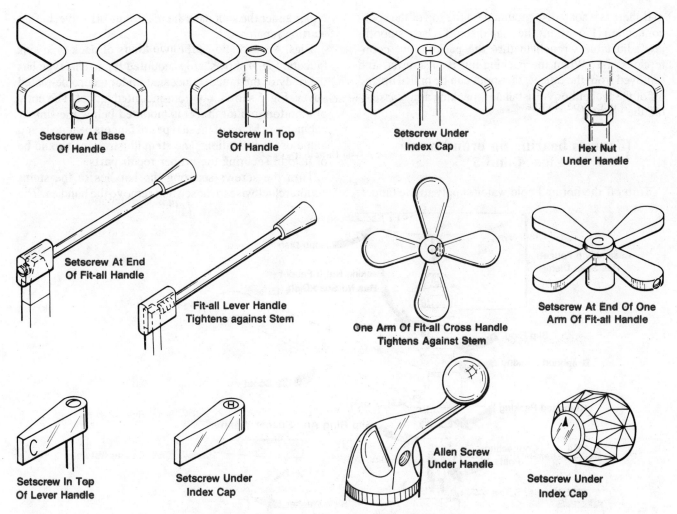

Fig. 2. Various methods of attaching faucet handles.

moved, the bibb washer seats can be inspected. Remove the screws securing the handles to the stems. Fig. 2 shows different methods of removing the handles. It may be necessary to tap lightly on the underside of the handles to remove them. Use an adjustable, smooth-jaw wrench (Fig. 4, Chapter 1), to turn the bonnet nut counterclockwise.

When the bonnet nuts are free from the faucet body, set the handles back on the stems and turn the stems in the direction of opening the faucet. When the stems are unscrewed from the faucet body, they can be lifted out. Slide the bonnet nuts off of the stem. The stems can then be compared with the full size stem illustrations to aid in determining which size and type of bibb washers are needed. The bibb washers are held in place, in a retainer or retainer cup, by a bibb screw. Some stems have the retainer made onto the stem, others have a separate retainer. If the retainer is damaged or corroded or one side broken or missing, the bibb washer will not be held in place. If the retainer is the loose type, it can be replaced. If the retainer is damaged, but is part of the stem,

it will be necessary to replace the stem. The bibb seat should be inspected, it may be necessary to shine the beam from a flashlight into the faucet to inspect the seats. If the seats are chipped or pitted, they should either be replaced or reseated—follow instructions at the end of this chapter on **Reseating Faucets.**

When the stem has been correctly matched, using the stem illustratlions, the correct size and number of the bibb washer seat can be purchased. If there is a leak around the faucet stems, add a single strand of graphited packing, 2 or 3 inches long, above the slip ring, between the brass slip ring and the bonnet packing. If the cone bonnet packing has completely deteriorated, replace it with a new bonnet packing. The faucet can now be reassembled.

If there is a leak at the swing spout, the packing in the spout should be replaced. Matching the stems with the illustrations (shown in the Appendix) should aid in learning the brand name of the faucet. When the faucet name is known, it should be helpful in choosing the right spout packing part number.

If there is a soap dish mounted on the top of the spout connection (Fig. 3), lift the soap dish off. Use a smooth-jaw adjustable wrench to turn the packing nut counterclockwise. When the packing nut is unscrewed and removed from the faucet, the spout can be pulled up and off of the faucet body. Install new packing and reassemble the spout.

(Faucets bearing no brand name Figs. 4 and 5)

Turn off the hot and cold water supply to the faucet, either under the sink or at the main shut-off valve, before starting repair work.

Figs. 4 and 5 shows the two kinds of deck-type sink faucets. Fig. 4 shows a top-mounted type. This type has the body of the faucet concealed under a chrome-plated cover. Fig. 5 shows a bottom-mounted faucet. The body of a bottom-mount faucet is mounted below the sink or cabinet top. Many of these types of faucets bear no brand name or identification. The stem illustrations should be of help in securing the proper repair parts.

Turn the screws securing the handles to the stems counterclockwise to loosen and remove the handles. The

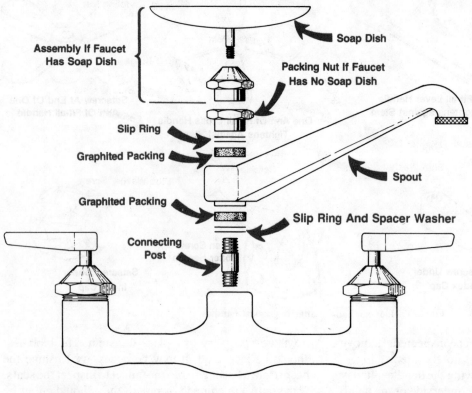

Fig. 3. Wall mounted sink faucets.

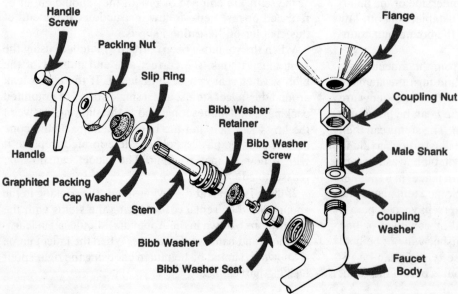

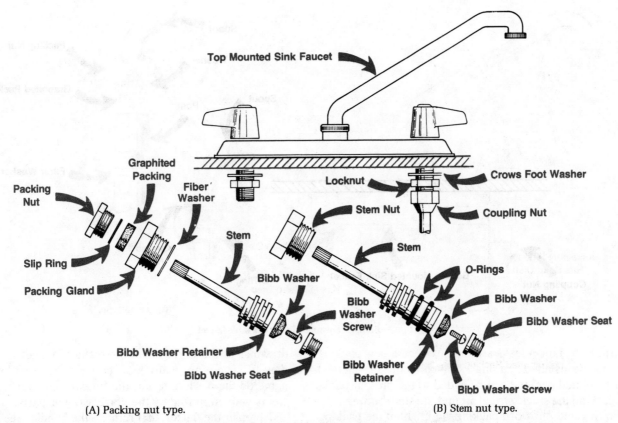

(A) Packing nut type.

(B) Stem nut type.

Fig. 4. Top mounted deck type sink faucet.

handles are often die-cast metal, corrosion between the die-cast metal and the brass stem may make the handles difficult to remove. Tap the handles lightly on the bottom side or pry up on the skirt at the bottom of the handles to loosen. If prying is necessary, exert very little pressure at each point and work all around the handle until it is loosened. Too much applied pressure when prying up on the skirt may cause the die-cast metal to crumble.

If the faucet has escutcheons, a packing nut may secure the escutcheons to the faucet body. Turn the packing nut counterclockwise to loosen and remove the escutcheons. If there is no visible nut securing the escutcheon, turning the escutcheon counterclockwise should loosen it so that it may be lifted off. If the stem has a packing nut and a packing gland, Fig. 4A, graphited packing is used as a water seal. If the stem has only a stem nut, Fig. 4B, then O-rings are used as a water seal. Use a smooth-jaw adjustable wrench, turn the packing gland or the stem nut counterclockwise to loosen and remove the gland or the nut. If the stem is similar to Fig. 4A and has a packing nut under the escutcheon, it is not necessary to loosen the packing nut. When the packing gland or the stem nut has been loosened, the stem can be unscrewed and lifted out of the faucet body.

When the stems have been removed, they can be compared with the full-size illustrations of stems to aid in selecting the proper repair parts. The correct bibb screw, seat washer and washer seat is shown for each stem. If a stem turns but will not tighten in the faucet body, the stem thread is probably stripped and a new replacement stem should be installed. Inspect the washer retainer—if the retainer is damaged a new retainer is needed. The retainer is the same size as the bibb washer. The bibb screw should be turned counterclockwise to loosen and remove it. If the bibb screw is corroded, it should be replaced. Inspect the bibb washer seats, if the seats are pitted or chipped, they should be replaced or reseated. Instructions for reseating a faucet are found later in this chapter on **Reseating Faucets.** if a new seat is to be installed, use a seat wrench shown in Fig. 25 Chapter 1. Turn the wrench counterclockwise to loosen and remove the worn seat.

If O-rings are used on the stem, new O-rings should be installed before replacing the stem. Compare the worn O-rings with the O-rings shown in the Appendix to aid in selecting the correct size. When new O-rings are installed on a stem, the O-rings should be coated lightly with a waterproof grease. The grease protects the O-ring when reassembling the faucet and lubricates the O-rings to make the stem turn easily when the faucet is used. If the faucet stems have a packing nut (Fig. 4A), the packing nut should be turned clockwise to slightly tighten it when reassembling the faucet. The packing nut should not be too tight, if the nut is too tight, the stem will be hard to

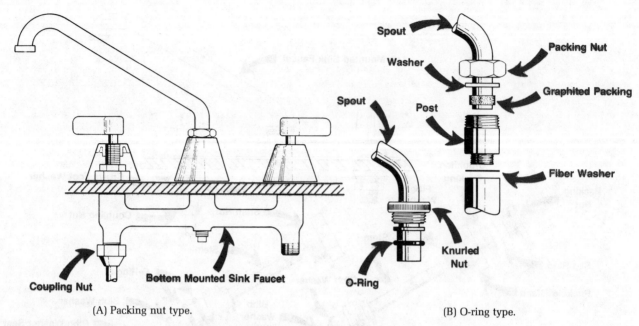

(A) Packing nut type.   (B) O-ring type.

Fig. 5. Bottom mounted deck type sink faucet.

turn. If the faucet leaks around the stem, a new packing should be installed or a single strand (about 2 or 3 inches) of graphited packing can be added to the present packing. Wind the single strand around the stem between the slip ring and the worn packing, and tighten the packing nut. Reassemble the faucet.

If the spout is secured to the faucet by a packing nut (Fig. 5A) turning the packing nut clockwise one quarter turn will often stop the leak. If the packing is badly worn, new packing should be installed. Swing spout packing is shown in Figs. 5A and 5B.

If the spout uses an O-ring as a water seal (Fig. 5B), the worn O-ring can be compared with the illustrations of O-rings shown in the Appendix to aid in selecting the correct replacement. If you were able to identify the brand name of the faucet by the stem illustrations, you may be able to cross check the O-ring by the number against the list of faucets shown in the O-ring illustration. Coat the O-ring lightly with waterproof grease before reassembling the spout.

Repairing Lavatory Faucets
(Faucets bearing no brand name)

Turn off the hot and cold water supply to the faucet, either under the sink or at the main shut-off valve, before starting repair work.

If a lavatory faucet which leaks and needs repairing has a name on the faucet, the correct repair parts should not be hard to find. In the Appendix which shows actual size stems, you will see that the stem illustration also lists the kind and size of bibb washer to use, and the correct replacement bibb washer seat. The Appendix shows the full size drawings of stem packings and the full size

drawings of the O-rings and the make of faucets which they fit. There are many faucets which have no brand name on them. Here again, the full size illustrations of stems will aid in finding the correct repair parts.

To repair the faucet, first remove the handle. See Fig. 2 which illustrates different types of handles and how the handles are secured to the faucet stem. Many handles are made of die-cast metal and corrosion between the die-cast metal and the brass stem may cause the handle to be difficult to remove. When the screw securing the handle to the stem has been removed, if the handle cannot be lifted off of the stem, tap the underside of the handle lightly with a small hammer. Some types of faucet handles have a skirt at the bottom and it may be necessary to pry up on the skirt with a screwdriver blade to loosen the handle. If prying is necessary, exert *very little pressure* and work all around the skirt, prying just a little at each point. Exerting too much pressure will cause the die-cast metal to crumble and break. If the faucet has no escutcheons, use an adjustable smooth- jaw wrench to turn the packing gland or the stem nut counterclockwise to loosen and remove it. Set the faucet handle back on the stem and turn the handle in the direction of opening the faucet to loosen and remove the stem. When the stem is removed, inspect the seat. (It may be necessary to use a flashlight to do this.) If the seat is pitted or chipped, use a seat wrench (Fig. 25 Chapter 1) to remove the old worn seat. Or if a seat is not readily obtainable, if you have a reseating tool, such as is shown in Fig. 26, the old worn seat can be restored to new condition by following the instructions given in the chapter on **Reseating Faucets.**

If water leaks around the stem when the faucet is turned on, install a new bonnet packing or a few strands of graphited packing can be added to the old bonnet packing. Often one strand of packing, 2 or 3 inches long, can be added on top of the brass slip ring. When the faucet is reassembled, the bonnet nut will compress the graphited bibb screw packing into shape. Quite often when the stem has been removed, the bibb screw will be badly corroded. If the corroded bibb screw can be removed, a new bibb screw should be installed. If the screw is twisted off, your local plumbing shop may be able to drill and tap the stem for a new screw. The bibb screw must be turned counterclockwise to loosen and remove. When replacing the bibb washer, use the correct size and type. Again, the correct size and type washers are shown for each stem in the Appendix section. With some types of faucets, the stem is threaded into the body of the faucet. With this type faucet, if the stem turns but will not tighten, the thread on the stem is probably stripped. A new stem will be necessary to correct the problem.

Certain types of single lever faucets use a cartridge element. It will be necessary to install a new cartridge when faucets drip. Repairs for *Delta* single lever faucets are covered in another chapter. When new O-rings are installed on a stem, the O-rings should be lightly coated with a waterproof grease. The grease protects the O-rings when reassembling the faucet and lubricates the O-rings to make the stem turn easily when the faucet is used.

Pop-up drains consist of a lift rod connected to a horizontal rod. The horizontal rod extends into the pop-up assembly. The drain plug is lifted or lowered by the combined action of these rods. The horizontal rod may have a ball seating against a nylon seat or it may be a cam-action type rod and depends on a graphited packing for a water seal. If a leak develops at this point, tightening the nut securing the horizontal rod will sometimes stop the leak. Replacement of the pop-up assembly is the recommended cure for this problem (see Fig. 6).

The following *problem* and *cure* examples are included on some of the most popular model faucets and valves. These examples are shown as a guide only in the disassembly sequence and general overall repair. Although your particular valve or faucet may not be included in these examples, the problem, cure and disassembly procedures are generally the same.

American Standard Single-Lever Faucet
R-4150-1 less hose and spray
R-4155-1 with hose and spray

PROBLEM: Faucet drips —will not shut off completely. (Fig. 7)
CURE: Replace parts 6-7-8-9 (2 each) Use repair kit No. 72495-07. The kit also contains item 3 (O-ring).

The spout (1) must be removed and the escutcheon (4) removed. Turn the spout nut 1(a) counterclockwise to loosen. When the nut is loose, lift off the spout. Lift off the escutcheon (4) to expose the faucet body. Turn the two plugs (10) counterclockwise to loosen, then remove. Remove the worn parts 6-7-8-9. Take one each of 6-7-8-9 from the repair kit and install them in the order shown in the exploded drawing (Fig. 7) in the hot (left) side of the faucet body and in the cold (right) side of the faucet body. Replace the two strainer plugs (10). Remove the worn O-ring (3) from the spout and install the new O-ring. Lightly coat the O-ring with waterproof grease to prevent damage when installing the spout. Replace the escutcheon (4) and insert the spout into the faucet body and tighten the spout nut 1(a). *Do not overtighten the spout nut.*

PROBLEM: Water leaks around the spout connection to the faucet.
CURE: Replace the O-ring (3) Part No. 55234-07. To replace the O-ring (3) remove spout and replace the worn O-ring as explained above. Even if the faucet does not drip, it would be wise to get the complete repair kit and replace items 6-7-8-9, since the spout has to be removed just to replace item (3).

PROBLEM: Spray head or spray hose leaks.
CURE: Replace complete hose and spray assembly, Part No. 63172-07. Hose is removed from underneath by turning coupling nut 25-(a) counterclockwise to loosen and remove hose.

PROBLEM: Spray does not operate properly.
CURE: Remove spout and replace diverter valve, 5, Part No. 12435-07. Grease O-ring as outlined previously.

Repairing American Standard Top-Mount Sink Faucet
(N-4110-1 less spray hose)
(N-4112-1 with spray hose)

PROBLEM: Faucet drips, will not shut off completely. (Fig. 8.)
CURE: Install 2 new aquaseal diaphragms (item 8), also inspect seats (item 10) and replace or resurface the seats if they are pitted or chipped.

Note that the stem nuts (item 4) and stem (item 5) are different. On the cold side (right) of the faucet, the stem nut and stem have L.H. (left hand) threads, while on the hot side (left) the stem nut and stem have R.H. (right hand) threads. It is best to take one side apart, repair that side and reassemble the parts, before starting on the other side.

Remove the screw (1) and handle (2). Turn the lock nut (3) counterclockwise to loosen and remove. Lift out the stem, stem nut, and aquaseal diaphragm. Inspect the seat, if the seat is pitted or chipped, use the seat wrench to remove the old seat. Insert the wrench into the seat and turn the wrench counterclockwise to loosen and remove. The seat can be resurfaced using the reseating tool without removing the seat. If a new seat is needed, install new seat, Part No. 862-14. Remove the old

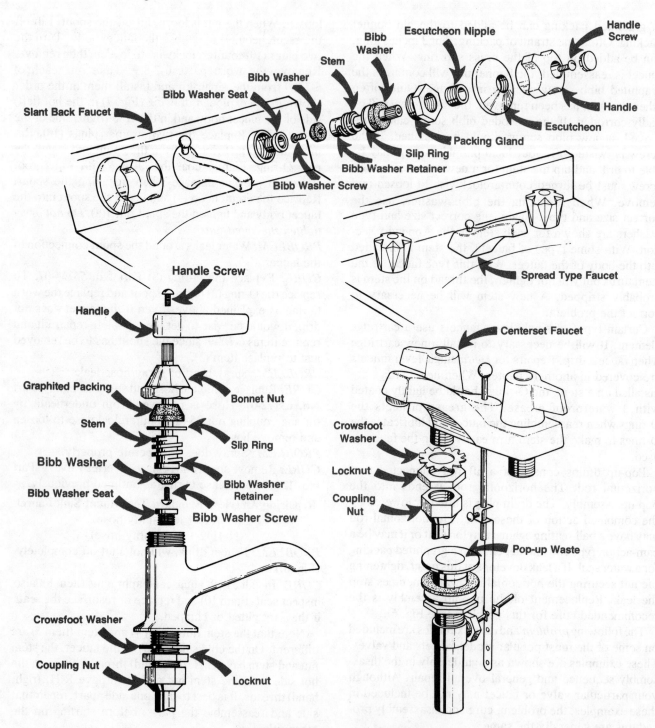

Fig. 6. Faucet and pop-up drain assemblies.

aquaseal diaphragm (Part No. 72940-7)—two are required, one for the hot water side and one for the cold water side and reassemble the faucet.

PROBLEM: Spray or spray hose leaks.
CURE: Replace the spray and hose, Part No. 63172-07.

The old hose is removed from underneath the sink by turning the coupling nut 24(a) counterclockwise to loosen and remove.

PROBLEM: Water leaks around the spout.
CURE: Replace the O-ring (item 20) Part No. 887-17.

Use a smooth jaw wrench to turn the spout nut counterclockwise to loosen and remove the spout. Remove the worn O-ring and install the new O-ring. Lightly coat the O-ring with waterproof grease to prevent damage when reassembling the spout.

PROBLEM: Hose and spray does not work properly.

Single Control Sink Fittings

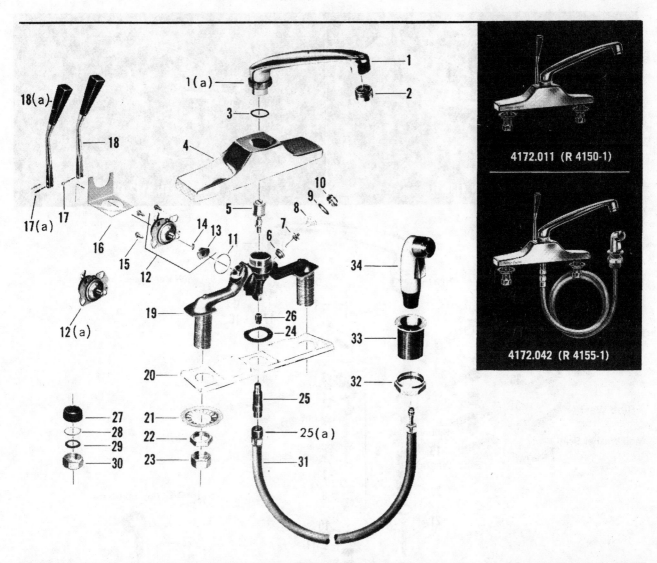

no.	description	part no.	R 4150-1	R 4155-1	no.	description	part no.	R 4150-1	R 4155-1
1	Spout S/A w/Aerator & "O" Ring	61011-02			19	Body S/A w/Hyseal Valve Seat	61056-07●		
2	Aerator	56135-02			20	Faucet Mounting Gasket	61013-07		
3	"O" Ring	55234-07			21	Shank Washer	754-17		
4	Escutcheon	61006-02			22	Jamb Nut	300-27		
5	Diverter Valve	12435-07			23	Swivel Nut	24220-07		
6	Hyseal Valve Seat	12002-07			24	Center Washer	61014-07●		
7	Hyseal Valve Stem	61428-07			25	Tube Connector	63418-07		
8	Conical Spring	72510-07			26	Pipe Plug	56519-09●		
9	Strainer Plug Gasket	12007-07			27	Slip Joint Gasket	1010-17		
10	Strainer Plug	12268-07			28	Slip Joint Washer	21391-07		
11	Control Mounting Gasket	12035-07			29	Slip Joint Washer	21545-07		
12	Control S/A w/Parts 13 & 14	12375-04			30	Slip Joint Nut	24906-07●		
12A	Control Sub Assembly	61226-04			31	Hose S/A	1178-17		
13	Cam	12390-07			32	Lock Nut	61300-06		
14	Cam Pin	12034-07			33	Spray Holder			
15	Control Mounting Screws	61020-04			34	Sprayhead S/A	63181-07		
16	Rear Closure	12067-07							
17	Handle Screw	12326-07				Sub-assemblies			
17A	Handle Pins (2)	63154-07				Spray & Hose S/A (Parts 31 & 34)	63172-07		
18	Control Lever S/A	61051-04				Control & Lever S/A (Parts 12A-17A-18A)	61232-04		
18A	Control Lever S/A w/ 2 Pin Openings	61228-04							

Repair Kit #72495-07 consisting of
((1) Part 3 and (2) each 6,7,8,9).

● NOT AVAILABLE

Courtesy American Standard.

Fig. 7. Single lever faucet.

Top Mount Sink Fittings

Heritage Trim—Lever Handles—Aquaseal

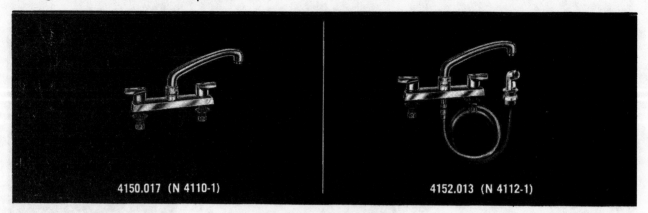

4150.017 (N 4110-1)   4152.013 (N 4112-1)

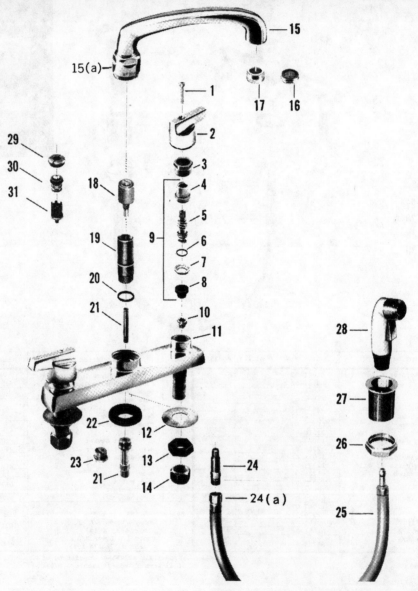

Fig. 8 Top mount

Top Mount Sink Fittings
Heritage Trim—Lever Handles—Aquaseal

no.	description	part no.	N 4100-1 N 4101-1	N 4102-1 N 4103-1	N 4110-1 N 4111-1	N 4112-1 N 4113-1
1	Handle Screw	690-31				
2	Handle - Specify Index	64074-02				
3	Lock Nut	855-17				
4	Stem Nut - R.H. Thread	72956-07				
	Stem Nut - L.H. Thread	72957-07				
5	Stem w/Swivel & Friction Ring - R.H. Thread	72952-07•				
	Stem w/Swivel & Friction Ring - L.H. Thread	72953-07•				
6	Friction Ring	246-37				
7	Stop Ring	72959-07				
8	Aquaseal Diaphragm (2)	72940-07				
9	Aquaseal Trim - R.H. Thread Stem	72950-17				
	Aquaseal Trim - L.H. Thread Stem	72951-17				
10	Seat	862-14				
11	Body	8171-02•				
12	Friction Washer	754-19				
13	Lock Nut	300-27				
14	Coupling Nut	24220-07				
15	Spout w/End Trim & "O" Ring	8099-02•				
	Spout w/Aerator & "O" Ring	8098-02				
16	End Trim	279-12•				
17	Aerator	56135-02				
18	Diverter	53828-07				
19	Post	1271-27				
20	"O" Ring	887-17				
21	Hose Connection Tube	1155-27				
22	Gasket	1153-27•				
23	Body Plug	875-17				
24	Hose Connector	63418-07				
25	Hose S/A	1178-17				
26	Lock Nut	61300-06				
27	Spray Holder					
28	Spray Head S/A	63181-07				
29	Cap w/washer	899-27				
30	Auto Spray (Upper)	696-17				
31	Auto Spray (Lower)	697-17				
	Sub-assembly					
	Spray & Hose S/A (Parts 25 & 28)	63172-07				

• NOT AVAILABLE

Note - If part #18 is not used, order parts 29, 30 & 31

Courtesy American Standard.

Top Mount Sink Fittings
Colony Trim

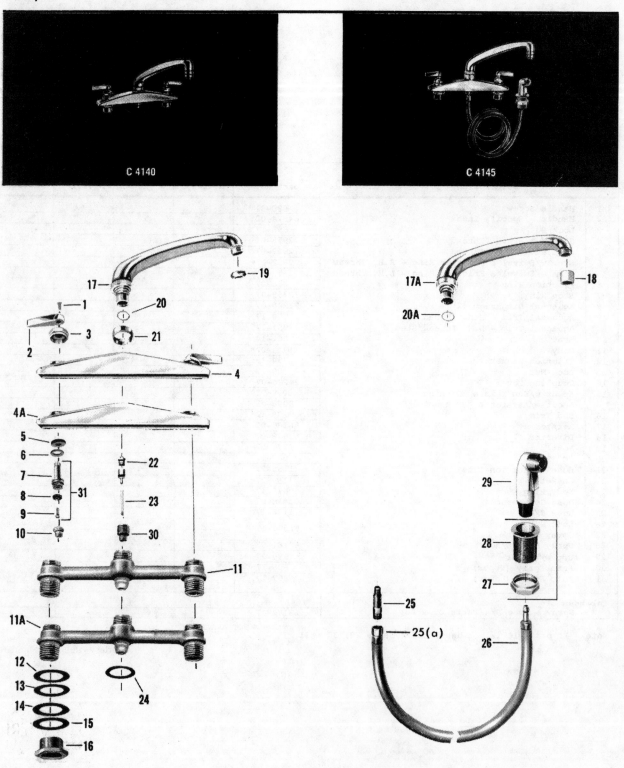

C 4140

C 4145

Fig. 9. Top mount

Top Mount Sink Fittings
Colony Trim

no.	description	part no.	C-4140	C-4141	C-4142	C-4143	C-4145	C-4146	C-4147	C-4148
1	Handle Screw	54868-01								
2	Handle - Specify Index	56725-02								
3	Packing Nut	56723-04								
4	Escutcheon	56499-02•								
4	Escutcheon Marked (Standard)	63223-02•								
5	Stem Packing	54111-07								
6	Stem Washer	54109-07								
7	Stem Hot	54106-04•								
	Stem Cold	54107-04•								
8	Seat Washer	57244-07								
9	Seat Washer Screw	56480-07								
10	Seat	56534-07								
11	Body w/seat	56356-07•								
11A	Body w/seat	56357-07•								
12	Shank Washer	63459-07•								
13	Shank Gasket	57542-07•								
14	Jamb Nut Gasket	56523-07•								
15	Jamb Nut Washer	54528-07								
16	Jamb Nut	56744-07								
	Jamb Nut	63200-07								
17	Spout w/Spout Nut	57166-02								
17A	Spout w/Spout Nut	57160-02								
18	Aerator	56135-02								
19	Stream Regulator	57364-04•								
20	"O" Ring I.D. 674	54212-07								
20A	"O" Ring I.D. 861	55234-07								
21	Escutcheon Nut	63219-04								
22	Diverter Valve	55275-07								
23	Diverter Tube	63218-07								
24	Center Washer	63226-07•								
25	Hose Connector	63418-07								
26	Hose S/A	1178-17								
27	Lock Nut	61300-06								
28	Spray Holder									
29	Sprayhead S/A	63181-07								
30	Adapter	63216-07								
31	Stem Hot w/washer & Screw	54156-04								
	Stem Cold w/washer & Screw	54157-04								
	Sub-assembly									
	Spray & Hose S/A - (Parts 26-29)	63172-07								

• NOT AVAILABLE

Courtesy American Standard.

sink faucet.

CURE: Install new diverter, (item 18), Part No. 53828-07. If diverter is type shown as 29-30-31 in illustration, order parts: Parts No. 899-27 (29), 696-17(30), 697-17(31).

Remove spout as outlined above; install new diverter spout. Grease O-ring as outlined above.

Top-Mount American Standard Sink Faucet
(C-4140 less spray)
(C-4145 with spray)

PROBLEM: Faucet drips, will not shut off completely. (Fig. 9.)
CURE: Replace seat washers (item 8) inspect seats (item 10). If seats are chipped or pitted, replace or resurface seats.

Remove screw (1) and handle (2). Use smooth jaw wrench and turn the packing nut (3) counterclockwise to loosen and remove. Set the handle back on the stem and turn until stem lifts it out of the faucet body. Inspect the seat (10) if the seat is chipped or pitted, use the seat wrench and turn counterclockwise to remove the worn seat. The seat can be resurfaced using the reseating tool without removing the seat. If the worn seat is not reusable, install a new seat, Part No. 56534-07. If the faucet leaks around the stem, install new stem packing, Part No. 54111-07. Install new seat washer and reassemble the faucet.
PROBLEM: Water leaks around the spout.
CURE: Replace the O-ring (20 or 20A). If the spout has a stream regulator, use O-ring, Part No. 54212-07. If the spout has an aerator, use O-ring, Part No. 55234-07.

Turn the spout nut (17 or 17A) counterclockwise to loosen the nut and remove. Remove the old O-ring and install the new O-ring. Lightly coat the O-ring with waterproof grease to prevent damage when replacing the spout.
PROBLEM: Hose or spray leaks.
CURE: Replace the hose and spray with Part No. 63172-07. Hose is removed from underneath the sink by turning the coupling nut (25a) counterclockwise to loosen and remove.
PROBLEM: Hose and spray unit does not work properly.
CURE: Remove the spout and install new diverter valve, (item 22) Part No. 55275-07. Apply grease to O-ring on spout as outlined above.

American Standard Bottom-Mount Sink Faucets
(N-4010-1 less spray)
(N-4012-1 with spray)

The repair procedure for this faucet is the same as for the N-4110-1 and N4112-1, listed earlier. The only basic difference is in the type of escutcheons used. See Fig. l0.

American Standard Bottom-Mount Sink Faucet
(C4040 less spray) (C4045 with spray)

PROBLEM: Faucet drips, will not shut off completely. (Fig. ll.)
CURE: Replace the seat washer, inspect the seat, replace the seat if seat is chipped or pitted.

Remove the handle screw and the handle. It may be necessary to tap the underside of the handle to loosen. Use pliers to grip the top of the escutcheon nipple and turn the nipple counterclockwise to loosen and remove. Lift off the escutcheon. Use an adjustable wrench to turn the stem packing nut counterclockwise to loosen and remove the packing nut. Set the handle back on the stem and unscrew the stem from the faucet body. Inspect the seat, if it is chipped or pitted, replace it with a new seat, Part No. 57285-07. Replace the old seat washer with a new washer. If the stem is stripped, or turns around and around without tightening, replace the stem. Note on the illustration that the stem may be either long or short, and that the stems are only available as stem subassemblies. It would be best to take the worn-out stem to the plumbing shop when ordering a new stem, in order to match the existing equipment.

Also note in the parts list that the hot stem and the cold stem have different part numbers. After the seat has been inspected and replaced, if needed, and new seat washer installed on stem, reassemble the faucet.
PROBLEM: Faucet leaks around the stem.
CURE: Add a strand of graphited packing between present packing and the stem packing nut, or remove old stem packing and replace it with new stem packing, Part No. 57283-07. Remove parts (5-6-7- 8-9) in illustration, as outlined above to repair or replace the stem packing.
PROBLEM: Water leaks around spout.
CURE: Replace the spout O-ring.

Use pliers and grip the spout nut tightly so that the pliers will not slip and mar the chrome nut, and turn the spout nut counterclockwise to loosen. When the nut is loose, remove spout. If the spout has an aerator (3), it should use the No. 2 O-ring, Part No. 55234-07. Remove the worn O-ring and install the new one. Coat the new O-ring lightly with waterproof grease to prevent damage when reassembling tbe spout to the faucet body. Do not overtighten the spout nut.

PROBLEM: Hose and spray do not work properly.
CURE: Replace the diverter valve (21), Part No. 55275-07. Remove the spout as outlined above. Lift out the old diverter valve and replace it with the new part. Anytime the spout is removed for this repair, a new O-ring should be installed as outlined above.

PROBLEM: Hose and spray leaks.
CURE: Replace the worn hose and spray with new hose and spray parts (25 and 26) Parts No. 63172-07.

The hose is removed from underneath the sink by turning the coupling nut 24a counterclockwise to loosen and remove.

Repairing American Standard
Wall-Mounted Sink Faucet
(R 4213)

PROBLEM: Faucet drips — will not shut off completely. (Fig. 12).

CURE: Replace seat washer, inspect seat, replace or resurface seat if it is chipped or pitted.

Remove the handle screw and remove the handle. Use a smooth jaw wrench and turn the cap (3) counterclockwise to loosen and remove. Set the handle back on the stem and turn the stem in the direction of opening the faucet (counterclockwise on the hot stem, clockwise on the cold stem) and unscrew the stem. Inspect the washer seat (10) if the seat is chipped or pitted use the seat wrench to turn the seat counterclockwise to loosen and remove. Replace the old seat, if needed with Part No. 174-14. The worn seat can be resurfaced without removing it, using reseating tool, install new seat washer (8) and reassemble the faucet.

PROBLEM: Water leaks around the faucet stem.

CURE: Add a strip of graphited packing to present packing (4) or install new stem packing.

Remove the handle and cap as outlined above and add a strip of graphited packing above present packing or install new packing Part No. 1311-07.

PROBLEM: Water leaks around the spout.

CURE: Install new spout packing.

Use smooth jaw wrench and turn spout packing nut counterclockwise to loosen. Lift off the soap dish, if used, and lift up and remove the spout. Remove the two worn spout packings, one below the spout and one above, with new packing, Parts No. 473-07. Reassemble the spout placing the lock washer and space washer above the upper packing as shown.

Kohler Rockford Ledge
(Top-Mount) Sink Faucet
(K-7827T with hose and spray)
(K-7825T less hose and spray)

PROBLEM: Faucet leaks — drips. (Fig. 13).

CURE: Replace the seat washer, inspect the renewable seat, (12) if the seat is chipped or pitted, replace the seat.

Remove the screw (1) and the handle (2). Use an adjustable wrench to turn the bonnet (3) counterclockwise to loosen and remove. When the stem (7) is screwed into the plunger (8), the stem and the plunger can be lifted out of the body of the faucet. Lift out the sleeve (11) and inspect the seat. If the seat is chipped or pitted, replace the worn seat with new part (12) Part No. 23004. The seat is pressed into the bottom of the sleeve. To remove the seat, insert a nail set through two opposite holes in the sleeve and tap the square head of the nail set lightly with a small hammer. The worn seat will be forced out of the sleeve. Insert the new seat into the sleeve and

tap it lightly with a small hammer to "set" it. Replace the O-rings, (4 and 5) on the stem with new O-rings, Part No. 34263 and 34264. Lightly coat the O-rings with waterproof grease to prevent damage when reassembling the faucet. Note that the sleeve is grooved to receive the plunger.

PROBLEM: Faucet leaks around the spout.

CURE: Replace the worn O-rings on the spout.

Use an adjustable wrench to turn the coupling nut, (16) counterclockwise to loosen and remove. Remove the spout. Replace the 2 worn O-rings (18) Parts No. 40933. Lightly coat the O-rings with waterproof grease to prevent damage when reassembling the spout.

PROBLEM: Spray Hose does not work properly.

CURE: Replace the auto spray unit, (21) Part No. 39931.

Remove the spout as outlined above. Turn the spout post counterclockwise to loosen and remove the post. Lift out the worn spray unit and install the new unit. Install new O-rings, (18) as outlined above, before reassembly of the spout.

Repairing Kohler Exilla
Single-Lever Faucet
(K - 7840 - K - 7842 Sink Fitting)

PROBLEM: Faucet Drips — Will not shut off completely. (Fig. 14)

CURE: Replace the Valve Unit (Item 4 - Part No. 35774)

Since it is necessary to partially disassemble the faucet to replace the valve unit, it would also be wise to install new O-rings on the valve body (Item (15) Part No. 38618-2 required) and also replace item (12) O-ring (Part No. 32657).

The numbered exploded-type drawing shows the steps for disassembly of this faucet. Remove the cap, handle, cover and skirt assembly. Remove the old valve assembly. Lift the spout off and remove the two O-rings (item 15) from the valve body. Remove the old O-ring (item 12) and install new O-ring in its place and install two new O-rings (item 15) on valve body. Lightly coat the outsides of the new O-rings with a waterproof grease. This will prevent damage when reassembling the faucet. Reassemble the faucet. The hose and spray are removed for replacement if necessary, from underneath the sink. The hose and spray unit are coupled to item (21) by a nut on the hose. Turn this nut counterclockwise to loosen and remove the old hose and spray unit. Replace the old unit with Part No. 38613, hose and spray.

PROBLEM: Hose and spray does not operate properly.

CURE: Disassemble faucet as outlined above. Install new director unit, item (14) Part No. 38636.

Repairing Kohler KYKA Single-Lever Faucet
(K-7830-K-7832 Sink Fitting)

PROBLEM: Faucet drips—Will not shut off completely. (Fig. 15)

Bottom Mount Sink Fittings

Heritage Trim—Lever Handles—Aquaseal

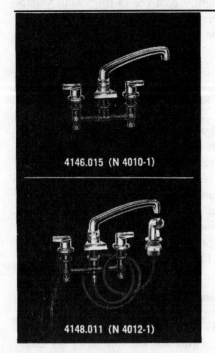

4146.015 (N 4010-1)

4148.011 (N 4012-1)

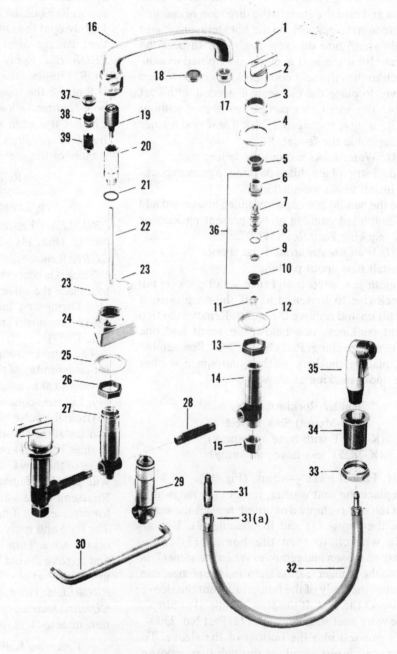

Fig. 10. Bottom

Bottom Mount Sink Fittings
Heritage Trim—Lever Handles—Aquaseal

no.	description	part no.	N 4020-1R N 4020-1L N 4021-1R N 4021-1L N 4000-1 N 4001-1	N 4022-1R N 4022-1L N 4023-1R N 4023-1L N 4002-1 N 4003-1	N 4030-1R N 4030-1L N 4031-1R N 4031-1L N 4010-1 N 4011-1	N 4032-1R N 4032-1L N 4033-1R N 4033-1L N 4012-1 N 4013-1
1	Handle Screw	852-11				
2	Handle - Specify Index	64071-02				
3	Escutcheon Holder	800-32				
4	Escutcheon	679-32				
5	Lock Nut	855-17				
6	Stem Nut - R.H. Threads	72956-07				
	Stem Nut - L.H. Threads	72957-07				
7	Stem w/swivel & Friction Ring (R.H. Threads)	72952-09●				
	Stem w/swivel & Friction Ring (L.H. Threads)	72953-07●				
8	Friction Ring	246-37				
9	Stop Ring	72959-07				
10	Aquaseal Diaphragm (2)	72940-07				
11	Seat	862-14				
12	Washer	1013-29				
13	Lock Nut	25045-07				
14	Body w/seat	8164-07●				
15	Coupling Nut	24220-07				
16	Spout w/end trim & "O" Ring	8099-02●				
	Spout w/aerator & "O" Ring	8098-02				
17	Aerator	56135-02				
18	End Trim	279-12				
19	Diverter	53828-07				
20	Post	964-17				
21	"O" Ring	887-17				
22	Tube	881-17				
23	Washer Retainer	401-27				
24	Flange	25893-02				
25	Washer	1013-29				
26	Lock Nut	27365-07				
27	Tee (Fittings Less Spray)	25584-07●				
	Tee (Fittings With Spray)	25582-07●				
28	Pipe	25140-07●				
29	Tee for Fittings w/3rd Water Valve Less Hose & Spray	25589-07●				
	Tee for Fittings w/3rd Water Valve with Hose & Spray	25590-07●				
30	Bent Pipe for All Fittings w/3rd Water Valve	63181-07				
31	Hose Connector	63418-07				
32	Hose S/A	1178-17				
33	Lock Nut	61300-06				
34	Spray Holder					
35	Sprayhead S/A	63181-07				
36	Aquaseal Trim Less Lock Nut (R.H. Threaded Stem)	72950-17				
	Aquaseal Trim Less Lock Nut (L.H. Threaded Stem)	72951-17				
37	Cap w/washer	899-27				
38	Auto Spray (Upper)	696-17				
39	Auto Spray (Lower)	697-17				
	Sub-assembly					
	Spray & Hose S/A - (Parts 32 & 35)	63172-07				

Note - If part #19 is not used, order parts 37, 38 and 39   ● NOT AVAILABLE

Courtesy American Standard.

mount sink faucet.

Bottom Mount Sink Fittings

Colony Trim

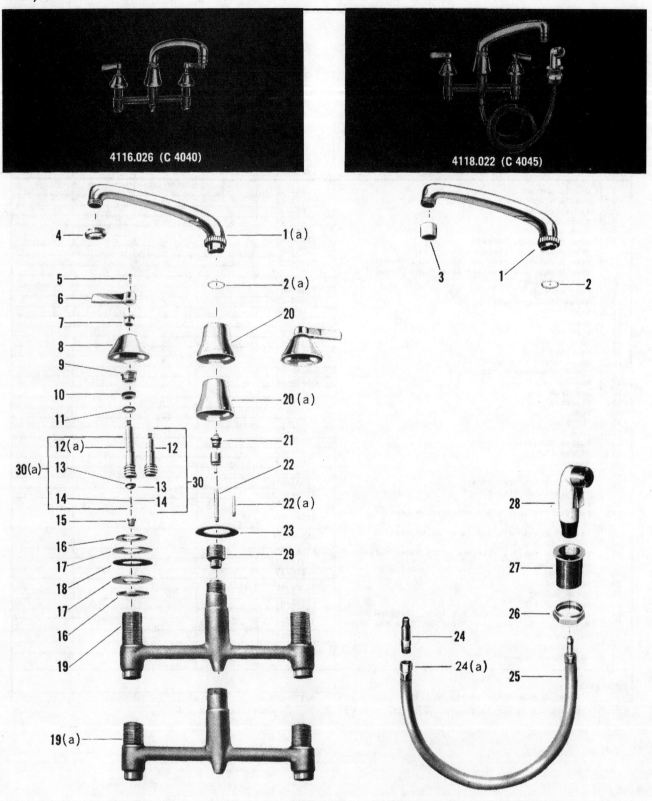

4116.026 (C 4040)

4118.022 (C 4045)

Fig. 11. Bottom mount

Bottom Mount Sink Fittings
Colony Trim

no.	description	part no.	C-4040	C-4041	C-4042	C-4043	C-4045	C-4046	C-4047	C-4048
1	Spout w/spout Nut	57165-02								
1A	Spout w/spout Nut	57160-02								
2	"O" Ring	56539-07								
2A	"O" Ring	55234-07								
3	Aerator	56135-02								
4	Stream Regulator	57364-04 •								
5	Handle Screw	56581-01								
6	Handle - Specify Index	57302-02								
7	Escutcheon Nipple	57309-04								
8	Escutcheon	55352-02								
9	Stem Packing Nut	57291-07								
10	Stem Packing	57283-07								
11	Stem Washer	57284-07								
12	Stem Hot	55815-04 •								
	Stem Cold	55816-04 •								
12A	Stem Hot	55822-04 •								
	Stem Cold	55823-04 •								
13	Seat Washer	858-17								
14	Seat Washer Screw	56480-07								
15	Seat (Noryl)	57285-07								
16	Jamb Nut	57296-09								
17	Jamb Nut Washer	55346-09 •								
18	Jamb Nut Gasket	55347-07 •								
19	Body	56355-07 •								
	Body	56354-07 •								
19A	Body	56359-07 •								
	Body	56358-07 •								
20	Spout Escutcheon Marked Standard	63221-02 •								
20A	Spout Escutcheon Marked MF Standard	63220-02 •								
21	Diverter Valve	55275-02								
22	Diverter Tube	63217-07								
22A	Diverter Tube	63231-07								
23	Center Washer	63238-07 •								
24	Hose Connector	63418-07								
25	Hose S/A	1178-17								
26	Lock Nut	61300-06								
27	Spray Holder									
28	Sprayhead S/A	63181-07								
29	Adapter	63216-07								
30	Stem Hot S/A w/washer & Screw	55838-04								
	Stem Cold S/A w/washer & Screw	55839-04								
30A	Stem Hot S/A w/washer & Screw	55788-04								
	Stem Cold S/A w/washer & Screw	55789-04								
	Sub-assembly									
	Spray & Hose S/A - (Parts 25 & 28)	63172-07								

• NOT AVAILABLE

sink faucet.

Courtesy American Standard.

Double Faucets
Renewable Seats

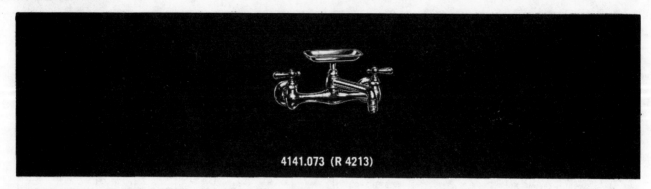

4141.073 (R 4213)

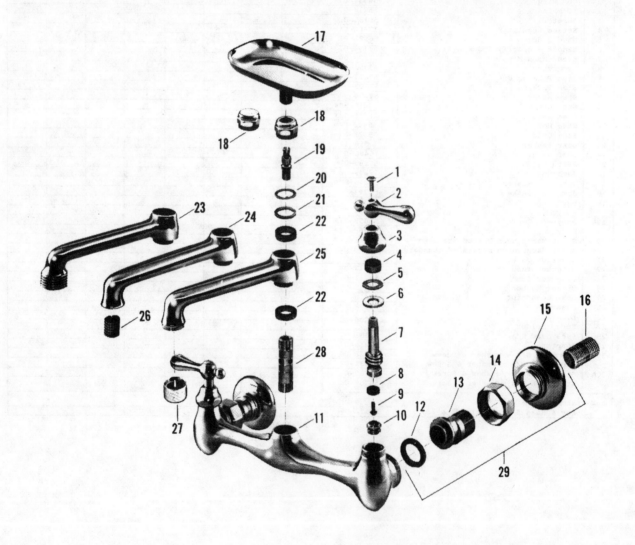

Fig. 12. Double

Double Faucets
Renewable Seats

no.	description	part no.	R 4200 R 4202 R 4220 R 4222 R 4240 R 4242 R 4250 R 4252	R 4210 R 4212 R 4230 R 4232	R 4201 R 4203 R 4221 R 4223 R 4241 R 4243 R 4251 R 4253	R 4211 R 4213 R 4231 R 4233
1	Handle Screw	173-01				
2	Handle - Specify Index	1185-12				
3	Cap	21436-02				
4	Stem Packing	1311-07				
5	Brass Washer	21625-07				
6	Cap Washer	125-07				
7	Stem w/washer & Screw - Cold - L.H. Thread	63375-04				
	Stem w/washer & Screw - Hot - R.H. Thread	63381-04				
8	Seat Washer	441-07				
9	Seat Washer Screw	159-17				
10	Seat	174-14				
11	Body w/seats	8190-02•				
12	Coupling Washer	456-07				
13	Female Shank	20427-02•				
14	Coupling Nut	1233-22				
15	Adjustable Flange	609-22				
16	Close Nipple	39525-07				
17	Soap Dish	315-22				
18	Spout Packing Nut	25014-02				
	Spout Packing Nut	25013-02				
19	Soap Dish Post	39786-05				
20	Lock Washer	21365-05				
21	Space Washer	305-17				
22	Spout Packing	473-07				
23	Spout (Hose End) w/regulator & Packings	8191-02				
	Spout (Hose End 8" Long) w/regulator & Packings	8192-02•				
24	Spout (Plain End) w/regulator & Packings	8193-02				
	Spout (Plain End 8" Long) w/regulator & Packings	8194-02•				
25	Spout w/aerator & Packings	8155-02				
	Spout (8" Long) w/aerator & Packings	8156-02				
26	Stream Regulator (Plain End Spout)	33952-07				
	Stream Regulator (Hose End Spout)	33953-07				
27	Aerator	56135-02				
28	Connecting Post	1162-17				
29	Male or Female Shank S/A - (Parts 12-13-14-15)	14319-02				

• NOT AVAILABLE

Note - If Male Shank needed, add Close Nipple, part 16 - #39525-07 to S/A part 29 - #14319-02

Courtesy American Standard.

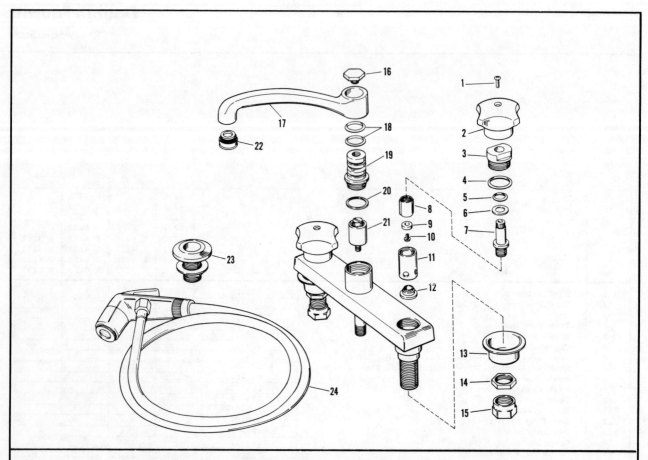

K-7827-T ROCKFORD LEDGE SINK FITTING PARTS
TRITON II SERIES
(K-7825-T Same as K-7827-T Except Less Hose and Spray)

ITEM NO.	PART NO.	DESCRIPTION	ITEM NO.	PART NO.	DESCRIPTION
1	53448	Screw	14	32010	Lock Nut
2	41980	Handle (Specify Index)	15	32751	Coupling Nut
3	34260	Bonnet	16	34312	Spout Cap Nut
4	34264	"O" Ring	17	38687	Spout w/Aerator
5	34263	"O" Ring	18	40933	"O" Ring
6	34265	Washer	19	38686	Spout Post
7	34320	Stem	20	40950	Gasket
8	22947	Plunger w/Seat Washer & Screw	21	39931	Auto Spray Unit
9	39541	Seat Washer	22	41056	Aerator
10	34848	Screw	23	34329	Hose Guide S.A.
11	34842	Sleeve	24	38613	Hose & Spray
12	23004	Renewable Seat		34570	Spray Only
13	34274	Spacer		38614	Hose Only

Items 5-12 Inclusive, Specify 22932 Valvet for Hot and Cold Valve.

Note: For fitting with soap dish, Specify No. 37766 Post and Nut Assembly in place of Item 16.   Soap dish for above, specify 37668.

Courtesy Kohler Co.

Fig. 13. Top mount sink faucet.

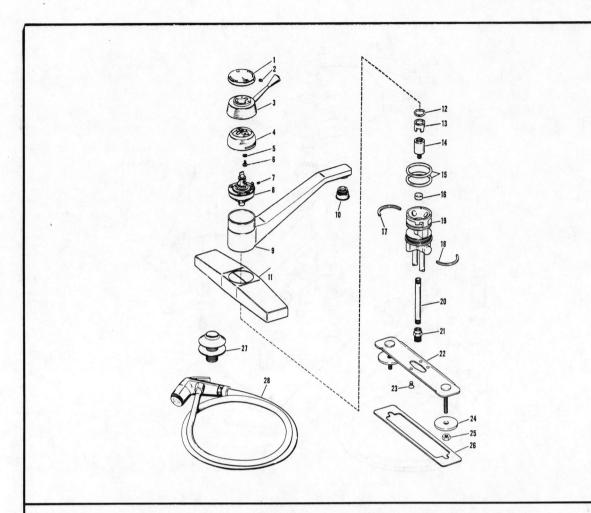

EXILLA ONE LEVER SINK SUPPLY FITTING PARTS
K-7840 and K-7842 Exilla Sink Fitting

ITEM NO.	PART NO.	DESCRIPTION		ITEM NO.	PART NO.	DESCRIPTION
1	38638	Cap		16	38701	Screen (2 Required)
2	35739	Screw		17	38620	Clamp
3	38622	Handle & Cover Assembly } #38623 handle, cover & skirt assembly		18		
				19		Valve Body (Not Available)
4	35762	Skirt		20	20667	Connection for Hose & Spray
5	35789	Washer		21	20666	Adaptor
6	35744	Screw		22	38628	Base Plate
7	35804	Set Screw		23	38635	Screw (3 Required)
8	35774	Valve Unit (RC-2)		24	38619	Washer
9	38750	Spout		25	51533	Nut
10	41056	Aerator		26	38630	Gasket
11	38616	Shroud		27	34329	Hose Guide Assembly
12	32657	"O" Ring		28	38613	Hose & Spray
13	38633	Spacer			38614	Hose Only
14	38636	Diverter Unit			34570	Spray Only
15	38618	"O" Ring (2 Required)				

Courtesy Kohler Co.

Fig. 14. Exilla single lever faucet.

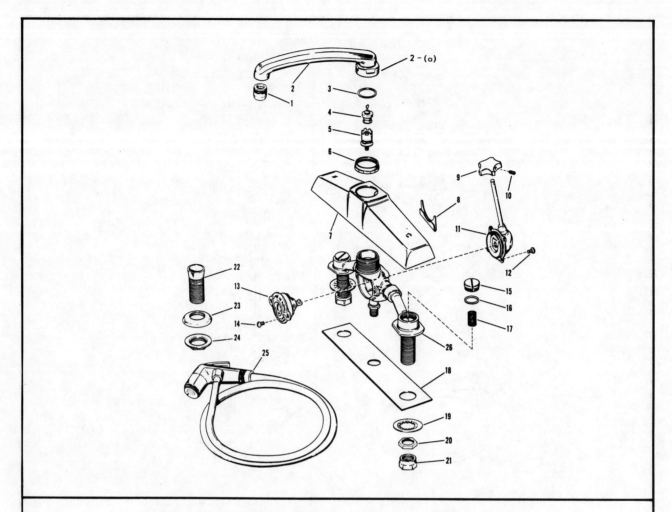

HYKA ONE LEVER SINK SUPPLY FITTING PARTS
K-7830 and K-7832 Hyka Sink Fitting

ITEM NO.	PART NO.	DESCRIPTION	ITEM NO.	PART NO.	DESCRIPTION
1	41056	Aerator	16	34300	"O" Ring
2	41431	Spout w/41434 Coupling Nut and 41433 Spout Post	17		Screen (Not Available)
			18		Gasket (Not Available)
3	29464	"O" Ring	19	33420	Washer
4			20	32010	Lock Nut
5	39867	Auto Spray Unit	21	32751	Coupling Nut
6	41435	Lock Ring	22		
7	41432	Cover	23	34329	Hose Guide Assembly
8	41443	Clip	24		
9	35118	Handle	25	38613	Hose and Spray
10	35176	Screw		38614	Hose only
11	41447	Stem & Cap Assembly		34570	Spray only
12	35744	Screw (4 Required)	26	41449	Valve Body w/Shanks for K-7832
13	39925	Cartridge (RC-1)		41441	Valve Body w/Shanks & 40630 Plug for K-7830
14	49931	Screw (2 Required)			
15	41437	Cap			

Courtesy Kohler Co.

Fig. 15. Sink single-lever faucet.

CURE: Replace Cartridge (Item 13 - Part No. 39925).
PROBLEM: Water leaks around spout connection to faucet body.
CURE: Replace O-ring (Item 3 - Part No. 29464).

Since it is necessary to remove the spout and cover in order to replace the cartridge, the old O-ring (item 3) should be replaced.

Use a smooth jaw wrench and turn nut 2 (a) counterclockwise to loosen. Lift off the spout and replace the old O-ring (item 3) with a new O-ring, Part No. 29464. Turn the lock ring (item 6) counterclockwise to loosen and remove. Lift off the cover (item 7), loosen and remove the four screws (item 12); the stem and cap assembly can be set aside for reassembly later. Remove the two screws (item 14) and remove the old cartridge (item 13). Install the new cartridge, RC - 1, (item 13 - Part No. 39925) replace the screws (item 14) reassemble the stem and cap assemble and replace the cover. Coat the new O-ring, (item 3) lightly with waterproof grease, and reassemble the spout. Do not over-tighten the nut on the spout 2(a).

The hose and spray assembly (item 25 - Part No. 38613) are removed for replacement if necessary from underneath the sink. The hose is connected to the valve body by a nut which should be turned counterclockwise to loosen and remove.

Kohler Clearwater Wall-Mounted
Sink Faucet
(K-7856)

PROBLEM: Faucet leaks — drips. (Fig. 16.)
CURE: Replace the worn seat washers (item 9). Inspect the renewable seats (item 12) replace the renewable seats if they are chipped or pitted.

Remove the handle screw (item 1) and the handle. Use an adjustable wrench to turn the bonnet (item 3) counterclockwise to loosen and remove. When the stem (item 7) is screwed into the plunger (item 8) the plunger can be lifted out of the faucet body. Lift the sleeve (item 11) out and inspect the washer seat. The washer seat should be replaced with a new seat, Part No. 23004. The washer seat should be replaced if it is chipped or pitted. The seat is pressed into the bottom of the sleeve. To remove the worn seat, insert a nail set through two opposite holes in the bottom of the sleeve and tap the square head of the nailset lightly with a smaller hammer. The worn seat will be forced out of the sleeve. Insert the new seat into the sleeve and tap it lightly to "set" it.

Replace the worn seat washer (item 9); the two O-rings (items 4 and 5) should also be replaced when installing new seat washers. When reassembling the faucet, note that the sleeve is grooved to receive the plunger. The stems and plungers on the hot-water side of the faucet are different from those on the cold-water side, therefore, it is best to repair and reassemble one side of the faucet

before repairing the other side, to avoid mixing the parts. Lightly coat the new O-rings with waterproof grease before reassembly.
PROBLEM: Water leaks around the spout.
CURE: Replace the worn O-rings (items 17 and 18).

Remove the soap dish and turn nut (item 15) counterclockwise to loosen and remove the nut and the post. Remove spout. Remove the two worn O-rings (items 17 and 18) and install two new O-rings, Parts No. 40933. Lightly coat the new O-rings with waterproof grease before reassembly.

Repairing Kohler Clearwater
Wall-Mount Sink Faucet
(K - 8655-A- K-8656-A- K-8657- K-8659A)

PROBLEM: Faucet drips — will not shut off completely. (Fig. 17.)
CURE: Replace seat washers (item 8) install new renewable seats (item 10) or resurface present seats, if seats are chipped or pitted.

Remove screw (item 1) and handle, (item 2). Use a smooth jaw wrench to turn the bonnet nut (item 3) counterclockwise to loosen and remove. Set the handle back on the stem and turn the stem in the direction of opening the faucet until the stem is loose; lift the stem out. Inspect the seat, if the seat is pitted or chipped, use a seat wrench to turn the seat counterclockwise to loosen and remove. The seat may be resurfaced without removing if a new seat is not available.

Remove the old seat washer (item 8) and install a new seat washer, Part No. 39541. Remove the old O-ring (item 4) from the stem and replace with a new ring, Part No. 34117. Lightly coat the new O-ring with waterproof grease before reassembly. Insert the stem into the faucet body and reassemble bonnet and handle. The above directions are for repair of the hot-water side of the faucet; the cold-water side is repaired in the same manner.
PROBLEM: Water leaks around spout connection to faucet body.
CURE: Replace the O-ring (item 18).

If the spout leaks at the connection to the faucet body, use a smooth jaw wrench to turn the nut (17-a) counterclockwise to loosen the nut. When the nut is free, the spout can be lifted out. Remove the old O-ring (item 18) and replace it with a new ring, Part No. 32725. Lightly coat the new O-ring with waterproof grease before reassembling the spout. Do not over-tighten the spout nut (item 17-a).

American Standard Centerset Lavatory Faucet
N-2001 with pop-up waste
N-2041 with chain and plug waste

PROBLEM: Faucet leaks — drips. (Fig. 18).
CURE: Replace aquaseal diaphragms, if the washer

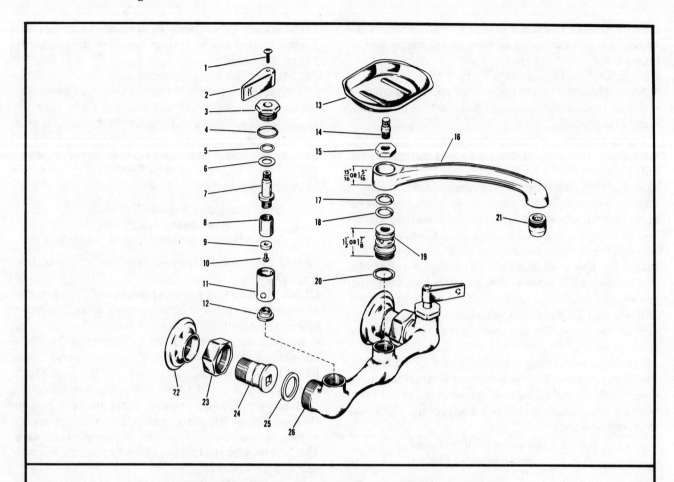

K-7856 CLEARWATER SINK FITTING PARTS

ITEM NO.	PART NO.	DESCRIPTION
1	33357	Screw
2	34064	Handle (Specify Index)
3	34365	Bonnet
4	34264	"O" Ring
5	34263	"O" Ring
6	34265	Washer
7	34319	Cold Stem
	34320	Hot Stem
8	22947	Cold Plunger w/seat washer & screw
	22948	Hot Plunger w/seat washer & screw
9	39541	Seat Washer
10	34848	Screw
11	34842	Sleeve
12	23004	Renewable Seat
13	37668	Soap Dish

ITEM NO.	PART NO.	DESCRIPTION
14	34532	Post 37766 Post & Nut
15	34533	Nut Sub-Assembly
16	37750	Spout with Aerator — 15⁄16"
	38687	Spout with Aerator — 15⁄16"
17	40933	"O" Ring
18	40933	"O" Ring
19	34430	Spout Post — 1½"
	38685	Spout Post — 1⅞"
20	32657	"O" Ring
21	41056	Aerator
22	39708	Flange
23	40748	Nut
24	40599	Shank Inside Thread
	40598	Shank Outside Thread
25	40718	Washer
26	34512	Body

Note: Valve parts for K-7857 and K-7856 same as above.
Items 5 through 12, Specify 22917 Valvet for cold valve, 22932 Valvet for hot valve.

Courtesy Kohler Co.

Fig. 16. Wall-mounted sink faucet.

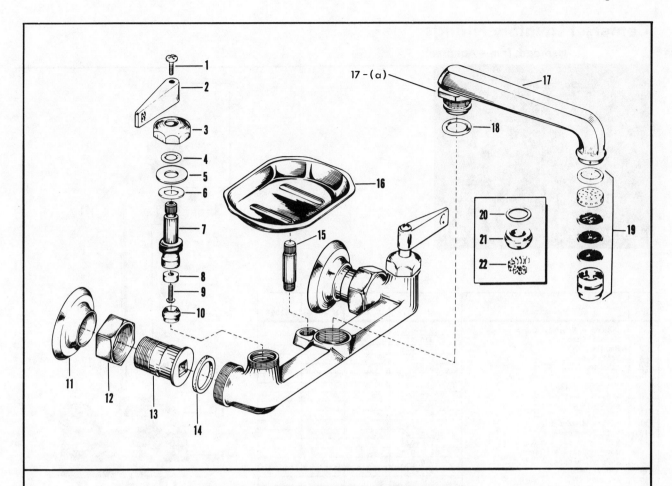

CLEARWATER SINK FITTING PARTS
For K-8655-A, K-8656-A, K-8657, K-8659-A

ITEM NO.	PART NO.	DESCRIPTION
1	33357	Screw
2	34064	Handle (Specify Index)
3	34074	Bonnet
4	34117	"O" Ring
5	39881	Gasket
6	39540	Washer
7	31871	Hot Stem with Washer and Screw
	31872	Cold Stem with Washer and Screw
8	39541	Seat Washer
9	31490	Screw
10	40602	Renewable Seat
11	39708	Flange
12	40748	Coupling Nut
13	40599	Shank — Female Connection
	40598	Shank — Male Connection
14	40718	Gasket

ITEM NO.	PART NO.	DESCRIPTION
15	39888	Post
16	37668	Soap Dish
17	37700	5" Spout with Streambreaker for K-8655 or K-8656
	37702	9" Spout with Streambreaker for K-8659
	37701	5" Spout with Hose Connection for K-8657
	37692	5" Spout with Aerator for K-8655-A or K-8656-A
	37693	9" Spout with Aerator for K-8659-A
18	32725	"O" Ring
19	37725	Aerator

Courtesy Kohler Co.

Fig. 17. Wall-mounted sink faucet.

Centerset Lavatory Fittings

Heritage Trim—Aquaseal

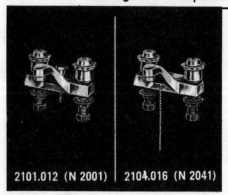

2101.012 (N 2001)   2104.016 (N 2041)

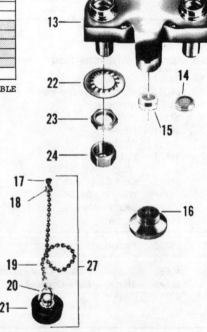

no.	description	part no.	N 2000 N 2001	N 2040 N 2041	N 2098
1	Button - Specify Index	1234-22			
2	Handle Screw	915-17			
3	Handle	5358-02●			
4	Lock Nut	855-17			
5	Stem Nut	72956-07			
6	Stem w/swivel & Friction Ring	72952-07●			
7	Friction Ring	246-37			
8	Stop Ring	72959-07			
9	Aquaseal Diaphragm	72960-07			
10	Seat	862-14			
11	Lift Rod w/knob	4358-02			
12	Lift Rod Guide	25845-02●			
13	Body w/seats, Shanks & Rod Guide	8176-02●			
	Body w/seats & Shanks	8177-02●			
	Body w/seats & Shanks	8078-02●			
14	End Trim	279-12			
15	Aerator	56135-02			
16	Spray Head S/A	5450-02●			
17	Sleeve for Chain	275-04			
18	Chainstay Nut	28182-04			
19	Chain	1112-02			
20	Link	1128-22			
21	Rubber Stopper	1012-12			
22	Friction Washer	54538-09			
23	Lock Nut	300-27			
24	Coupling Nut	24220-07			
25	Assembled Handle Specify Index	6458-02			
26	Aquaseal Trim Less Lock Nut	72950-17			
27	Chain & Stopper S/A	8174-02			

● NOT AVAILABLE

Note - For Pop-up Drain parts refer to 2420.016 (N 2542)
       For Chain & Stopper Drain parts refer to 2440.014 (N 2522)

Courtesy American Standard.

Fig. 18. Centerset lavatory faucet.

seats are chipped or pitted, replace or resurface the washer seats.

Insert a small screwdriver blade under the edge of the index cap (1) and lift. Remove the handle screw (2) and the handle. Use an adjustable wrench to turn the lock nut counterclockwise to loosen and remove the lock nut. Lift the stem (6) and the worn aquaseal diaphragm out. Inspect the seat (10) if the seat is chipped or pitted, replace it with a new seat, Part No. 862-14. Remove the damaged seat with the seat wrench.

The damaged seat can be resurfaced in place, using the reseating tool.

Install new aquaseal diaphragms, (Part No. 72940-07), on the ends of the stems and reassemble the faucet.

American Standard Single-Lever Lavatory Faucet
N - 2055 with pop-up waste.

PROBLEM: Faucet leaks — drips. (Fig. 19.)
CURE: Replace parts 7, 8, 9, 10, using repair kit No. 12417-07.

No. 1, bolt, is not used on this faucet. Turn the control knob, (2), counterclockwise to unscrew it from the waste lever. Turn the waste control bushing (3), counterclockwise to loosen and remove it. Remove the escutcheon (4). Use a screwdriver to turn the valve plug (6), counterclockwise to remove the plug. Remove worn parts, (7, 8, 9, 10). The repair kit has two each of parts (7, 8, 9, 10). Install one new set of these parts in each side of the faucet and reassemble the faucet.

American Standard (8") Spread Lavatory Faucet

PROBLEM: Faucet leaks — drips.
CURE: Replace the aquaseal diaphragms; if the washer seats are chipped or pitted, replace or resurface the washer seats.

The repair procedure is the same for this faucet as for the N - 2001 or N - 2041 faucets. See instructions for these faucets. The part numbers for the aquaseal diaphragms and the seats are also the same as for the N - 2002 and N - 2041 faucets.

Kohler Constellation and Galaxy Centerset Lavatory Faucets
(K - 7400 K - 6960)

PROBLEM: Faucet drips. (Fig. 20.)
CURE: Replace the worn seat washers, item 11; inspect and replace the renewable seats if they are chipped or pitted.

Remove the handle screw (item 1) and the handle. Use an adjustable wrench to turn the bonnet (item 3) counterclockwise to loosen and remove. When the stem (item 7) is screwed into the plunger (item 8), the plunger can be lifted out of the faucet body. Lift the sleeve (item 11) out and inspect the washer seat. The washer seat should be replaced if it is chipped or pitted. The seat is pressed into the bottom of the sleeve. To remove the worn seat, insert

a nail set through two opposite holes in the bottom of the sleeve and tap the square head of the nail set lightly with a small hammer. The worn seat will be forced out of the sleeve. Insert the new seat (item 12) Part No. 23004, into the sleeve and tap it lightly to "set" it. Replace the worn seat washers (item 9). The two O-rings (items 4 and 5) Parts Nos. 34263 and 34264, should also be replaced when installing new seat washers. When reassembling the faucet, note that the sleeve is grooved to receive the plunger. Lightly coat the new O-rings with waterproof grease to prevent damage when reassembling the faucet.

Kohler Triton Bancroft Lavatory Faucets
(K - 7436 K- 7437 K - 7439)

PROBLEM: Faucet drips. (Fig. 21.)
CURE: Replace the worn seat washers, item 11; inspect the renewable seats if they are chipped or pitted.

Remove the screw (1) and the handle. Use an adjustable wrench to turn the bonnet (item 5) counterclockwise to loosen and remove. When the stem (item 9) is screwed into the plunger (item 10), the plunger can be lifted out of the faucet body. Lift the sleeve (item 13) out and inspect the washer seat. The washer seat should be replaced with a new seat, Part No. 23004, if it is chipped or pitted. The seat is pressed into the bottom of the sleeve. To remove the worn seat, insert a nail set through two opposite holes in the bottom of the sleeve and tap the square head of the nail set lightly with a small hammer. The worn seat will be forced out of the sleeve. Insert the new seat into the sleeve and tap it lightly to "set" it. Replace the worn seat washer (item 11). The two O-rings (items 6 and 7) Parts No. 34263 and 34300, should also be replaced when installing new seat washers. When reassembling the faucet note that the sleeve is grooved to receive the plunger. The stems and plungers on the hot-water side of the faucet are different from those on the cold side, therefore, it is best to repair and reassemble one side of the faucet before repairing the other side to avoid mixing the parts. Lightly coat the new O-rings with waterproof grease before reassembly.

Kohler Triton Shelf Back Lavatory Faucets
K - 8040

PROBLEM: Faucet drips. (Fig. 22.)
CURE: Replace the seat washers, inspect the renewable seats, if the seats are chipped or pitted, replace the seats.

Remove the screw (1) and the handle. Use an adjustable wrench to turn the bonnet (3) counterclockwise to loosen and remove. When the stem (7) is screwed into the plunger (8) the stem and the plunger can be pulled out of the faucet body. Pull the sleeve (11) out and inspect the renewable seat. If the seat is chipped or pitted, replace the seat with Part No. 23004. The seat is pressed into the bottom of the sleeve. To remove the worn seat, insert a nail set through two holes in the bottom of the sleeve and tap the square head of the nail set lightly with a small

Single Control Lavatory Fitting

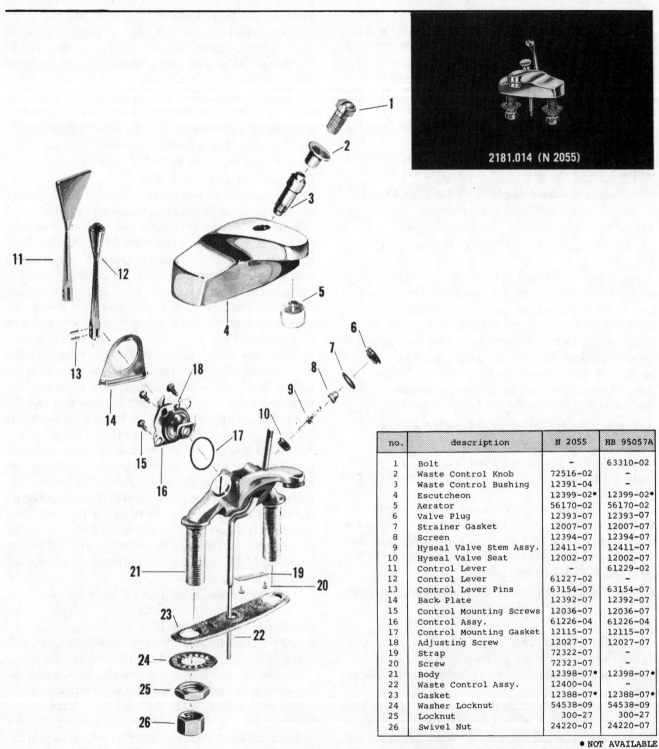

2181.014 (N 2055)

no.	description	N 2055	HB 95057A
1	Bolt	–	63310-02
2	Waste Control Knob	72516-02	–
3	Waste Control Bushing	12391-04	–
4	Escutcheon	12399-02•	12399-02•
5	Aerator	56170-02	56170-02
6	Valve Plug	12393-07	12393-07
7	Strainer Gasket	12007-07	12007-07
8	Screen	12394-07	12394-07
9	Hyseal Valve Stem Assy.	12411-07	12411-07
10	Hyseal Valve Seat	12002-07	12002-07
11	Control Lever	–	61229-02
12	Control Lever	61227-02	–
13	Control Lever Pins	63154-07	63154-07
14	Back Plate	12392-07	12392-07
15	Control Mounting Screws	12036-07	12036-07
16	Control Assy.	61226-04	61226-04
17	Control Mounting Gasket	12115-07	12115-07
18	Adjusting Screw	12027-07	12027-07
19	Strap	72322-07	–
20	Screw	72323-07	–
21	Body	12398-07•	12398-07•
22	Waste Control Assy.	12400-04	–
23	Gasket	12388-07•	12388-07•
24	Washer Locknut	54538-09	54538-09
25	Locknut	300-27	300-27
26	Swivel Nut	24220-07	24220-07

• NOT AVAILABLE

Note – Repair Kit consisting of ((2) each
       parts 7, 8, 9, 10, – #12417-07

Courtesy American Standard.

Fig. 19. Single-lever lavatory faucet.

CONSTELLATION AND GALAXY CENTRA LAVATORY FITTINGS PARTS
K-7400 & K-6960 FITTINGS

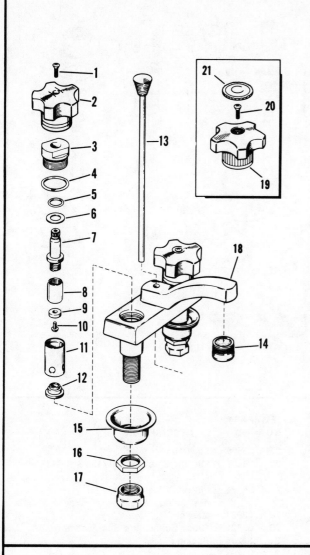

ITEM NO.	PART NO.	DESCRIPTION
1	22859	Screw
2	34255	Handle (Specify Index)
3	34260	Bonnet
4	34264	"O" Ring
5	34263	"O" Ring
6	34265	Washer
7	34320	Stem
8	22947	Plunger w/Seat Washer & Screw
9	39541	Seat Washer
10	34848	Screw
11	34842	Sleeve
12	23004	Renewable Seat
13	51196	Lift Rod
14	41056	Aerator
15	34274	Spacer
16	32010	Lock Nut
17	32751	Coupling Nut
18	34281	Body & 34285 Spout
19	29658	Galaxy Handle
20	51677	Screw
21	34249	Index Button (Specify Index)

Items 5-12 Inclusive, Specify 22932 Valvet Unit.

Note: Internal valve parts as shown above are also used in following Lavatory fittings:

K-7405   K-7406   K-7409

Courtesy Kohler Co.

Fig. 20. Constellation and galaxy lavatory faucet.

TRITON BANCROFT LAVATORY FITTING PARTS WITH POP-UP DRAIN
K-7436, K-7437 & K-7439 Bancroft

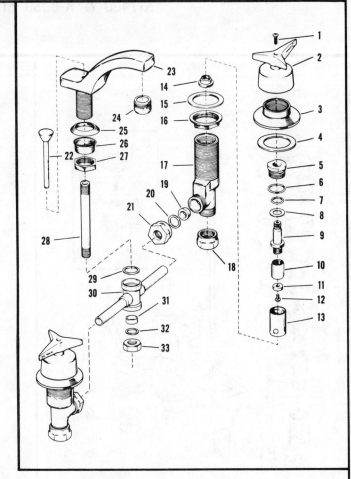

ITEM NO.	PART NO.	DESCRIPTION
1	41110	Screw (Round Head)
	53448	Screw (Countersunk)
2	41107	Triton Handle (Specify Index)
3	34304	Escutcheon
4	35715	Washer
5	20475	Bonnet
6	34300	"O" Ring
7	34263	"O" Ring
8	34265	Washer
9	34320	Hot Stem
	34319	Cold Stem
10	22947	Hot Plunger w/Seat Washer & Screw
	22948	Cold Plunger w/Seat Washer & Screw
11	39541	Seat Washer
12	34848	Screw
13	34842	Sleeve
14	23004	Renewable Seat
15	32060	Washer
16	33907	Lock Nut
17	34303	Valve Body for K-7436 & K-7439 — 4¼" Overall
	20480	Valve Body for K-7437 — 5" Overall
18	32751	Coupling Nut
19	20564	Slip Joint Washer
20	20565	Friction Ring
21	32751	Coupling Nut
22	51196	Lift Rod
23	35800	Spout w/Lift Rod Sleeve for K-7436 Spout Dimension 3⅝" c-c, Spout Shank 1¾"
	35795	Spout w/Lift Rod Sleeve for K-7437 Spout Dimension 4⅝" c-c, Spout Shank 2½"
	35798	Spout w/Lift Rod Sleeve for K-7439 Spout Dimension 4⅝" c-c, Spout Shank 1¾"
24	41007	Aerator
25	33504	Washer
26	34296	Spacer

ITEM NO.	PART NO.	DESCRIPTION
27	33505	Lock Nut
28	34307	Lift Rod Sleeve for K-7436 & K-7439 — 4" Long
	34971	Lift Rod Sleeve for K-7437 — 4¾" Long
29	33031	Gasket
30	35821	Cross Connection Sub-Assembly
31	34294	Packing
32	34295	Friction Ring
33	34309	Lock Nut

Items 5-17, For K-7436 and K-7439, Specify 34342 Hot Valve, 34343 Cold Valve.

Items 5-17, For K-7437, Specify 20481 Hot Valve, 20684 Cold Valve.

Items 7-14 Inclusive, Specify 22932 Valvet for Hot Valve, 22917 Valvet for Cold Valve.

Note: Internal valve parts as shown above are also used in following Lavatory fittings:

| K-7444 | K-7445 | K-7446 | K-7447 |

Courtesy Kohler Co.

Fig. 21. Triton bancraft lavatory faucet.

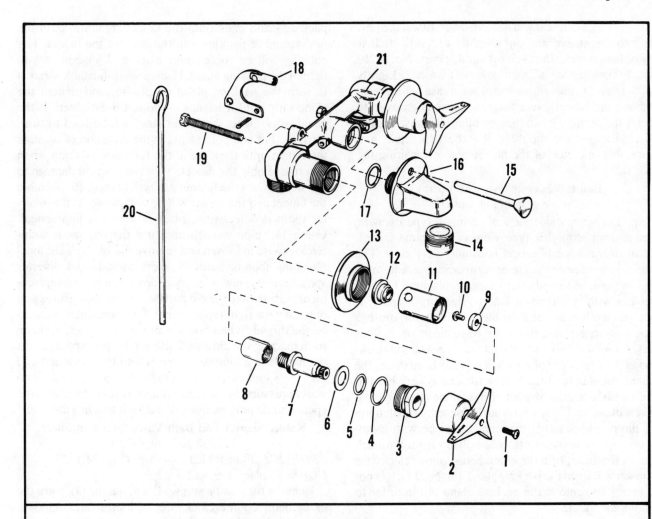

TRITON SHELF LAVATORY FITTING PARTS
For K-8040 Fitting
For Gramercy, Juneau, Strand, Rockport, Hampton, Taunton, Marston Lavatories

ITEM NO.	PART NO.	DESCRIPTION	ITEM NO.	PART NO.	DESCRIPTION
1	41110	Screw (Round Head)	10	34848	Screw
	53448	Screw (Countersunk)	11	34842	Sleeve
2	41105	Triton Handle (Specify Index)	12	23004	Renewable Seat
3	20475	Bonnet	13	34301	Flange
4	34300	"O" Ring	14	41007	Aerator
5	34263	"O" Ring	15	34349	Push Rod
6	34265	Washer	16	34435	Spout w/Aerator & "O" Ring
7	34320	Cold Stem	17	40933	"O" Ring
	34319	Hot Stem	18	34898	Trip Lever with #33799 Cotter Key
8	22947	Cold Plunger w/Seat Washer & Screw	19	35227	Screw
	22948	Hot Plunger w/Seat Washer & Screw	20	51198	Lift Rod
9	39541	Seat Washer	21	34298	Valve Body Only

Items 5-12 Inclusive Specify 22932 Valvet for Cold Valve, 22917 Valvet for Hot Valve.

Courtesy Kohler Co.

Fig. 22. Triton shelf back lavatory faucet.

hammer. The worn seat will be forced out. Insert the new seat into the sleeve and tap it lightly to "set" it. It is advisable to replace the O-rings (4 and 5) Parts No. 34263 and 34300, when installing a new seat washer. Lightly coat the new O-rings with waterproof grease before reassembling the faucet. Note that the sleeve is grooved to receive the plunger when reassembling. Because the hot and cold stems are different, it is best to repair and reassemble one side of the faucet before repairing the other side.

Bathtub Faucets (over-rim fillers)
(bearing no brand name)

Fig. 23 shows a side view of common type over-rim filler and an exploded type view of the same faucet. Repair of this type of faucet is usually very easy. The seats rarely need replacing or resurfacing; usually only the bibb washer needs occasional replacing. The big problem with this type of faucet is getting it apart. It becomes even more of a problem if the bathroom has been remodeled and the walls have been tiled. This causes the faucet to be recessed into the wall the thickness of the tile. Very often when this has been done, the bonnet nut is behind the wall or tile and is very hard to reach with a regular wrench. First, remove the handle and escutcheon. If there is no nut holding the escutcheon on, the nipple is either made in one piece with escutcheon, or it screws into the escutcheon from the back side. In this case, turn the escutcheon counterclockwise to unscrew it from the packing gland nut. NOTE: It is not necessary to remove the packing gland nut in order to remove the stem.

If the bonnet nut is accessible from the front of the faucet, use an adjustable wrench and turn the nut counterclockwise to unscrew it. If the bonnet nut is behind the wall, try to reach it through the access door (if there is one) in the wall behind the tub. If this is impossible, carefully enlarge the hole in the wall around the stem and under the escutcheon and carefully work the basin wrench, shown in Fig. 18, Chapter 1, through the hole until you can lock the jaws of the basin wrench around the bonnet nut.

Turn the wrench counterclockwise to remove the nut. Turn the stem in the same direction as opening the faucet to remove the stem and install a new bibb washer. Inspect the seat; in the event the seat needs replacing or resurfacing, proceed as outlined in chapter on **Reseating Faucets.**

It is very unlikely that the faucet stems will require any new packing. If there is a water leak around the stem, tighten the packing gland by turning the gland ¼ turn clockwise. If you wish to add more packing, turn the packing gland counterclockwise to loosen and remove the gland. Wind a strand of graphited packing approximately 1 ½" long around the stem of the faucet, slide the packing gland back onto the faucet stem and push the new strand of packing into the body of the faucet. The packing will be forced into place and shaped as you tighten the packing gland. Use a small adjustable wrench to turn the packing gland clockwise, and tighten the gland only until it is snug — do not overtighten. If the gland is too tight, the faucet stem will be hard to turn. Turn the hot and cold stems in the direction of opening the faucet until they are in the full open position when you reassemble the faucet. This will prevent damage to the stems when the bonnet nut is tightened. Reassemble the faucet and the repair will be completed. If the diverter spout is defective, replace it, repair is impractical. Use a 14" pipe wrench and turn the old spout counterclockwise to loosen and remove the spout. Your local plumbing shop or hardware store should stock diverter spouts and it is always a good idea when buying replacement parts to take the old parts with you; this insures you of getting the right replacement. Use pipe thread cement on the threads when you install the new spout. To keep from marring the chrome finish on the new spout, insert the handle of a long screwdriver into the spout end and turn the spout clockwise to tighten it. Since there is no water pressure on the water running out of the spout, the spout should only be tightened until it touches the wall.

Kohler Shower and Bath Valve (over-rim filler)
K-7032 K-7100 K-7210

PROBLEM: Faucets leak — drip. (Fig. 24.)
CURE: Replace the seat washer.

Remove the handle screw (1) and handle (2). Turn the escutcheon counterclockwise to loosen and remove. Use an adjustable wrench to turn the bonnet nut (7) counterclockwise to loosen and remove. Place the handle back on the stem and turn the stem in the direction of opening the faucet. The stem will come out of the faucet body. Replace the seat washer (12) with a new washer. If the seat (Remov Unit) is chipped or pitted, replace it with new part (item 14) Part No. 32462.

If the faucet leaks around the stem it is usually only necessary to tighten the packing nut (5), turning it ½ to 1 turn clockwise to stop the leak. If the packing is badly worn, replace it with Part No. 32005. Reassemble the faucet.

Kohler Triton Diverter Shower and Bath Faucet
K -7014

PROBLEM: Faucet leaks. (Fig. 25.)
CURE: Replace the seat washers, inspect the seats. If the seats are chipped or pitted, replace the seats.

Remove the screw (1) and the handle (2). Use an adjustable wrench to turn the bonnet counterclockwise to loosen and remove. When the plunger is screwed onto the stem, the stem and plunger can be pulled out together. Pull the sleeve (12) out and inspect the seat. If the seat is chipped or pitted, replace it with new part (13) Part

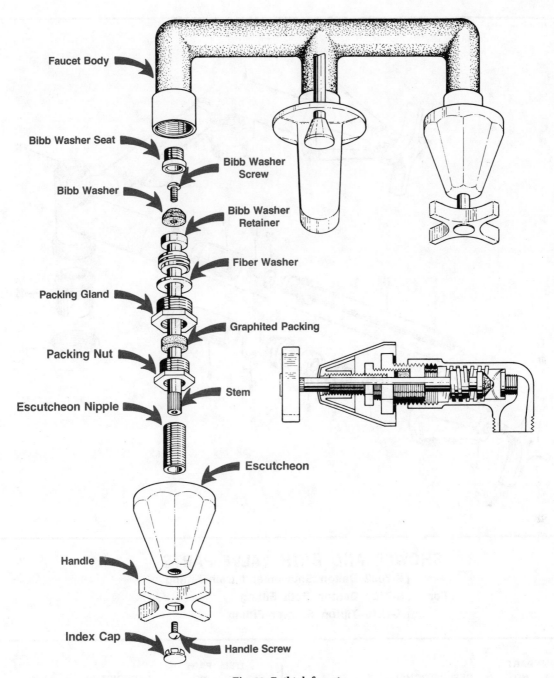

Fig. 23. Bathtub faucets.

No. 23004. Insert a nail set through two opposite holes in the sleeve, tap the square head of the nail set lightly with a small hammer and the worn seat will be forced out of the sleeve. Insert the new seat into the sleeve and tap it lightly with a small hammer to "set" it. Replace the washer (10). Replace the O-rings (5 and 6) Parts No. 34263 and 34264. Lightly coat the O-rings with waterproof grease to prevent damage when reassembling the faucet. It is unlikely that the diverter assembly will require repair. However, if this is necessary, refer to the illustration; start with the screw, (14) and disassemble the diverter assembly. Replace the worn parts, if needed, and grease the O-rings, 18 and 19 when reassembling the diverter.

How to Reseat a Faucet

Refer to Fig. 26 showing reseating tool. At this point you have already disassembled the faucet and determined the need for resurfacing the bibb washer seat. The next step is to assemble the reseating tool to fit your faucet. Note that the female threads in the adapter collar (e) and adapter collar (f) are different sizes and one of

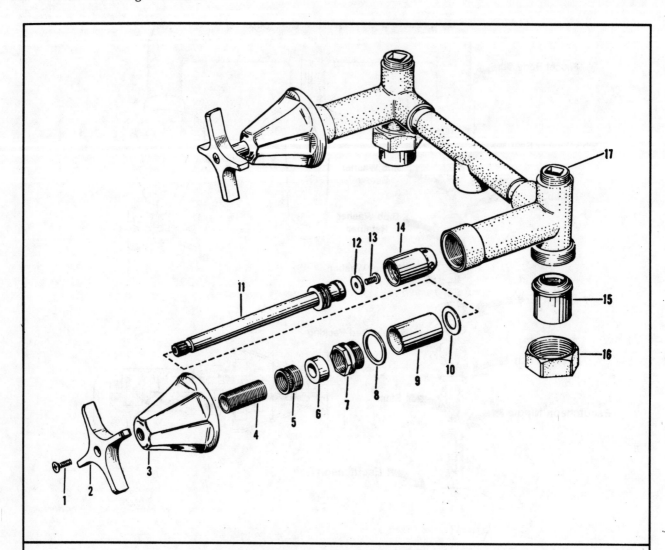

SHOWER AND BATH VALVE PARTS

For { K-7032 Dalton Shower and Bath Fitting
      K-7100 Denton Bath Fitting
      K-7210 Tipton Shower Fitting

ITEM NO.	PART NO.	DESCRIPTION		ITEM NO.	PART NO.	DESCRIPTION
1	33357	Handle Screw		11	20242	Stem w/Washer & Screw
2	20419	Handle (Specify Index)		12	39541	Seat Washer
3	34116	Escutcheon		13	31490	Screw
4	22511	Adjusting Sleeve		14	32462	Remov Unit
5	32008	Packing Nut	Adjusting Sleeve &	15	24015	Union Joint for ½" I.P.
6	32005	Stem Packing	Packing Nut Assembly		24899	Union Joint for ⅝" O.D. Copper Tube
7	32002	Bonnet Nut	No. 20241	16	24003	Nut
8	32057	Gasket		17	34191	Plug
9	34190	Spacer			K-9800	Socket Wrench for #32002
10	32470	Washer (Not required with new style #34190 spacer)				Bonnet Nut (Not Shown)

Fig. 24. Shower and bath faucet.

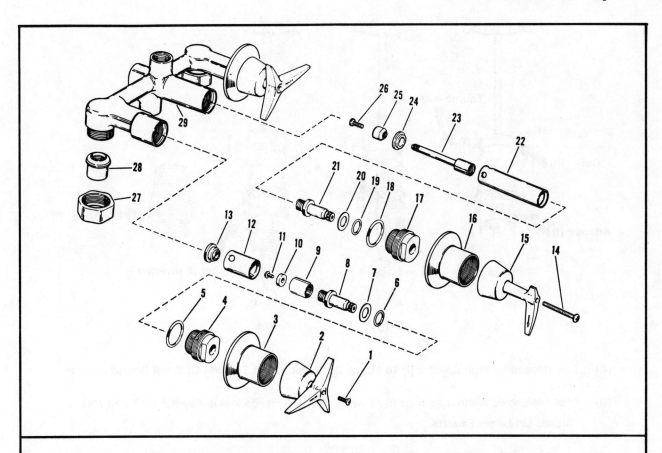

TRITON DIVERTER SHOWER AND BATH VALVE PARTS
K-7014 Dalney Shower and Bath Fitting

ITEM NO.	PART NO.	DESCRIPTION
1	41110	Screw (Round Head)
	53448	Screw (Countersunk)
2	41099	Triton Handle (Specify Index)
*3	22904	Escutcheon
4	34253	Bonnet
5	34264	"O" Ring
6	34263	"O" Ring
7	34265	Washer
8	34320	Cold Stem
	34319	Hot Stem
9	22947	Cold Plunger w/Seat Washer & Screw
	22948	Hot Plunger w/Seat Washer & Screw
10	39541	Seat Washer
11	34848	Screw
12	34842	Sleeve
13	23004	Renewable Seat

ITEM NO.	PART NO.	DESCRIPTION
14	41117	Screw
15	41119	Diverter Handle
16	22904	Escutcheon
17	34253	Bonnet
18	34264	"O" Ring
19	34263	"O" Ring
20	34265	Washer
21	34320	Stem
22	22924	Sleeve with 23057 sleeve
23	22967	Stem
24	22927	Renewable Seat
25	22942	Diverter Seat
26	31490	Screw with 22941 Washer
27	24003	Union Nut
28	24015	Union for ½" I.P.
	24899	Union for ⅝" O.D. Copper Tube
29	22928	Valve Body Only

Items 6-13, Specify 22932 Valvet for Cold Valve, 22917 Valvet for Hot Valve.

* Fitting with Screwdriver Stops, Specify #29516 Escutcheon and #29514 Sleeve.

Courtesy Kohler Co.

Fig. 25. Shower and bath faucet.

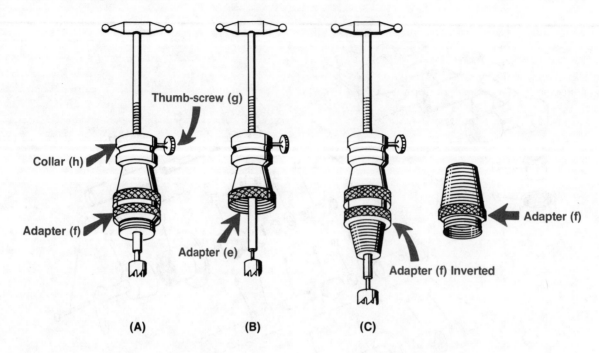

Thumb-screw (g)

Collar (h)

Adapter (f)

Adapter (e)

Adapter (f)

Adapter (f) Inverted

(A)   (B)   (C)

(A)   Tool Assembled With Adapter (f) To Mount On Outside Male Threads Of Small Bodied Faucets

(B)   Tool Assembled Without Adapter (f) Allowing Female Threads Inside Adapter (e) To Fit Male
       Threads On Larger Faucets

(C)   Tool Assembled With Adapter (f) Inverted To Thread Into Body Of Faucet

Fig. 26. Reseating tool.

them will fit the bonnet threads of most faucets. In the event they do not fit, then the adapter (f) can be inverted and threaded onto the stem of the resurfacing tool and the male threads of the adapter can be threaded into the faucet body. Remove the old bibb washer from the washer retainer or the faucet stem and using the bibb washer retainer as a guide select the proper cutter. Thread the cutter on the stem of the reseating tool, turning it clockwise to tighten it. The reseating tool is now assembled for use. Turn the stem of the reseating tool counterclockwise until the top of the cutting tool is resting against the bottom of the adapter. Insert the tool into the top of the faucet and turning the adapter clockwise, tighten the adapter hand-tight on the faucet body. Grasp the Tee handle on the top of the tool and turn the stem of the tool clockwise until you feel the cutter just touch the seat. Tighten the thumb-screw (g) on the collar (h) finger-tight. Then with one hand on the Tee handle and the other grasping the collar (h), turn them both at the same time, clockwise, until you can no longer feel the cutter touching metal. Loosen the thumbscrew (g) and turn the Tee handle approximately 1/16 of a turn clockwise; tighten the thumbscrew and again, as before, turn them both together clockwise until you can no

longer feel the cutter touching metal. It is very important that both of these parts, the Tee handle and the collar, be turned together. This prevents the cutter from cutting too deeply into the metal of the bibb washer seat.

Remove the tool from the faucet body and using the beam of a flashlight inspect the washer seat. The seat should be bright, round and polished. If any pits still show on the washer seat, repeat the operation.

When all pits, chips etc., have been removed and the washer seat presents a bright, round and unblemished surface, the seat repair is completed. Pour water into the top of the faucet to flush out all the metallic cuttings. It is very important that all metal cuttings be flushed out of the faucet; if these cuttings are left in the faucet, they will embed themselves in the soft rubber bibb washer and the faucet will leak. When the faucet has been thoroughly flushed out, install a new bibb washer in the bibb washer retainer on the faucet stem and reassemble the faucet. Repair is now complete.

Aerators

Aerators on faucet spouts serve a dual purpose. Aerators direct the flow of water from the spout to eliminate splashing and they screen the flow of water to trap sand or other foreign substances.

Faucet spouts which are equipped with aerators may have either internal or external threads on the aerator end of the spout. An aerator can be added to a faucet spout which is not threaded. A universal ''clip-adapter'' type aerator can be used.

If the nylon or copper screens in an aerator become obstructed, the water flow through the aerator will be decreased. Remove the aerator body from the spout. Use slip joint pliers to grasp the body of the aerator firmly, turn the aerator clockwise (viewed from the front and above) to loosen and remove the aerator body. Remove the nylon and copper screens and wash out the foreign material caught in the screens. When removing the screens note the order in which the screens are assembled in the aerator body. The screens and washers must be reinstalled in the aerator in the same order as they were removed, in order for the aerator to work properly.

Faucets for the LAVATORY

4" CENTER SETS

DELTA

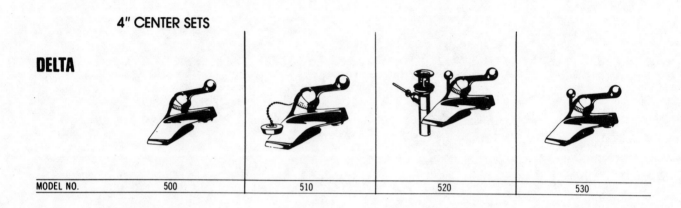

| MODEL NO. | 500 | 510 | 520 | 530 |

SINGLE HOLE WITH 4" SWING SPOUT

DELTA

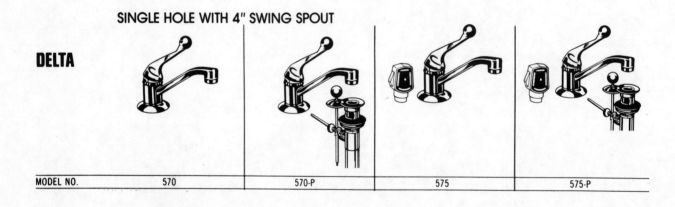

| MODEL NO. | 570 | 570-P | 575 | 575-P |

4" CENTERSETS

DELTA

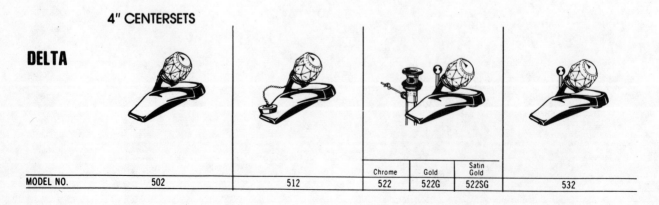

| MODEL NO. | 502 | 512 | Chrome 522 | Gold 522G | Satin Gold 522SG | 532 |

4" CENTERSETS

DELEX

MODEL NO.	2500	2502	2510	2512

710

**IMPORTANT INFORMATION ON SPECIAL ADAPTATIONS
AND PRICING APPLYING TO:**

DELTA...SINGLE LEVER AND CRYSTAL HANDLE
SERIES FOR THE LAVATORY.

½" I.P.S. Adapters—Available on all models. Specify by adding "W/F" to model number.

Satin Finish—Available in all models except 710. Specify by adding SF to model number.

WIDESPREAD 6" TO 16" CENTERS

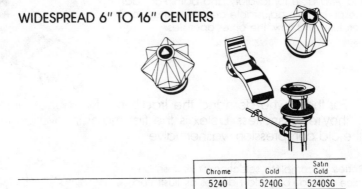

Chrome	Gold	Satin Gold
5240	5240G	5240SG

WIDESPREAD—6" TO 16" CENTERS

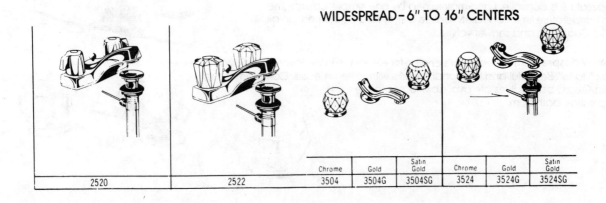

					Chrome	Gold	Satin Gold	Chrome	Gold	Satin Gold
2520	2522				3504	3504G	3504SG	3524	3524G	3524SG

INDEX

Illustrated guide for easy selection.

Selection of the faucets just right for your applications is made easy with this "Quick Reference Illustrated Price List." There is a section for each application—kitchen, lavatory and bath. And within each section, the various series (Delta Single Lever, Delta Crystal Handle and Delex Crystal and Blade Handle) are featured.

With Delta single control faucets, temperature and water flow are controlled by one convenient handle. Credited is the patented rotating ball...the only moving part inside the faucet. Delta single control faucets are the first choice of professional decorators, plumbers, builders and homeowners. The same internal design concept is utilized in a wide range of single control faucets.

Single Lever Series: Traditionally Delta. Proven in over 17 years of service. This series offers the widest range of styles, making the right selection of a Delta faucet most convenient. The ease of control, requiring minimum effort of a single hand or elbow, has made the single lever series the most preferred by home-makers, builders and architects. Available for the kitchen, lavatory and bath.

Crystal Handle Series: Created for those discriminating buyers who demand the ultimate in beauty, dependability and utility. Incorporating all the advantages of a Delta single lever series, plus the added convenience of dialing the exact temperature. What's more the temperature control can be pre-set to turn off the flow with a flick of one hand. Available for lavatory and bath. For added luxury, choose the Delta widespread with adjustable centers from 6" to 16". One handle controls temperature and flow; the other operates the drain assembly.

For those who demand the traditional two handle installation, Delex breaks tradition...they're Washerless. Delex is the first major innovation in two handle faucets since the old compression washer valve.

Crystal or Blade Handle Series: A complete selection of kitchen and lavatory faucets offers a choice of blade or crystal handles as illustrated. Washerless, too, are the Delex Scald-Guard bath mixing valves. Scald-Guard is the ultimate in protection against bodily injury. Desired temperature is controlled and maintained by one simple control. The high temperature limit stop safety feature is standard equipment on all Scald-Guard tub and shower valves.

Delex Widespread: The decorator's choice for elegance in the lavatory. Adjustable centers from 6" to 16". Beautiful and functional, it works with extreme ease. Complementing Scald-Guard bath valves are available to complete the prestige bathroom.

Delex Two Handle BATH VALVES

8" CENTERS, RIGID CASTING
Crystal Handles, Eliptical Escutcheon, Standard Showerhead

DELEX

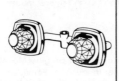    

MODEL NO.	2600-C	2610-C	2620-C		2640-C

8" CENTERS, RIGID CASTING
Larger Crystal Handles, Deluxe Escutcheon and Deluxe Adjustable Showerhead

DELEX

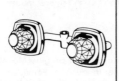    

FINISH	Chrome	Chrome	Satin Gold	Chrome	Satin Gold	Satin Gold	Chrome
MODEL NO.	2602-C	2612-C	2612-C-SG	2622-C	2622-C-SG	2632-C-SG	2642-C

▲

Delex Glamour and Step-Up Deluxe

Delex offers two levels of luxury in two handle rigid casting 8" center bath valves...all feature famous Delex washerless rotating cylinder design. Choose either the Delex trim or upgrade to deluxe.

IMPORTANT INFORMATION ON SPECIAL ADAPTATIONS AND PRICING APPLYING TO:

DELEX...Two Handle Bath Valves.

Small Showerhead Deletions—Shower models available less small head, arm and flange when specified.

Deluxe Showerhead Deletion—Shower models available less Deluxe head, arm and flange when specified.

Integral Screwdriver Stops—Available on all Delex two handle bath valves except model 2704. Specify by adding "With Stops" to model number.

Brass Diverter Spout—On models 2640-C, 2642-C, 2742 and 2746 specify by adding 472 to model number.

Volume Control—Available on all shower valves. Specify by adding 2955 to model number.

Note—All Delex Two Handle Bath Valves supplied with sweat inlets. Not available with IPS connections.

BATH VALVES *Continued*

COMBINATION VALVE, DELUXE ESCUTCHEON, DELUXE ADJUSTABLE SHOWERHEAD

DELEX

	FLEXIBLE MOUNT	FLEXIBLE MOUNT		FLEXIBLE MOUNT		FLEXIBLE MOUNT	FLEXIBLE MOUNT
FINISH	Chrome	Chrome	Satin Gold	Chrome	Satin Gold	Satin Gold	Chrome
MODEL NO.	2700	2712	2712-SG	2722	2722-SG	2732-SG	2742

DELEX ### OVER THE RIM SPOUT, LARGE DELUXE HANDLES

FINISH	Chrome	Satin Gold
MODEL NO.	2704	2704-SG

Widespread, Flexible Mount

This innovation in bath valves permits the location of each valve almost anywhere on the wall. Mounts directly to wall framing via a handy bracket. Tee is furnished with tub or shower models. Ejector cross supplied with combination tub and shower models.

Widespread, Fiberglass Mount

Yet another innovation. Fiberglass mount valves are especially designed to mount directly to the shower and/or tub enclosure panels in almost any position.

DELEX ## COMBINATION VALVE, DELUXE ESCUTCHEON, DELUXE ADJUSTABLE SHOWERHEAD FOR MOUNTING DIRECT TO BATH ENCLOSURE

	FIBERGLASS MOUNT	FIBERGLASS MOUNT		FIBERGLASS MOUNT		FIBERGLASS MOUNT	FIBERGLASS MOUNT
FINISH	Chrome	Chrome	Satin Gold	Chrome	Satin Gold	Satin Gold	Chrome
MODEL NO.	2702	2716	2716-SG	2726	2726-SG	2736-SG	2746

BATH VALVES *Continued*

PRESSURE BALANCE & CHECK STOPS

DELTA

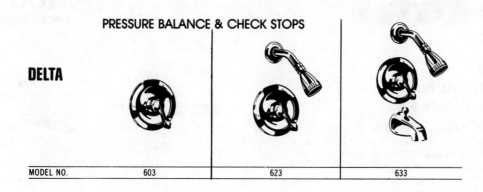

MODEL NO.	603	623	633

PRESSURE BALANCE & CHECK STOPS

DELTA

FINISH	Chrome	Gold	Satin Gold	Chrome	Gold	Satin Gold	Chrome	Gold	Satin Gold
MODEL NO.	604	604G	604SG	624	624G	624SG	634	634G	634SG

DELEX PRESSURE BALANCE & CHECK STOPS

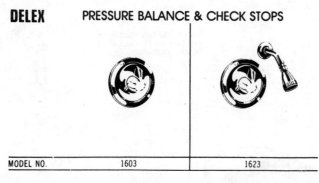

MODEL NO.	1603	1623

PRESSURE BALANCE & CHECK STOPS

DELEX

FINISH	Chrome	Gold	Satin Gold	Chrome	Gold	Satin Gold
MODEL NO.	1604	1604G	1604SG	1624	1624G	1624SG

IMPORTANT INFORMATION ON SPECIAL ADAPTATIONS AND PRICING APPLYING TO:

DELTA...SINGLE LEVER AND CRYSTAL HANDLE SERIES BATH VALVES.

Sweat Inlets—Available on all chrome bath valves. Specify by adding "C" to model number. Not available on special finish models.

Integral Screwdriver Stops—Available on all conventional mixing valves. Specify by adding "WITH STOPS" to model number.

Less Shower Head Assembly—Shower models available less head, arm and flange when specified.

Satin Finish—Available in all models. Specify by adding SF to model number.

DELEX...MIXING VALVES

Sweat Inlets—Available on all chrome bath valves. Specify by adding "C" to model number. Not available on special finish models.

Integral Screwdriver Stops—Available on all non-pressure balance mixing valves. Specify by adding "WITH STOPS" to model number

Less Shower Head Assembly—Shower models available less head, arm and flange when specified.

delta

Single Lever Kitchen Deck Faucets For 3-Hole Sinks—8″ Centers

SUGGESTED SPECIFICATION: Single handle mixing faucets, kitchen deck, for exposed mounting, 8″ center to center with swing spout and aerator. Lever type handle. Control mechanism shall be of the rotating ball type with replaceable non-metallic seats operating in stainless steel lined sockets. Control handle shall always return to neutral position when valve is turned off. All exposed surfaces shall be chrome plated. Delta Models 100, 110, 100-D, 100-DX or equal.

Model 100: Single Lever Kitchen Deck Faucet

Model 110: Single Lever Kitchen Deck Faucet With Spray 'Round Aerator

Model 100-D: Single Lever Kitchen Deck Faucet With Dispenser

Model 100-DX: Single Lever Kitchen Deck Faucet With Double Dispenser

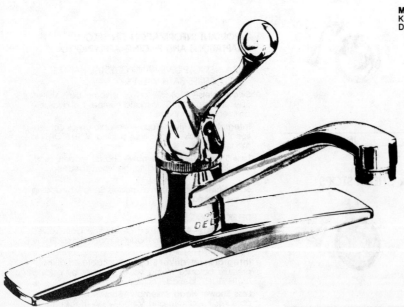

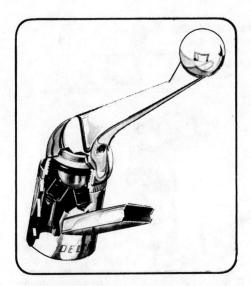

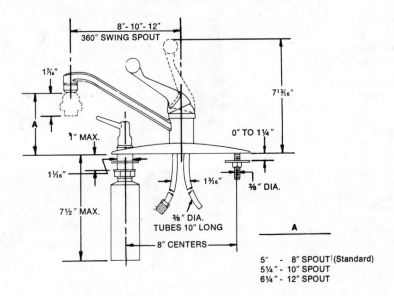

Model 100: Single Lever Kitchen Deck Faucet

Model 110: Single Lever Kitchen Deck Faucet With Spray 'Round Aerator (not available in satin chrome)

Model 100-D: Single Lever Kitchen Deck Faucet With Dispenser

Model 100-DX: Single Lever Kitchen Deck Faucet With Double Dispenser

Special Finish Available: Satin Chrome

Meets federal specifications WW-P-541/5A Type 1 Oct. 1971

½" I.P.S. adapters available. Specify by adding "W/F" to model number.

The Rotating Ball is the heart of the single control faucet. The patented design is the most successful and trouble-free single handle faucet on the market today. The ball has ports opening into the center of the ball. Water flow and temperature are controlled by simply lining up the ports in the ball with hot and cold inlet holes in the brass body. Pushing the handle back, controls the amount of flow. Move the handle to the right, you get cold water; move the handle to the left, you get hot water. There's an infinite control of temperature and flow in between. Both operations, flow and temperature, are controlled simultaneously.

Delta SINGLE LEVER KITCHEN FAUCETS Series 100, 300, 350, 400, 450

Installation Guide
WITH OR WITHOUT SPRAY AND/OR DISPENSER

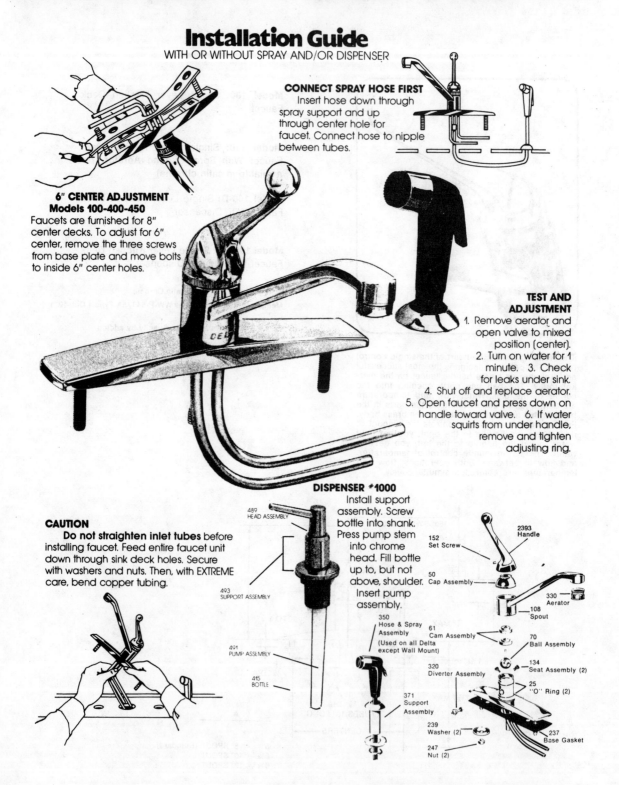

6" CENTER ADJUSTMENT
Models 100-400-450
Faucets are furnished for 8" center decks. To adjust for 6" center, remove the three screws from base plate and move bolts to inside 6" center holes.

CONNECT SPRAY HOSE FIRST
Insert hose down through spray support and up through center hole for faucet. Connect hose to nipple between tubes.

TEST AND ADJUSTMENT
1. Remove aerator and open valve to mixed position (center).
2. Turn on water for 1 minute. 3. Check for leaks under sink.
4. Shut off and replace aerator.
5. Open faucet and press down on handle toward valve. 6. If water squirts from under handle, remove and tighten adjusting ring.

CAUTION
   Do not straighten inlet tubes before installing faucet. Feed entire faucet unit down through sink deck holes. Secure with washers and nuts. Then, with EXTREME care, bend copper tubing.

DISPENSER #1000
Install support assembly. Screw bottle into shank. Press pump stem into chrome head. Fill bottle up to, but not above, shoulder. Insert pump assembly.

489 HEAD ASSEMBLY
493 SUPPORT ASSEMBLY
491 PUMP ASSEMBLY
415 BOTTLE

350 Hose & Spray Assembly (Used on all Delta except Wall Mount)

320 Diverter Assembly

371 Support Assembly

239 Washer (2)

247 Nut (2)

152 Set Screw
2393 Handle
50 Cap Assembly
61 Cam Assembly
330 Aerator
108 Spout
70 Ball Assembly
134 Seat Assembly (2)
25 "O" Ring (2)
237 Base Gasket

Routine Maintenance Instructions

Delta faucets have earned the reputation of superiority in design, engineering, performance and durability in millions of home installations. Routine faucet maintenance to compensate for foreign materials in varying water conditions, will restore like new performance and extend the life of the faucet. Ease and simplicity of service and maintenance is another advantage of having Delta faucets throughout the home.

A. If you should have a leak under handle—tighten adjusting ring by following steps 1, 14, 15 and 16.

B. If you should have a leak from spout—replace seats and springs by following steps 1, 2, 3, 4 and 10. Then reassemble with steps 11, 12, 13, 14, 15 and 16.

C. If you should have a leak from around spout collar—replace "O" rings by following steps 1, 2, 5, 6 and 9. Then reassemble by following steps 13, 14, 15 and 16.

D. If spray does not work properly—check hose for wear or cuts. Clean or replace diverter part No. 320 by following steps 1, 2, 5, 7, 8 and 9A. Then reassemble, following steps 13, 14, 15 and 16.

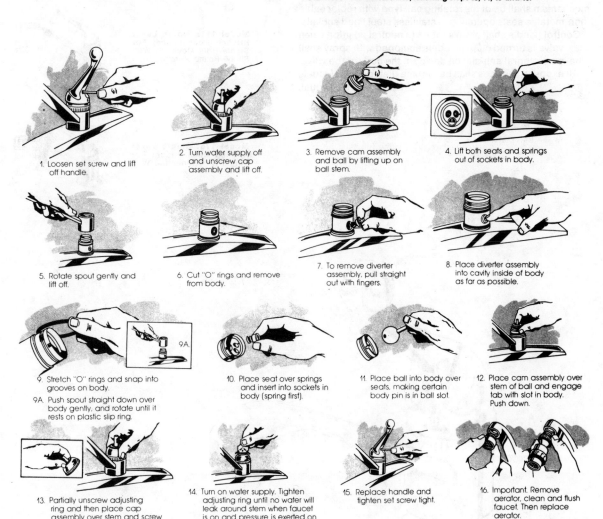

1. Loosen set screw and lift off handle.

2. Turn water supply off and unscrew cap assembly and lift off.

3. Remove cam assembly and ball by lifting up on ball stem.

4. Lift both seats and springs out of sockets in body.

5. Rotate spout gently and lift off.

6. Cut "O" rings and remove from body.

7. To remove diverter assembly, pull straight out with fingers.

8. Place diverter assembly into cavity inside of body as far as possible.

9. Stretch "O" rings and snap into grooves on body.

9A. Push spout straight down over body gently, and rotate until it rests on plastic slip ring.

10. Place seat over springs and insert into sockets in body (spring first).

11. Place ball into body over seats, making certain body pin is in ball slot.

12. Place cam assembly over stem of ball and engage tab with slot in body. Push down.

13. Partially unscrew adjusting ring and then place cap assembly over stem and screw down tight onto body.

14. Turn on water supply. Tighten adjusting ring until no water will leak around stem when faucet is on and pressure is exerted on handle to force ball into socket.

15. Replace handle and tighten set screw tight.

16. Important. Remove aerator, clean and flush faucet. Then replace aerator.

Single Lever Kitchen Deck Faucets For Single Hole Mounting

SUGGESTED SPECIFICATION: Single handle mixing faucets, kitchen deck, for exposed single hole mounting, 8″ swing spout and aerator. Lever type handle. Control mechanism shall be of the rotating ball type with replaceable non-metallic seats operating in stainless steel lined sockets. Control handle shall always return to neutral position when valve is turned off. All models equipped with spray shall have an integral anti-siphon device in the body of the valve. All exposed surfaces shall be chrome plated. Delta Models 101, 111, 175 or equal.

Model 101: Single Lever Kitchen Deck Faucet For Single Hole Mounting

Model 111: Single Lever Kitchen Deck Faucet For Single Hole Mounting With Spray 'Round Aerator

Model 175: Single Lever Kitchen Deck Faucet For Single Hole Mounting With Spray

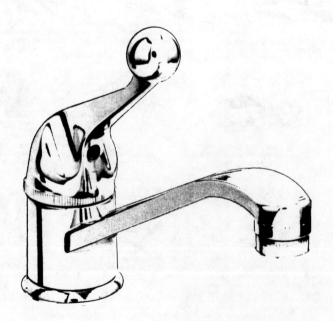

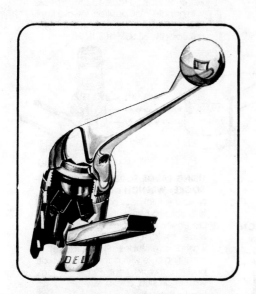

Model 101: Single Lever Kitchen Deck Faucet For Single Hole Mounting

Model 111: Single Lever Kitchen Deck Faucet For Single Hole Mounting With Spray 'Round Aerator (not available in satin chrome)

Model 175: Single Lever Kitchen Deck Faucet For Single Hole Mounting With Spray

Special Finish Available: Satin Chrome
Meets federal specifications WW-P-541/5A Type 1 Oct. 1971

½" I.P.S. adapters available. Specify by adding "W/F" to model number.

The Rotating Ball is the heart of the single control faucet. The patented design is the most successful and trouble-free single handle faucet on the market today. The ball has ports opening into the center of the ball. Water flow and temperature are controlled by simply lining up the ports in the ball with hot and cold inlet holes in the brass body. Pushing the handle back, controls the amount of flow. Move the handle to the right, you get cold water; move the handle to the left, you get hot water. There's an infinite control of temperature and flow in between. Both operations, flow and temperature, are controlled simultaneously.

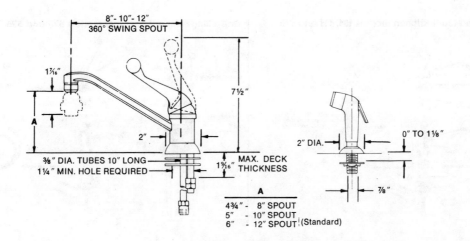

Delta Series

SINGLE MOUNT, KITCHEN AND LAVATORY SINGLE LEVER FAUCETS

101, 111, 175 Kitchen and 570 and 575 Lavatory

Installation Guide

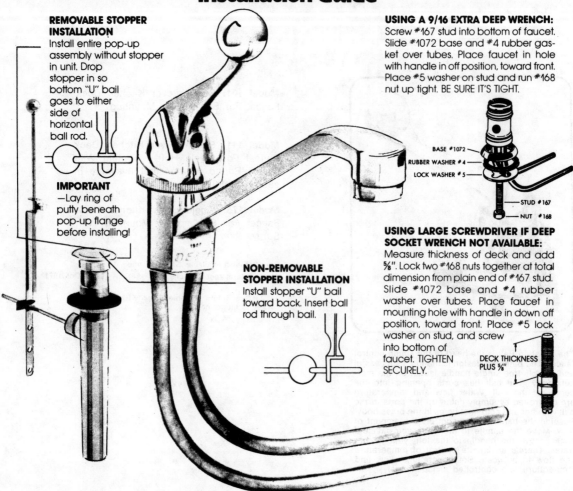

REMOVABLE STOPPER INSTALLATION
Install entire pop-up assembly without stopper in unit. Drop stopper in so bottom "U" bail goes to either side of horizontal ball rod.

IMPORTANT
—Lay ring of putty beneath pop-up flange before installing!

NON-REMOVABLE STOPPER INSTALLATION
Install stopper "U" bail toward back. Insert ball rod through bail.

USING A 9/16 EXTRA DEEP WRENCH:
Screw #167 stud into bottom of faucet. Slide #1072 base and #4 rubber gasket over tubes. Place faucet in hole with handle in off position, toward front. Place #5 washer on stud and run #168 nut up tight. BE SURE IT'S TIGHT.

BASE #1072
RUBBER WASHER # 4
LOCK WASHER # 5
STUD #167
NUT #168

USING LARGE SCREWDRIVER IF DEEP SOCKET WRENCH NOT AVAILABLE:
Measure thickness of deck and add ⅝". Lock two #168 nuts together at total dimension from plain end of #167 stud. Slide #1072 base and #4 rubber washer over tubes. Place faucet in mounting hole with handle in down off position, toward front. Place #5 lock washer on stud, and screw into bottom of faucet. TIGHTEN SECURELY.

DECK THICKNESS PLUS ⅝"

Delta Single Mount Kitchen models 101, 111 and 175

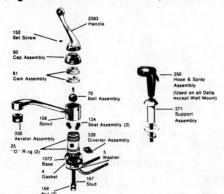

152 Set Screw
2393 Handle
50 Cap Assembly
61 Cam Assembly
70 Ball Assembly
108 Spout
330 Aerator Assembly
134 Seat Assembly (2)
25 "O" Ring (2)
1072 Base
4 Gasket
5 Washer
320 Diverter Assembly
167 Stud
168 Nut (2)

350 Hose & Spray Assembly (Used on all Delta except Wall Mount)
371 Support Assembly

Delta Single Mount Lavatory models 570 and 575

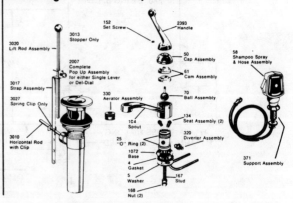

3020 Lift Rod Assembly
3013 Stopper Only
2007 Complete Pop Up Assembly for either Single Lever or Del-Dial
3017 Strap Assembly
3027 Spring Clip Only
3010 Horizontal Rod with Clip
152 Set Screw
2393 Handle
50 Cap Assembly
61 Cam Assembly
70 Ball Assembly
330 Aerator Assembly
104 Spout
134 Seat Assembly (2)
25 "O" Ring (2)
320 Diverter Assembly
1072 Base
4 Gasket
5 Washer
168 Nut (2)
167 Stud

58 Shampoo Spray & Hose Assembly
371 Support Assembly

Routine Maintenance Instructions

Delta faucets have earned the reputation of superiority in design, engineering, performance and durability in millions of home installations. Routine faucet maintenance to compensate for foreign materials in varying water conditions, will restore like new performance and extend the life of the faucet. Ease and simplicity of service and maintenance is another advantage of having Delta faucets throughout the home.

A. If you should have a leak under handle—tighten adjusting ring by following steps 1, 14, 15 and 16.

B. If you should have a leak from spout—replace seats and springs by following steps 1, 2, 3, 4 and 10. Then reassemble with steps 11, 12, 13, 14, 15 and 16.

C. If you should have a leak from around spout collar—replace "O" rings by following steps 1, 2, 5, 6 and 9. Then reassemble by following steps 13, 14, 15 and 16.

D. If spray does not work properly—check hose for wear or cuts. Clean or replace diverter part No. 320 by following steps 1, 2, 5, 7, 8 and 9A. Then reassemble, following steps 13, 14, 15 and 16.

1. Loosen set screw and lift off handle.

2. Turn water supply off and unscrew cap assembly and lift off.

3. Remove cam assembly and ball by lifting up on ball stem.

4. Lift both seats and springs out of sockets in body.

5. Rotate spout gently and lift off.

6. Cut "O" rings and remove from body.

7. To remove diverter assembly, pull straight out with fingers.

8. Place diverter assembly into cavity inside of body as far as possible.

9. Stretch "O" rings and snap into grooves on body.
9A. Push spout straight down over body gently, and rotate until it rests on plastic slip ring.

10. Place seat over springs and insert into sockets in body (spring first).

11. Place ball into body over seats, making certain body pin is in ball slot.

12. Place cam assembly over stem of ball and engage tab with slot in body. Push down.

13. Partially unscrew adjusting ring and then place cap assembly over stem and screw down tight onto body.

14. Turn on water supply. Tighten adjusting ring until no water will leak around stem when faucet is on and pressure is exerted on handle to force ball into socket.

15. Replace handle and tighten set screw tight.

16. Important. Remove aerator, clean and flush faucet. Then replace aerator.

SERIES
200
WALL MOUNT

Model 200: Single Lever Kitchen Wall Mount Faucet

Model 250: Single Lever Kitchen Wall Mount Faucet With Dishwashing Attachment

Model 275: Single Lever Kitchen Wall Mount Faucet With Spray Attachment

Single Lever Wall Mount Faucets 8" Centers

SUGGESTED SPECIFICATION: Single handle mixing faucets, kitchen, for exposed wall mounting, 8" center to center with swing spout and aerator. Lever type handle. Control mechanism shall be of the rotating ball type with replaceable non-metallic seats operating in stainless steel lined sockets. Control handle shall always return to neutral position when valve is turned off. All models equipped with spray shall have an integral anti-siphon device in the body of the valve. All exposed surfaces shall be chrome plated. Delta Models 200, 250, 275 or equal.

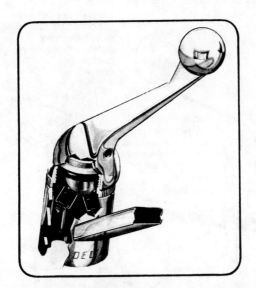

Model 200: Single Lever Kitchen Wall Mount Faucet

Model 250: Single Lever Kitchen Wall Mount Faucet With Dishwashing Attachment

Model 275: Single Lever Kitchen Wall Mount Faucet With Spray Attachment

Bright chrome finish only

The Rotating Ball is the heart of the single control faucet. The patented design is the most successful and trouble-free single handle faucet on the market today. The ball has ports opening into the center of the ball. Water flow and temperature are controlled by simply lining up the ports in the ball with hot and cold inlet holes in the brass body. Pushing the handle back, controls the amount of flow. Move the handle to the right, you get cold water; move the handle to the left, you get hot water. There's an infinite control of temperature and flow in between. Both operations, flow and temperature, are controlled simultaneously.

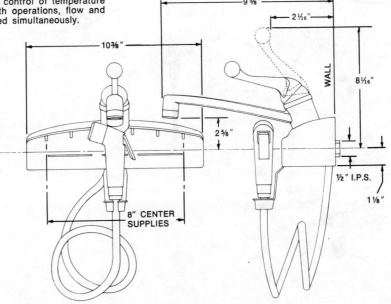

Delta
KITCHEN
WALL
MOUNT
FAUCET
Series 200, 250, 275

Installation Guide
(With or without spray or dishwashing attachment.)

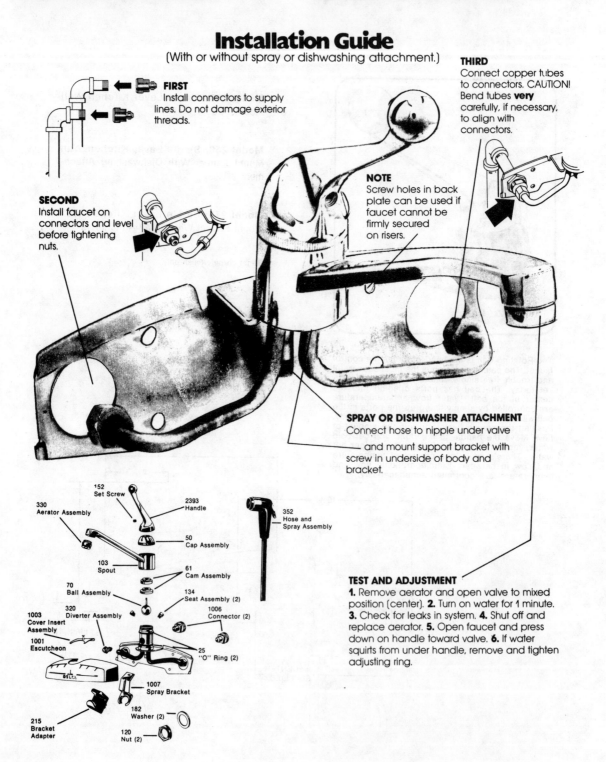

THIRD
Connect copper tubes to connectors. CAUTION! Bend tubes **very** carefully, if necessary, to align with connectors.

FIRST
Install connectors to supply lines. Do not damage exterior threads.

SECOND
Install faucet on connectors and level before tightening nuts.

NOTE
Screw holes in back plate can be used if faucet cannot be firmly secured on risers.

SPRAY OR DISHWASHER ATTACHMENT
Connect hose to nipple under valve and mount support bracket with screw in underside of body and bracket.

330
Aerator Assembly

152
Set Screw

2393
Handle

352
Hose and
Spray Assembly

50
Cap Assembly

103
Spout

61
Cam Assembly

70
Ball Assembly

134
Seat Assembly (2)

1003
Cover Insert
Assembly

320
Diverter Assembly

1006
Connector (2)

1001
Escutcheon

25
"O" Ring (2)

1007
Spray Bracket

215
Bracket
Adapter

182
Washer (2)

120
Nut (2)

TEST AND ADJUSTMENT
1. Remove aerator and open valve to mixed position (center). **2.** Turn on water for 1 minute. **3.** Check for leaks in system. **4.** Shut off and replace aerator. **5.** Open faucet and press down on handle toward valve. **6.** If water squirts from under handle, remove and tighten adjusting ring.

Routine Maintenance Instructions

Delta faucets have earned the reputation of superiority in design, engineering, performance and durability in millions of home installations. Routine faucet maintenance to compensate for foreign materials in varying water conditions, will restore like new performance and extend the life of the faucet. Ease and simplicity of service and maintenance is another advantage of having Delta faucets throughout the home.

A. If you should have a leak under handle—tighten adjusting ring by following steps 1, 14, 15 and 16.

B. If you should have a leak from spout—replace seats and springs by following steps 1, 2, 3, 4 and 10. Then reassemble with steps 11, 12, 13, 14, 15 and 16.

C. If you should have a leak from around spout collar—replace "O" rings by following steps 1, 2, 5, 6 and 9. Then reassemble by following steps 13, 14, 15 and 16.

D. If spray does not work properly—check hose for wear or cuts. Clean or replace diverter part No. 320 by following steps 1, 2, 5, 7, 8 and 9A. Then reassemble, following steps 13, 14, 15 and 16.

1. Loosen set screw and lift off handle.

2. Turn water supply off and unscrew cap assembly and lift off.

3. Remove cam assembly and ball by lifting up on ball stem.

4. Lift both seats and springs out of sockets in body.

5. Rotate spout gently and lift off.

6. Cut "O" rings and remove from body.

7. To remove diverter assembly, pull straight out with fingers.

8. Place diverter assembly into cavity inside of body as far as possible.

9. Stretch "O" rings and snap into grooves on body.

9A. Push spout straight down over body gently, and rotate until it rests on plastic slip ring.

10. Place seat over springs and insert into sockets in body (spring first).

11. Place ball into body over seats, making certain body pin is in ball slot.

12. Place cam assembly over stem of ball and engage tab with slot in body. Push down.

13. Partially unscrew adjusting ring and then place cap assembly over stem and screw down tight onto body.

14. Turn on water supply. Tighten adjusting ring until no water will leak around stem when faucet is on and pressure is exerted on handle to force ball into socket.

15. Replace handle and tighten set screw tight.

16. Important. Remove aerator, clean and flush faucet. Then replace aerator.

Single Lever Kitchen Deck Faucets For 3-Hole Sinks—8" Centers

SUGGESTED SPECIFICATION: Single handle mixing faucets, kitchen deck, for exposed mounting, 8" center to center with swing spout and aerator. Lever type handle. Control mechanism shall be of the rotating ball type with replaceable non-metallic seats operating in stainless steel lined sockets. Control handle shall always return to neutral position when valve is turned off. All models equipped with hose and spray shall have an integral anti-siphon device in the body of the valve. All exposed surfaces shall be chrome plated. Delta Models 300, 300-D, 350, 350-D or equal.

Model 300: Single Lever Kitchen Deck Faucet With Spray

Model 300-D: Single Lever Kitchen Deck Faucet With Spray and Dispenser

Model 350: Single Lever Kitchen Deck Faucet With Dishwashing Attachment

Model 350-D: Single Lever Kitchen Deck Faucet With Dishwashing Attachment and Dispenser

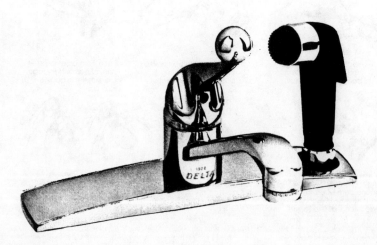

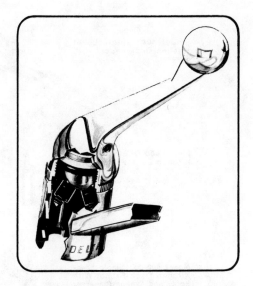

Model 300: Single Lever Kitchen Deck Faucet With Spray

Model 300-D: Single Lever Kitchen Deck Faucet With Spray and Dispenser

Model 350: Single Lever Kitchen Deck Faucet With Dishwashing Attachment

Model 350-D: Single Lever Kitchen Deck Faucet With Dishwashing Attachment and Dispenser

Special Finish Available: Satin Chrome

Meets federal specifications WW-P-541/5A Type 1 Oct. 1971

½″ I.P.S. adapters available. Specify by adding ''W/F'' to model number.

The Rotating Ball is the heart of the single control faucet. The patented design is the most successful and trouble-free single handle faucet on the market today. The ball has ports opening into the center of the ball. Water flow and temperature are controlled by simply lining up the ports in the ball with hot and cold inlet holes in the brass body. Pushing the handle back, controls the amount of flow. Move the handle to the right, you get cold water; move the handle to the left, you get hot water. There's an infinite control of temperature and flow in between. Both operations, flow and temperature, are controlled simultaneously.

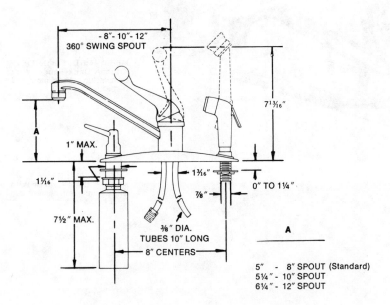

5″ - 8″ SPOUT (Standard)
5¼″ - 10″ SPOUT
6¼″ - 12″ SPOUT

Single Lever Kitchen Deck Faucets For 4-Hole Sinks—8″ Centers

SUGGESTED SPECIFICATION: Single handle mixing faucets, kitchen deck, for exposed mounting, 8″ center to center with swing spout and aerator. Lever type handle. Control mechanism shall be of the rotating ball type with replaceable non-metallic seats operating in stainless steel lined sockets. Control handle shall always return to neutral position when valve is turned off. All models equipped with hose and spray shall have an integral anti-siphon device in the body of the valve. All exposed surfaces shall be chrome plated. Delta Models 400, 400-D, 400-DX, 450, 450-D, 450-DX or equal.

Model 400: Single Lever Kitchen Deck Faucet With Spray

Model 400-D: Single Lever Kitchen Deck Faucet With Spray and Dispenser

Model 400-DX: Single Lever Kitchen Deck Faucet With Spray and Double Dispenser

Model 450: Single Lever Kitchen Deck Faucet With Dishwashing Attachment

Model 450-D: Single Lever Kitchen Deck Faucet With Dishwashing Attachment and Dispenser

Model 450-DX: Single Lever Kitchen Deck Faucet With Dishwashing Attachment and Double Dispenser

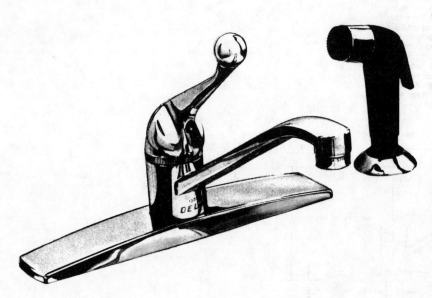

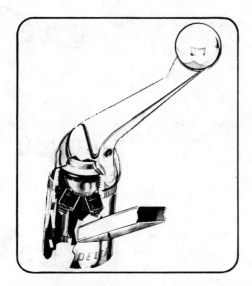

The Rotating Ball is the heart of the single control faucet. The patented design is the most successful and trouble-free single handle faucet on the market today. The ball has ports opening into the center of the ball. Water flow and temperature are controlled by simply lining up the ports in the ball with hot and cold inlet holes in the brass body. Pushing the handle back, controls the amount of flow. Move the handle to the right, you get cold water; move the handle to the left, you get hot water. There's an infinite control of temperature and flow in between. Both operations, flow and temperature, are controlled simultaneously.

Model 400: Single Lever Kitchen Deck Faucet With Spray

Model 400-D: Single Lever Kitchen Deck Faucet With Spray and Dispenser

Model 400-DX: Single Lever Kitchen Deck Faucet With Spray and Double Dispenser

Model 450: Single Lever Kitchen Deck Faucet With Dishwashing Attachment

Model 450-D: Single Lever Kitchen Deck Faucet With Dishwashing Attachment and Dispenser

Model 450-DX: Single Lever Kitchen Deck Faucet With Dishwashing Attachment and Double Dispenser

Special Finish Available: Satin Chrome

Meets federal specifications WW-P-541/5A   Type 1 Oct. 1971

½" I.P.S. adapters available. Specify by adding "W/F" to model number.

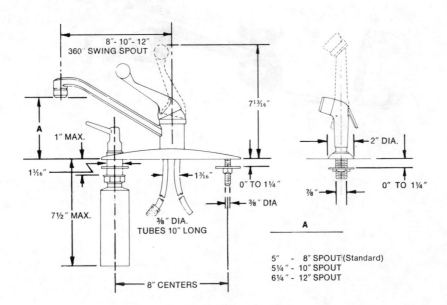

5"   - 8" SPOUT (Standard)
5¼" - 10" SPOUT
6¼" - 12" SPOUT

delex by delta

Two Handle Kitchen Deck Faucets With Blade Handles— 8″ Centers

SUGGESTED SPECIFICATION: Kitchen deck faucet shall be for "exposed" mounting on 8″ centers with 8″ swing spout, aerator and blade handles. Models with spray attachment shall have approved automatic diverter and hose and spray assembly. Valve mechanisms shall be of the rotating cylinder type with 180° rotation. Hot and cold stem units shall be interchangeable. All operating parts must be replaceable from above the deck. Valves shall be equipped with replaceable non-metallic seats which are contained in stainless steel lined sockets. All models equipped with hose and spray shall have an anti-siphon device. All exposed surfaces shall be chrome plated. Delta/Delex Models 2100, 2110, 2400 or equal.

Model 2100: Two Handle Kitchen Deck Faucet—Blade Handles

Model 2110: Two Handle Kitchen Deck Faucet With Spray 'Round Aerator—Blade Handles

Model 2400: Two Handle Kitchen Deck Faucet With Spray — Blade Handles (For 4-Hole Sinks)

Two Handle Kitchen Deck Faucets With Blade Handles— 8" Centers

Model 2100: Two Handle Kitchen Deck Faucet—Blade Handles

Model 2110: Two Handle Kitchen Deck Faucet With Spray 'Round Aerator—Blade Handles

Model 2400: Two Handle Kitchen Deck Faucet With Spray—Blade Handles (For 4-Hole Sinks)

Meets federal specifications WW-P-541/5A Type 1 Oct. 1971

Bright chrome finish only

Here's the heart of the revolutionary Delex two handle faucet showing the simplicity of the rotating cylinder valve that ends washer headaches forever. There are no compression valves or washers to crush, tear or replace—the rotating cylinder valve simply shears off the water with no tightening down, no grinding, no force.

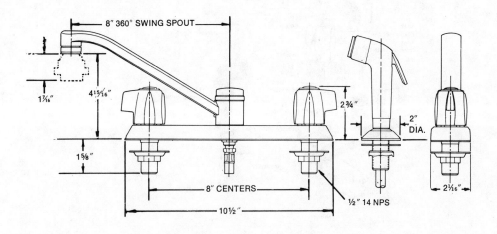

Delex Series
WASHERLESS KITCHEN FAUCETS
BY DELTA
2100 and 2400

Installation Guide
with or without spray

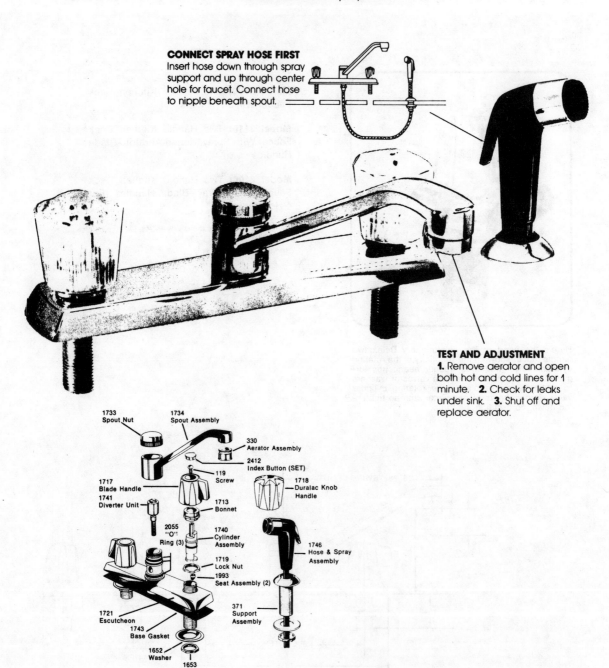

CONNECT SPRAY HOSE FIRST
Insert hose down through spray support and up through center hole for faucet. Connect hose to nipple beneath spout.

TEST AND ADJUSTMENT
1. Remove aerator and open both hot and cold lines for 1 minute.   **2.** Check for leaks under sink.   **3.** Shut off and replace aerator.

1733 Spout Nut
1734 Spout Assembly
330 Aerator Assembly
2412 Index Button (SET)
119 Screw
1717 Blade Handle
1741 Diverter Unit
1718 Duralac Knob Handle
1713 Bonnet
2055 "O" Ring (3)
1740 Cylinder Assembly
1719 Lock Nut
1746 Hose & Spray Assembly
1993 Seat Assembly (2)
1721 Escutcheon
1743 Base Gasket
371 Support Assembly
1652 Washer
1653 Nut

Routine Maintenance Instructions

Delex faucets have earned the reputation of superiority in design, engineering, performance and durability in millions of home installations. Routine faucet maintenance, to compensate for foreign materials in varying water conditions, will restore like new performance and extend the life of the faucet. Ease and simplicity of service and maintenance is another advantage of having Delex faucets throughout the home.

A. If faucet leaks from under handle or from spout outlet, replace stem unit and/or seat as in steps 1, 2, 3, 4, 5. Reassemble following steps 12, 13 and 16.

B. If leak occurs at top or bottom of spout body, replace "O" Rings as shown in steps 6, 7, 8, 9, 14 and 15.

C. If volume of water from spray models decreases or spray ceases to function, follow procedure in steps 6, 10,11 and 15.

SHUT OFF WATER SUPPLY

1. Pry off index button, remove screw and lift off handle.

2. Unscrew bonnet.

3. Pull stem straight up and out.

4. Lift seat and spring out of body.

5. Place new seat over new spring and push into socket in body.

6. Unscrew spout nut.

7. Lift spout off by rotating back and forth, and pull up gently.

8. Cut "O" rings and remove from body.

9. Stretch new "O" rings and snap into grooves on body.

10. Unscrew diverter using screwdriver. Wash off diverter, and clean and flush out socket of body.

11. Screw diverter tightly in spout socket.

12. IMPORTANT—STEM MUST BE REPLACED PROPERLY!

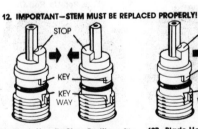

12A. Knob Handle Stem Position—Slip stem unit into body, aligning key with key way, so "stop" on both stems point toward spout.

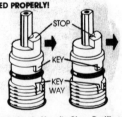

12B. Blade Handle Stem Position—As you face faucet, slip stem units into body, aligning keys in key way so "stops" on both units point to right.

13. Replace bonnet. Tighten securely.

14. Push spout straight down over body, rotating back and forth gently until it rests on faucet base.

15. Screw spout nut on snugly.

16. Replace handle. Tighten screw. Press index button in position.

Use only genuine Delta replacement parts.
Kit 1740—Cylinder Assembly
Kit 1933—Seat Assembly (2)
Kit 2055—"O" Ring (3)

Single Control, Lever Handle, Lavatory Faucets— 4" Centersets

SUGGESTED SPECIFICATION: Single handle mixing faucets, lavatory 4" centerset exposed deck mounting, with aerator, and pop-up type drain fitting or chain and stopper as required. Lever type handle. Control mechanism shall be of the rotating ball type with replaceable non-metallic seats operating in stainless steel lined sockets. Control handle shall always return to neutral position when valve is turned off. All exposed surfaces shall be chrome pated. Delta Models 500, 510, 520, 530 or equal.

Model 500: Single Control, Lever Handle, Lavatory Faucet

Model 510: Single Control, Lever Handle, Lavatory Faucet With Chain and Stopper

Model 520: Single Control, Lever Handle, Lavatory Faucet With Complete Pop-Up Drain Assembly

Model 530: Single Control, Lever Handle, Lavatory Faucet With Pop-Up Rod Only

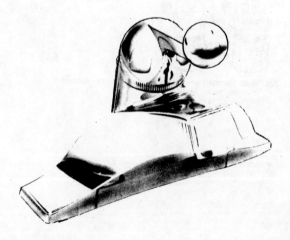

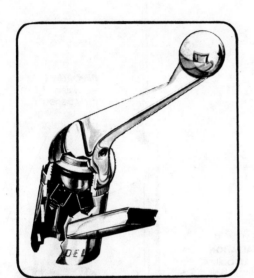

Single Control, Lever Handle, Lavatory Faucets— 4" Centersets

Model 500: Single Control, Lever Handle, Lavatory Faucet

Model 510: Single Control, Lever Handle, Lavatory Faucet With Chain and Stopper

Model 520: Single Control, Lever Handle, Lavatory Faucet With Complete Pop-Up Drain Assembly

Model 530: Single Control, Lever Handle, Lavatory Faucet With Pop-Up Rod Only

Special Finish Available: Satin Chrome

Meets federal specification WW-P-541/4—Type 4—Oct. 1971

½″ I.P.S. adapters available. Specify by adding "W/F" to model number.

The Rotating Ball is the heart of the single control faucet. The patented design is the most successful and trouble-free single handle faucet on the market today. The ball has ports opening into the center of the ball. Water flow and temperature are controlled by simply lining up the ports in the ball with hot and cold inlet holes in the brass body. Pushing the handle back, controls the amount of flow. Move the handle to the right, you get cold water; move the handle to the left, you get hot water. There's an infinite control of temperature and flow in between. Both operations, flow and temperature, are controlled simultaneously.

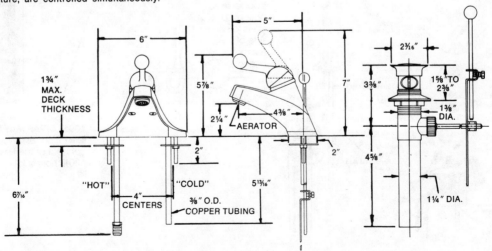

Delta SINGLE CONTROL LEVER LAVATORY FAUCET Series 500, 510, 520, 530

Installation Guide

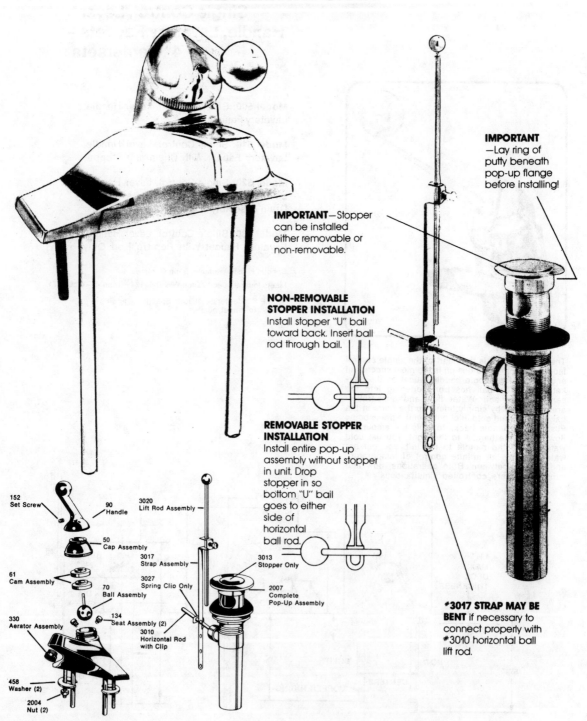

IMPORTANT—Stopper can be installed either removable or non-removable.

NON-REMOVABLE STOPPER INSTALLATION
Install stopper "U" bail toward back. Insert ball rod through bail.

REMOVABLE STOPPER INSTALLATION
Install entire pop-up assembly without stopper in unit. Drop stopper in so bottom "U" bail goes to either side of horizontal ball rod.

IMPORTANT
—Lay ring of putty beneath pop-up flange before installing!

#3017 STRAP MAY BE BENT if necessary to connect properly with #3010 horizontal ball lift rod.

152 Set Screw
90 Handle
50 Cap Assembly
61 Cam Assembly
70 Ball Assembly
330 Aerator Assembly
458 Washer (2)
2004 Nut (2)
134 Seat Assembly (2)
3020 Lift Rod Assembly
3017 Strap Assembly
3027 Spring Clip Only
3010 Horizontal Rod with Clip
3013 Stopper Only
2007 Complete Pop-Up Assembly

Routine Maintenance Instructions

Delta faucets have earned the reputation of superiority in design, engineering, performace and durability in millions of home installations. Routine faucet maintenance to compensate for foreign materials in varying water conditions, will restore like new performance and extend the life of the faucet. Ease and simplicity of service and maintenance is another advantage of having Delta faucets throughout the home.

A. If you should have a leak under handle—tighten adjusting ring following steps 1 and 9. Reassemble as in step 10.

B. If you should have a leak from spout—shut off water supply, and follow steps 1, 2, 3, 4 and 5. Reassemble as in steps 6, 7 and 8. Set adjusting ring as in 9. Replace handle as in 10.

1. Pry off handle button, remove screw and lift off handle.

2. Unscrew cap assembly and lift off.

3. Remove cam assembly and ball by lifting up on ball stem.

4. Lift both seats and springs out of sockets in body.

5. Place seat over springs and insert into sockets in body (spring first).

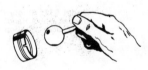

6. Place ball into body over seats.

7. Place cam assembly over stem of ball and engage tab with slot in body. Push down.

8. Partially unscrew adjusting ring and then place cap assembly over ball stem and screw down tight onto body.

9. Tighten adjusting ring until no water will leak around stem when faucet is on and pressure is exerted on handle to force ball into socket.

10. Replace handle. Tighten handle screw—tight. Replace handle button.

Single Control, Duralac Knob Handle, Lavatory Faucets—4″ Centersets

SUGGESTED SPECIFICATION: Single handle mixing faucets, lavatory 4″ centerset exposed deck mounting, with aerator, pop-up type drain fitting or chain and stopper as required. Alcohol resistant Duralac knob handle. Control mechanism shall be of the rotating ball type with replaceable non-metallic seats operating in stainless steel lined sockets. All exposed surfaces shall be chrome plated. Delta Models 502, 512, 522, 532 or equal.

Model 502: Single Control, Duralac Knob Handle, Lavatory Faucet

Model 512: Single Control, Duralac Knob Handle, Lavatory Faucet With Chain and Stopper

Model 522: Single Control, Duralac Knob Handle, Lavatory Faucet With Pop-Up Assembly

Model 532: Single Control, Duralac Knob Handle, Lavatory Faucet With Pop-Up Rod Only

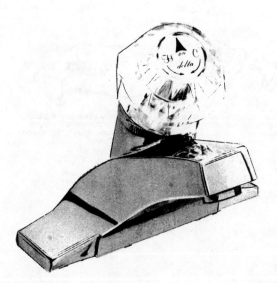

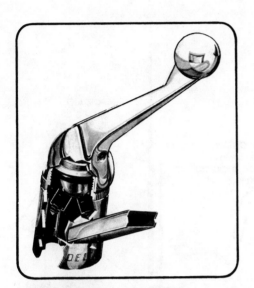

The Rotating Ball is the heart of the single control faucet. The patented design is the most successful and trouble-free single handle faucet on the market today. The ball has ports opening into the center of the ball. Water flow and temperature are controlled by simply lining up the ports in the ball with hot and cold inlet holes in the brass body. Pushing the handle back, controls the amount of flow. Move the handle to the right, you get cold water; move the handle to the left, you get hot water. There's an infinite control of temperature and flow in between. Both operations, flow and temperature, are controlled simultaneously.

Model 502: Single Control, Duralac Knob Handle, Lavatory Faucet

Model 512: Single Control, Duralac Knob Handle, Lavatory Faucet With Chain and Stopper

Model 522: Single Control, Duralac Knob Handle, Lavatory Faucet With Pop-Up Assembly

Model 532: Single Control, Duralac Knob Handle, Lavatory Faucet With Pop-Up Rod Only

Special Finish Available: Satin Chrome. Model 522 also available in Bright Gold and Satin Gold.

Meets federal specification WW-P-541/4—Type 4 — Oct. 1971

½″ I.P.S. adapters available. Specify by adding "W/F" to model number.

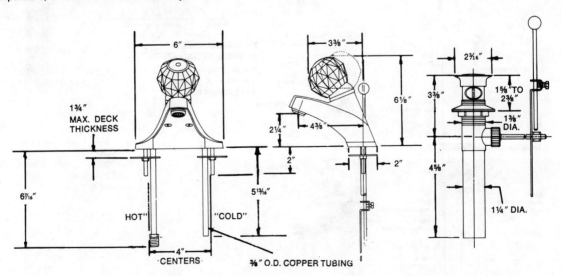

Delta Series 502, 512, 522, 532

SINGLE CONTROL,
KNOB HANDLE
LAVATORY
FAUCET

Installation Guide

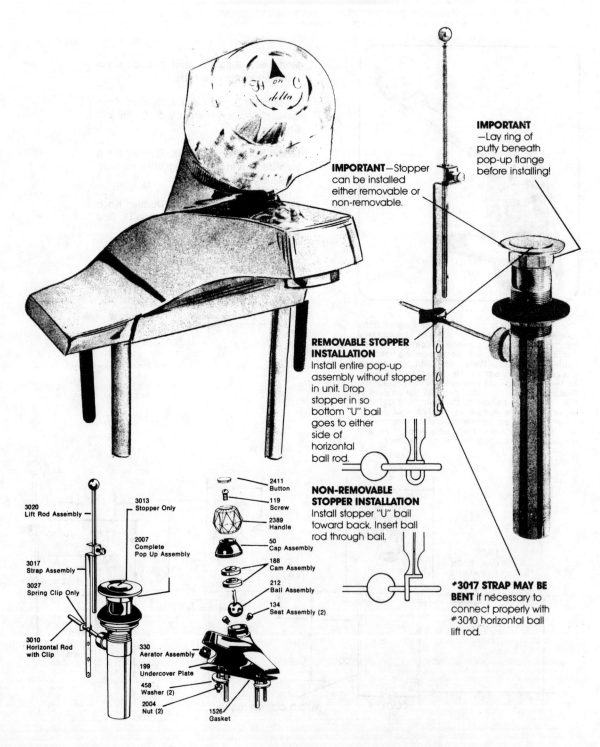

IMPORTANT—Stopper can be installed either removable or non-removable.

IMPORTANT—Lay ring of putty beneath pop-up flange before installing!

REMOVABLE STOPPER INSTALLATION
Install entire pop-up assembly without stopper in unit. Drop stopper in so bottom "U" bail goes to either side of horizontal ball rod.

NON-REMOVABLE STOPPER INSTALLATION
Install stopper "U" bail toward back. Insert ball rod through bail.

#3017 STRAP MAY BE BENT if necessary to connect properly with #3010 horizontal ball lift rod.

3020
Lift Rod Assembly

3013
Stopper Only

2007
Complete
Pop Up Assembly

3017
Strap Assembly

3027
Spring Clip Only

3010
Horizontal Rod
with Clip

2411
Button

119
Screw

2389
Handle

50
Cap Assembly

188
Cam Assembly

212
Ball Assembly

134
Seat Assembly (2)

330
Aerator Assembly

199
Undercover Plate

458
Washer (2)

2004
Nut (2)

1526
Gasket

Routine Maintenance Instructions

Delta faucets have earned the reputation of superiority in design, engineering, performace and durability in millions of home installations. Routine faucet maintenance to compensate for foreign materials in varying water conditions, will restore like new performance and extend the life of the faucet. Ease and simplicity of service and maintenance is another advantage of having Delta faucets throughout the home.

A. If you should have a leak under handle—tighten adjusting ring following steps 1 and 9. Reassemble as in step 10.

B. If you should have a leak from spout—shut off water supply, and follow steps 1, 2, 3, 4 and 5. Reassemble as in steps 6, 7 and 8. Set adjusting ring as in 9. Replace handle as in 10.

1. Pry off handle button, remove screw and lift off handle.

2. Unscrew cap assembly and lift off.

3. Remove cam assembly and ball by lifting up on ball stem.

4. Lift both seats and springs out of sockets in body.

5. Place seat over springs and insert into sockets in body (spring first).

6. Place ball into body over seats.

7. Place cam assembly over stem of ball and engage tab with slot in body. Push down.

8. Partially unscrew adjusting ring and then place cap assembly over ball stem and screw down tight onto body.

9. Tighten adjusting ring until no water will leak around stem when faucet is on and pressure is exerted on handle to force ball into socket.

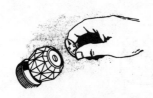

10. Replace handle. Tighten handle screw—tight. Replace handle butto with "ON" arrow pointing to rear of lavatory faucet and pointing up on shower or tub and shower faucet.

Single Control, Lever Handle, Lavatory Faucets for Single Hole Mounting

SUGGESTED SPECIFICATION: Single handle mixing faucets, lavatory deck, for single hole exposed mounting, 4″ swing spout and aerator. Lever type handle. Control mechanism shall be of the rotating ball type with replaceable non-metallic seats operating in stainless steel lined sockets. Control handle shall always return to the neutral position when valve is turned off. All models equipped with hose and spray shall have an integral anti-siphon device in the body of the valve. All exposed surfaces shall be chrome plated. Delta Models 570, 570-P, 575, 575-P or equal.

Model 570: Single Control, Lever Handle, Lavatory Faucet For Single Hole Mounting

Model 570-P: Single Control, Lever Handle, Lavatory Faucet For Single Hole Mounting With Pop-Up Assembly

Model 575: Single Control, Lever Handle, Lavatory Faucet For Single Hole Mounting With Shampoo Spray

Model 575-P: Single Control, Lever Handle, Lavatory Faucet For Single Hole Mounting With Shampoo Spray and Pop-Up Assembly

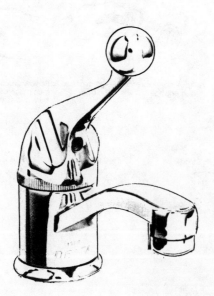

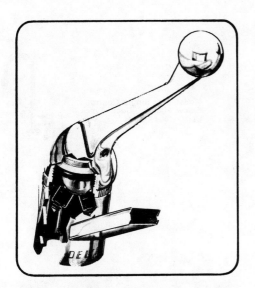

The Rotating Ball is the heart of the single control faucet. The patented design is the most successful and trouble-free single handle faucet on the market today. The ball has ports opening into the center of the ball. Water flow and temperature are controlled by simply lining up the ports in the ball with hot and cold inlet holes in the brass body. Pushing the handle back, controls the amount of flow. Move the handle to the right, you get cold water; move the handle to the left, you get hot water. There's an infinite control of temperature and flow in between. Both operations, flow and temperature, are controlled simultaneously.

Model 570: Single Control, Lever Handle, Lavatory Faucet For Single Hole Mounting

Model 570-P: Single Control, Lever Handle, Lavatory Faucet For Single Hole Mounting With Pop-Up Assembly

Model 575: Single Control, Lever Handle, Lavatory Faucet For Single Hole Mounting With Shampoo Spray

Model 575-P: Single Control, Lever Handle, Lavatory Faucet For Single Hole Mounting With Shampoo Spray and Pop-Up Assembly

Special Finish Available: Satin Chrome

½" I.P.S. adapters available. Specify by adding "W/F" to model number.

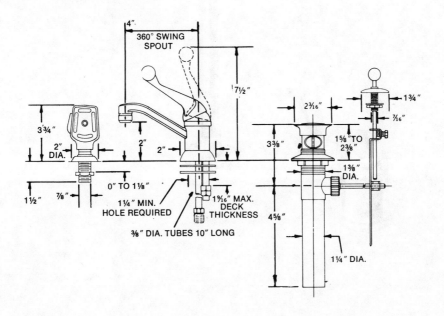

Single Control, Elbow Handle, Deck Type Commercial Faucets

SUGGESTED SPECIFICATION: Single handle mixing faucet, deck type exposed mounting 8″ center to center with 11″ gooseneck and aerator. Institutional type 6″ elbow handle. Control mechanism shall be of the rotating ball type with replaceable non-metallic seats operating in stainless steel lined sockets. Control handle shall always return to the neutral position when valve is turned off. All exposed surfaces shall be chrome plated. Delta Model 710 or equal.

Model 710: Single Control, Elbow Handle, Deck Type Commercial Faucet

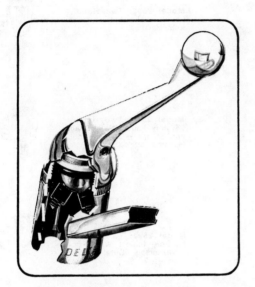

Model 710: Single Control, Elbow Handle, Deck Type Commercial Faucet

Meets federal specification WW-P-00541/9 May 1971 and WW-P-541/4A Oct. 1971

Bright chrome finish only

½" I.P.S. adapters available. Specify by adding "W/F" to model number.

The Rotating Ball is the heart of the single control faucet. The patented design is the most successful and trouble-free single handle faucet on the market today. The ball has ports opening into the center of the ball. Water flow and temperature are controlled by simply lining up the ports in the ball with hot and cold inlet holes in the brass body. Pushing the handle back, controls the amount of flow. Move the handle to the right, you get cold water; move the handle to the left, you get hot water. There's an infinite control of temperature and flow in between. Both operations, flow and temperature, are controlled simultaneously.

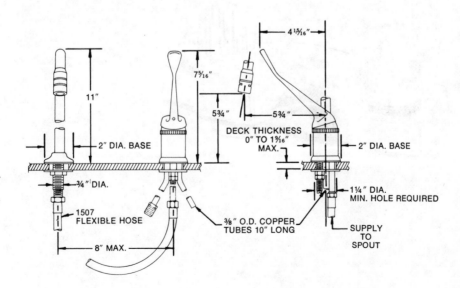

delex by delta

Two Handle Lavatory Faucets With Duralac Knob Handles— 4" Centersets

SUGGESTED SPECIFICATION: All lavatory faucets shall be top-mounting 4″ centersets with alcohol resistant Duralac knob handles (available as faucet only, with chain and stopper or with pop-up waste). Valve mechanisms shall be of the rotating cylinder type with 180° rotation. Hot and cold stem units shall be interchangeable. All operating parts must be replaceable from above the deck. Valves shall be equipped with replaceable non-metallic seats which are contained in stainless steel lined sockets. All exposed metal surfaces shall be chrome plated. Delta/Delex Models 2502, 2512, 2522 or equal.

Model 2502: Two Handle Lavatory Faucet—Duralac Knob Handles

Model 2512: Two Handle Lavatory Faucet With Chain and Stopper—Duralac Knob Handles

Model 2522: Two Handle Lavatory Faucet With Pop-Up Assembly—Duralac Knob Handles

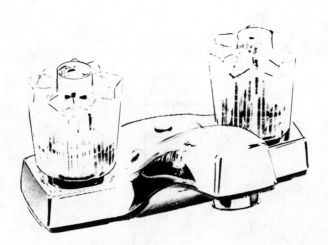

Here's the heart of the revolutionary Delex two handle faucet showing the simplicity of the rotating cylinder valve that ends washer headaches forever. There are no compression valves or washers to crush, tear or replace—the rotating cylinder valve simply shears off the water with no tightening down, no grinding, no force.

Two Handle Lavatory Faucets With Duralac Knob Handles— 4" Centers

Model 2502: Two Handle Lavatory Faucet —Duralac Knob Handles

Model 2512: Two Handle Lavatory Faucet With Chain and Stopper—Duralac Knob Handles

Model 2522: Two Handle Lavatory Faucet With Pop-Up Assembly—Duralac Knob Handles

Meets federal specifications WW-P-541/4A   Type 1 Oct. 1971

Bright chrome finish only

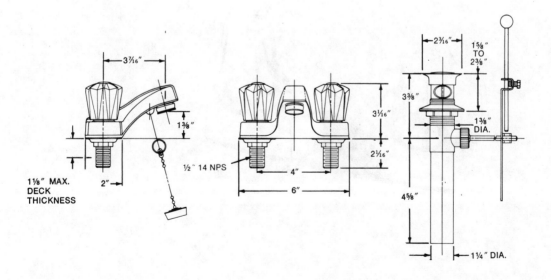

delex by delta

Two Handle Lavatory Faucets With Blade Handles— 4" Centersets

SUGGESTED SPECIFICATION: All lavatory faucets shall be top mounting 4" centersets with blade handle (available as faucet only, with chain and stopper or with pop-up waste.) Valve mechanisms shall be of the rotating cylinder type with 180° rotation. Hot and cold stem units shall be interchangeable. All operating parts must be replaceable from above the deck. Valves shall be equipped with replaceable non-metallic seats which are contained in stainless steel lined sockets. All exposed metal surfaces shall be chrome plated. Delta / Delex Models 2500, 2510, 2520 or equal.

Model 2500: Two Handle Lavatory Faucet — Blade Handles

Model 2510: Two Handle Lavatory Faucet With Chain and Stopper—Blade Handles

Model 2520: Two Handle Lavatory Faucet With Pop-Up Assembly—Blade Handles

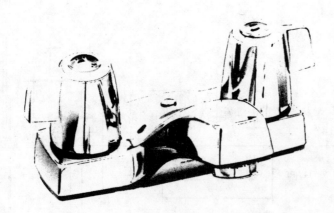

Here's the heart of the revolutionary Delex two handle faucet showing the simplicity of the rotating cylinder valve that ends washer headaches forever. There are no compression valves or washers to crush, tear or replace—the rotating cylinder valve simply shears off the water with no tightening down, no grinding, no force.

Two Handle Lavatory Faucets With Blade Handles— 4" Centersets

Model 2500: Two Handle Lavatory Faucet—Blade Handles

Model 2510: Two Handle Lavatory Faucet With Chain and Stopper—Blade Handles

Model 2520: Two Handle Lavatory Faucet With Pop-Up Assembly—Blade Handles

Meets federal specifications WW-P-541/4A   Type 1 Oct. 1971

Bright chrome finish only

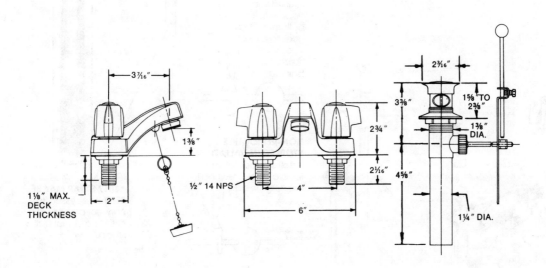

Delex Series

TWO HANDLE LAVATORY

BY DELTA

2500,

Installation Guide

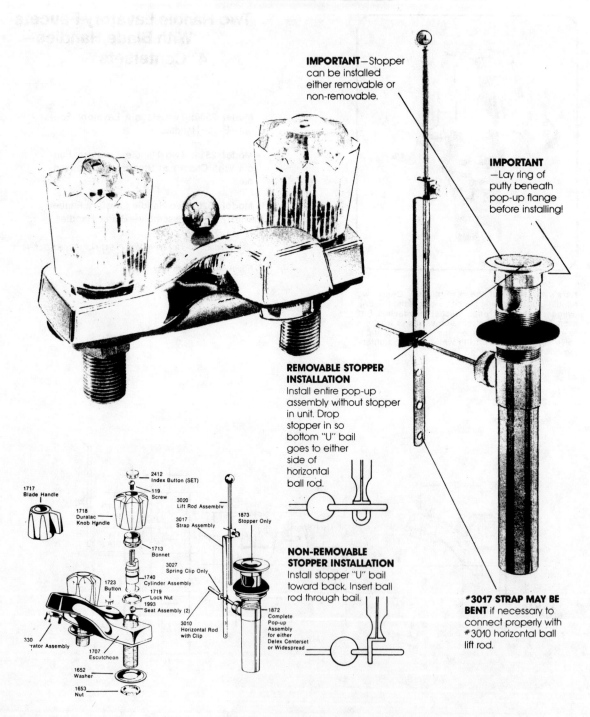

IMPORTANT—Stopper can be installed either removable or non-removable.

IMPORTANT —Lay ring of putty beneath pop-up flange before installing!

REMOVABLE STOPPER INSTALLATION
Install entire pop-up assembly without stopper in unit. Drop stopper in so bottom "U" bail goes to either side of horizontal ball rod.

NON-REMOVABLE STOPPER INSTALLATION
Install stopper "U" bail toward back. Insert ball rod through bail.

#3017 STRAP MAY BE BENT if necessary to connect properly with #3010 horizontal ball lift rod.

1717 Blade Handle

2412 Index Button (SET)

119 Screw

1718 Duralac Knob Handle

3020 Lift Rod Assembly

3017 Strap Assembly

1873 Stopper Only

1713 Bonnet

3027 Spring Clip Only

1723 Button

1740 Cylinder Assembly

1719 Lock Nut

1993 Seat Assembly (2)

3010 Horizontal Rod with Clip

1872 Complete Pop-up Assembly for either Delex Centerset or Widespread

330 Aerator Assembly

1707 Escutcheon

1652 Washer

1653 Nut

Routine Maintenance Instructions

Delex faucets have earned the reputation of superiority in design, engineering, performance and durability in rhillions of home installations. Routine faucet maintenance to compensate for foreign materials in varying water conditions, will restore like new performance and extend the life of the faucet. Ease and simplicity of service and maintenance is another advantage of having Delex faucets throughout the home.

A. If faucet leaks from under handle or from spout outlet, replace stem unit and/or seat as in steps 1, 2, 3, 4, 5. Reassemble following steps 6, 7 and 8.

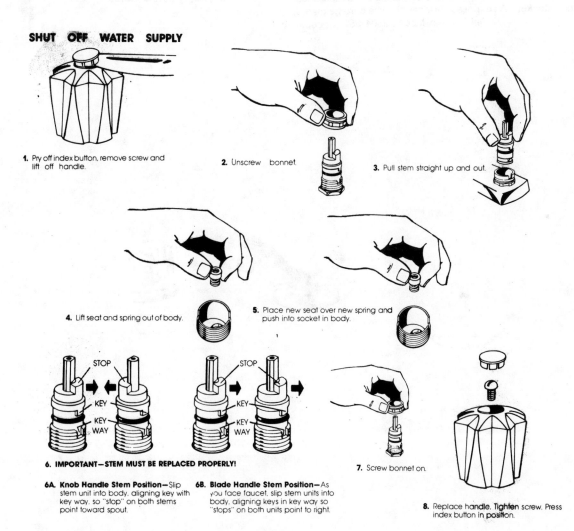

SHUT OFF WATER SUPPLY

1. Pry off index button, remove screw and lift off handle.

2. Unscrew bonnet.

3. Pull stem straight up and out.

4. Lift seat and spring out of body.

5. Place new seat over new spring and push into socket in body.

STOP   KEY   KEY WAY

STOP   KEY   KEY WAY

6. IMPORTANT—STEM MUST BE REPLACED PROPERLY!

6A. Knob Handle Stem Position—Slip stem unit into body, aligning key with key way, so "stop" on both stems point toward spout.

6B. Blade Handle Stem Position—As you face faucet, slip stem units into body, aligning keys in key way so "stops" on both units point to right.

7. Screw bonnet on.

8. Replace handle. Tighten screw. Press index button in position.

delex by delta

Two Handle Lavatory Faucets—Widespread Duralac Knob Handles

SUGGESTED SPECIFICATION: Lavatory faucets with alcohol resistant Duralac handles for concealed mounting from 6" to 16" centers. Valve mechanisms shall be the rotating cylinder type with 180° rotation. Hot and cold stem units shall be interchangeable. All operating parts must be replaceable from above the deck. Valves shall be equipped with replaceable non-metallic seats which are contained in stainless steel lined stockets. All exposed surfaces shall be chrome plated. Delta/Delex Models 3504, 3524 or equal.

Model 3504: Two Handle Lavatory Faucet — Widespread — Duralac Knob Handles

Model 3524: Two Handle Lavatory Faucet — Widespread — With Pop-Up Assembly Duralac Knob Handles

Here's the heart of the revolutionary Delex two handle faucet showing the simplicity of the rotating cylinder valve that ends washer headaches forever. There are no compression valves or washers to crush, tear or replace—the rotating cylinder valve simply shears off the water with no tightening down, no grinding, no force.

Two Handle Lavatory Faucets— Widespread Duralac Knob Handles

Model 3504: Two Handle Lavatory Faucet —Widespread—Duralac Knob Handles

Model 3524: Two Handle Lavatory Faucet — Widespread — With Pop-Up Assembly Duralac Knob Handles

Special Finishes Available: Satin Chrome, Bright Gold, or Satin Gold

½" I.P.S. adapters available. Specify by adding "W/F" to model number.

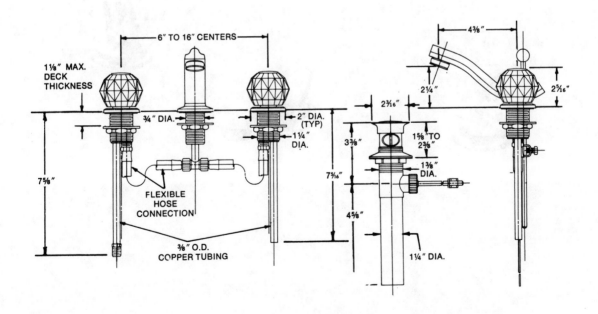

Single Control, Lavatory Faucet —Widespread Duralac Knob Handles—Both Faucet and Drain

SUGGESTED SPECIFICATION: Single control mixing lavatory faucet, for concealed mounting from 6 to 16" centers. Alcohol resistant Duralac handles. One handle controls both temperature and water flow. Second actuates pop-up type drain. Sculptured spout shall be solid brass with aerator. Control mechanism shall be of the rotating ball type with replaceable non-metallic seats operating in stainess steel lined sockets. All exposed surfaces shall be chrome plated. Delta Model 5240 or equal.

Model 5240: Single Control Lavatory Faucet—Widespread —Duralac Knob Handles— Both Faucet and Pop-Up Assembly

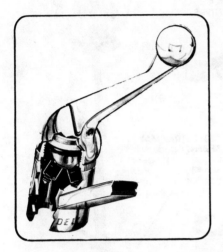

Here's the heart of the revolutionary Delex two handle faucet showing the simplicity of the rotating cylinder valve that ends washer headaches forever. There are no compression valves or washers to crush, tear or replace — the rotating cylinder valve simply shears off the water with no tightening down, no grinding, no force.

Single Control, Lavatory Faucet — Widespread Duralac Knob Handles Both Faucet and Drain

Model 5240: Single Control Lavatory Faucet—Widespread—Duralac Knob Handles —Both Faucet and Pop-Up Assembly

Special Finishes Available: Satin Chrome, Bright Gold, or Satin Gold

½" I.P.S. adapters available. Specify by adding "W/F" to model number.

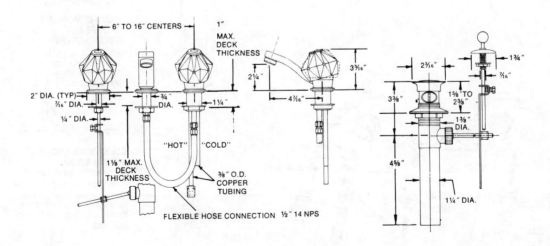

Delta SINGLE CONTROL WIDESPREAD FAUCETS Series 5240

Installation Guide

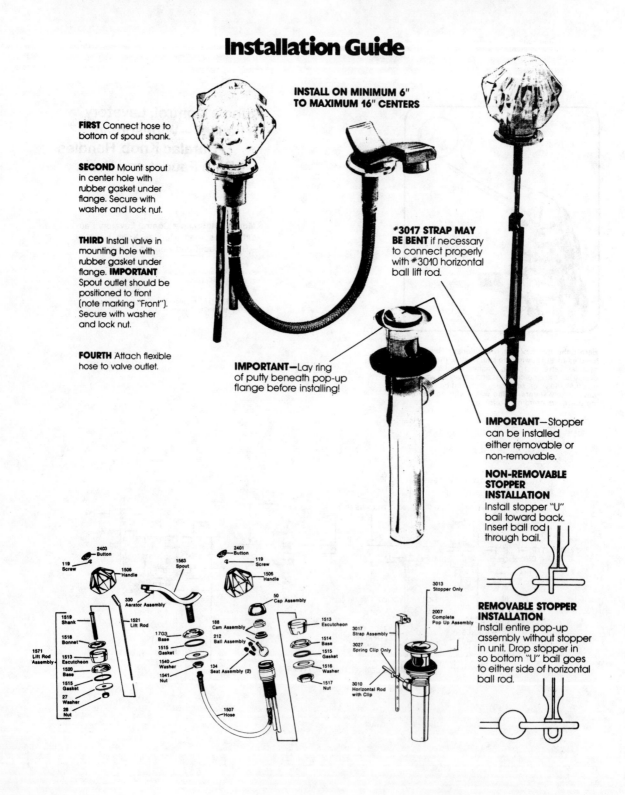

FIRST Connect hose to bottom of spout shank.

SECOND Mount spout in center hole with rubber gasket under flange. Secure with washer and lock nut.

THIRD Install valve in mounting hole with rubber gasket under flange. **IMPORTANT** Spout outlet should be positioned to front (note marking "Front"). Secure with washer and lock nut.

FOURTH Attach flexible hose to valve outlet.

INSTALL ON MINIMUM 6" TO MAXIMUM 16" CENTERS

#3017 STRAP MAY BE BENT if necessary to connect properly with #3010 horizontal ball lift rod.

IMPORTANT—Lay ring of putty beneath pop-up flange before installing!

IMPORTANT—Stopper can be installed either removable or non-removable.

NON-REMOVABLE STOPPER INSTALLATION
Install stopper "U" bail toward back. Insert ball rod through bail.

REMOVABLE STOPPER INSTALLATION
Install entire pop-up assembly without stopper in unit. Drop stopper in so bottom "U" bail goes to either side of horizontal ball rod.

Routine Maintenance Instructions

Delta faucets have earned the reputation of superiority in design, engineering, performance and durability in millions of home installations. Routine faucet maintenance to compensate for foreign materials in varying water conditions, will restore like new performance and extend the life of the faucet. Ease and simplicity of service and maintenance is another advantage of having Delta faucets throughout the home.

A. If you should have a leak under handle—tighten adjusting ring following steps 1 and 9. Reassemble as in step 10.

B. If you should have a leak from spout—shut off water supply, and follow steps 1, 2, 3, 4 and 5. Reassemble as in steps 6, 7 and 8. Set adjusting ring as in 9. Replace handle as in 10.

1. Pry off handle button, remove screw and lift off handle.

2. Unscrew cap assembly and lift off.

3. Remove cam assembly and ball by lifting up on ball stem.

4. Lift both seats and springs out of sockets in body.

5. Place seat over springs and insert into sockets in body (spring first).

6. Place ball into body over seats.

7. Place cam assembly over stem of ball and engage tab with slot in body. Push down.

8. Partially unscrew adjusting ring and then place cap assembly over ball stem and screw down tight onto body.

9. Tighten adjusting ring until no water will leak around stem when faucet is on and pressure is exerted on handle to force ball into socket.

10. Replace handle. Tighten handle screw—tight. Replace handle button.

delta

SERIES
600
**LEVER HANDLE
PRESSURE BALANCE
PUSH BUTTON
DIVERTER CHECK
STOPS**

Single Control, Lever Handle Bath Mixing Valves—Pressure Balance, Check Stops & Push Button Diverters

SUGGESTED SPECIFICATION: Single handle bath mixing valve pressure balancing concealed type for tub and shower. Lever type handle. Control mechanism shall be of the rotating ball type with replaceable non-metallic seats operating in stainless steel lined sockets. Control handle shall always return to the neutral position when valve is turned off. Valves shall be equipped with pressure balance spool and sleeve device of 302 stainless steel which will maintain a pre-set mix of hot and cold water and comply with Federal Specification WW-P-541. Valve shall be equipped with check stops to prevent cross flow. All parts shall be replaceable from the front of the valve. All exposed surfaces shall be chrome plated. Delta Models 607, 637 or equal.

Model 607: Single Control, Lever Handle Bath Mixing Valve—Pressure Balance, Check Stops & Push Button Diverter

Model 637: Single Control, Lever Handle Bath Mixing Valve—Pressure Balance, Check Stops & Push Button Diverter With Tub & Shower Assembly

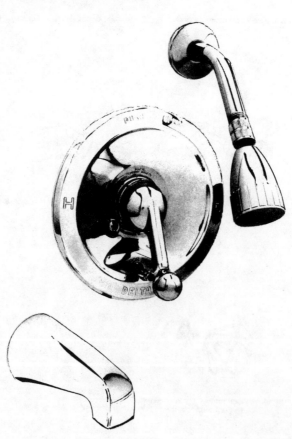

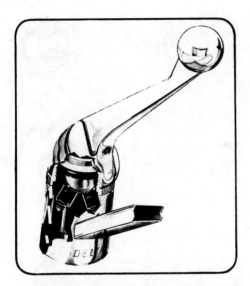

The Rotating Ball is the heart of the single control faucet. The patented design is the most successful and trouble-free single handle faucet on the market today. The ball has ports opening into the center of the ball. Water flow and temperature are controlled by simply lining up the ports in the ball with hot and cold inlet holes in the brass body. Pushing the handle back, controls the amount of flow. Move the handle to the right, you get cold water; move the handle to the left, you get hot water. There's an infinite control of temperature and flow in between. Both operations, flow and temperature, are controlled simultaneously.

Single Control, Lever Handle Bath Mixing Valves—Pressure Balance, Check Stops & Push Button Diverters

Model 607: Single Control, Lever Handle Bath Mixing Valve — Pressure Balance, Check Stops and Push Button Diverter

Model 637: Single Control, Lever Handle Bath Mixing Valve — Pressure Balance, Check Stops and Push Button Diverter With Tub and Shower Assembly

Special Finish Available: Satin Chrome, DH Models Only

Meets federal specification WW-P-541/7A Type P Class 2 Oct. 1972

Sweat inlets available on all chrome bath valves. Specify by adding "C" to model number. Not available on special finish models.

Deluxe showerhead featuring cam control provides variable shower patterns. Available on all shower models. Specify by adding "DH" to model number.

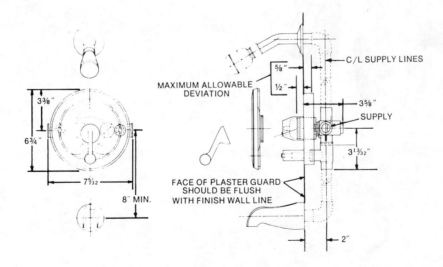

delta

SERIES
600
LEVER HANDLE
PRESSURE
BALANCE
CHECK STOPS

Single Control, Lever Handle Bath Mixing Valves—Pressure Balance, & Check Stops

SUGGESTED SPECIFICATION: Single handle bath mixing valve pressure balancing concealed type for tub and shower. Lever type handle. Control mechanism shall be of the rotating ball type with replaceable non-metallic seats operating in stainless steel lined sockets. Control handle shall always return to the neutral position when valve is turned off. Valves shall be equipped with pressure balance spool and sleeve device of 302 stainless steel which will maintain a pre-set mix of hot and cold water and comply with Federal Specification WW-P-541. Valve shall be equipped with check stops to prevent cross flow. All parts shall be replaceable from front of the valve. All exposed surfaces shall be chrome plated. Delta Models 603, 623, 633, 643 or equal.

Model 603: Single Control, Lever Handle Bath Mixing Valve—Pressure Balance & Check Stops

Model 623: Single Control, Lever Handle Bath Mixing Valve—Pressure Balance & Check Stops With Shower Assembly

Model 633: Single Control, Lever Handle Bath Mixing Valve—Pressure Balance & Check Stops With Shower Assembly and Brass Diverter Spout.

Model 643: Single Control, Lever Handle Bath Mixing Valve—Pressure Balance & Check Stops With Tub and Shower Assembly.

Single Control, Lever Handle Bath Mixing Valves—Pressure Balance, & Check Stops

Model 603: Single Control, Lever Handle Bath Mixing Valve—Pressure Balance and Check Stops

Model 623: Single Control, Lever Handle Bath Mixing Valve—Pressure Balance and Check Stops With Shower Assembly

Model 633: Single Control, Lever Handle Bath Mixing Valve—Pressure Balance & Check Stops With Shower Assembly and Brass Diverter Spout.

Model 643: Single Control, Lever Handle Bath Mixing Valve—Presure Balance & Check Stops With Tub and Shower Assembly.   Bright Chrome Only

Special Finish Available: Satin Chrome, DH Models Only

Meets federal specification WW-P-541/7A Type P Class 2 Oct. 1972

Sweat inlets available on all chrome bath valves. Specify by adding "C" to model number. Not available on special finish models.

Deluxe showerhead featuring cam control provides variable shower patterns. Available on all shower models. Specify by adding "DH" to model number.

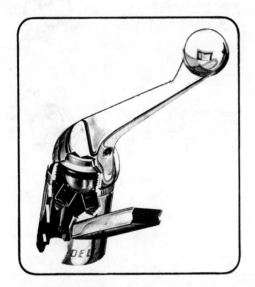

The Rotating Ball is the heart of the single control faucet. The patented design is the most successful and trouble-free single handle faucet on the market today. The ball has ports opening into the center of the ball. Water flow and temperature are controlled by simply lining up the ports in the ball with hot and cold inlet holes in the brass body. Pushing the handle back, controls the amount of flow. Move the handle to the right, you get cold water; move the handle to the left, you get hot water. There's an infinite control of temperature and flow in between. Both operations, flow and temperature, are controlled simultaneously.

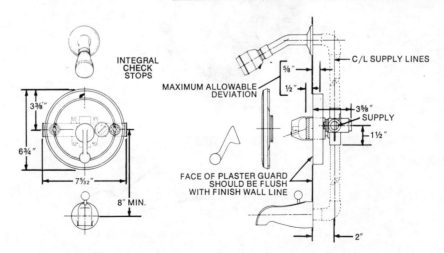

delta

Single Control, Lever Handle Bath Mixing Valves— Push Button Diverters

SUGGESTED SPECIFICATION: Single handle bath mixing valve, concealed type for tub and shower. Lever type handle. Control mechanism shall be of the rotating ball type with replaceable non-metallic seats operating in stainless steel lined sockets. Control handle shall always return to the neutral position when valve is turned off. All parts shall be replaceable from the front of the valve. All exposed surfaces shall be chrome plated. Delta Models 605, 635 or equal.

Model 605: Single Control, Lever Handle, Bath Mixing Valve—Push Button Diverter

Model 635: Single Control, Lever Handle, Bath Mixing Valve—Push Button Diverter With Tub and Shower Assembly

The Rotating Ball is the heart of the single control faucet. The patented design is the most successful and trouble-free single handle faucet on the market today. The ball has ports opening into the center of the ball. Water flow and temperature are controlled by simply lining up the ports in the ball with hot and cold inlet holes in the brass body. Pushing the handle back, controls the amount of flow. Move the handle to the right, you get cold water; move the handle to the left, you get hot water. There's an infinite control of temperature and flow in between. Both operations, flow and temperature, are controlled simultaneously.

Single Control, Lever Handle Bath Mixing Valves— Push Button Diverters

Model 605: Single Control, Lever Handle, Bath Mixing Valve—Push Button Diverter

Model 635: Single Control, Lever Handle, Bath Mixing Valve—Push Button Diverter With Tub and Shower Assembly

Special Finish Available: Satin Chrome   DH Models Only

Meets federal specification WW-P-541/7A Type M Class 2 Oct. 1972

Sweat inlets available on all chrome bath valves. Specify by adding "C" to model number. Not available on special finish models.

Integral screw driver stops available on all models. Specify by adding "With Stops" to model numbers.

Deluxe showerhead featuring cam control provides variable shower patterns. Available on all shower models. Specify by adding "DH" to model number.

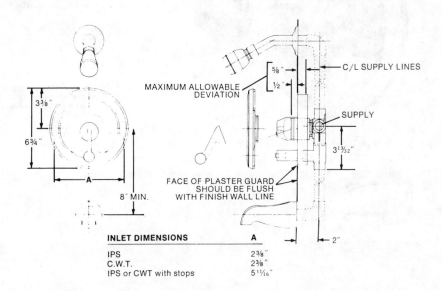

INLET DIMENSIONS	A
IPS	2⅜"
C.W.T.	2⅜"
IPS or CWT with stops	5¹¹⁄₁₆"

Single Control, Lever Handle Bath Mixing Valves

SUGGESTED SPECIFICATION: Single handle bath mixing valve, concealed type for tub and shower. Lever type handle. Control mechanism shall be of the rotating ball type with replaceable non-metallic seats operating in stainless steel lined sockets. Control handle shall always return to the neutral position when valve is turned off. All parts shall be replaceable from the front of the valve. All exposed surfaces shall be chrome plated. Delta Models 601, 611, 621, 631, 641 or equal.

Model 601: Single Control, Lever Handle, Bath Mixing Valve

Model 611: Single Control, Lever Handle, Bath Mixing Valve With Tub Spout

Model 621: Single Control, Lever Handle, Bath Mixing Valve With Shower Assembly

Model 631: Single Control, Lever Handle, Bath Mixing Valve With Shower Assembly and Brass Diverter Spout.

Model 641: Single Control, Lever Handle, Bath Mixing Valve With Tub and Shower Assembly.

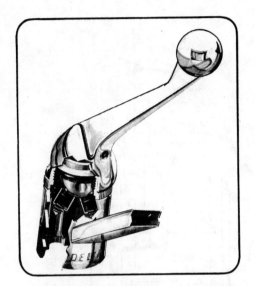

The Rotating Ball is the heart of the single control faucet. The patented design is the most successful and trouble-free single handle faucet on the market today. The ball has ports opening into the center of the ball. Water flow and temperature are controlled by simply lining up the ports in the ball with hot and cold inlet holes in the brass body. Pushing the handle back, controls the amount of flow. Move the handle to the right, you get cold water; move the handle to the left, you get hot water. There's an infinite control of temperature and flow in between. Both operations, flow and temperature, are controlled simultaneously.

Single Control, Lever Handle Bath Mixing Valves

Model 601: Single Control, Lever Handle, Bath Mixing Valve

Model 611: Single Control, Lever Handle, Bath Mixing Valve With Tub Spout

Model 621: Single Control, Lever Handle, Bath Mixing Valve With Shower Assembly

Model 631: Single Control, Lever Handle, Bath Mixing Valve With Shower Assembly and Brass Diverter Spout.

Model 641: Single Control, Lever Handle, Bath Mixing Valve With Tub and Shower Assembly.

Special Finish Available: Satin Chrome

Meets federal specification WW-P-541/7A Type M Class 2 Oct. 1972

Sweat inlets available on all chrome bath valves. Specify by adding "C" to model number. Not available on special finish models.

Integral screw driver stops available on all models. Specify by adding "With Stops" to model numbers.

Deluxe showerhead featuring cam control provides variable shower patterns. Available on all shower models. Specify by adding "DH" to model number.

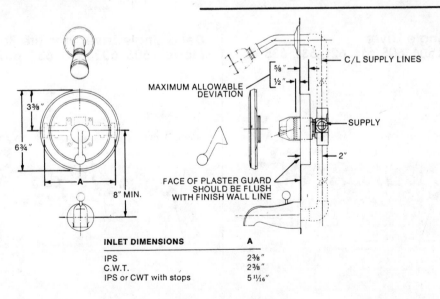

MAXIMUM ALLOWABLE DEVIATION

C/L SUPPLY LINES

5/8"

1/2"

SUPPLY

2"

3⅜"

6¾"

A

8" MIN.

FACE OF PLASTER GUARD SHOULD BE FLUSH WITH FINISH WALL LINE

INLET DIMENSIONS	A
IPS	2⅜"
C.W.T.	2⅜"
IPS or CWT with stops	5¹¹⁄₁₆"

Delta SINGLE LEVER BATH VALVES Series
Single lever with or without pressure balance

Installation Guide
(Pressure Balance Model Illustrated)

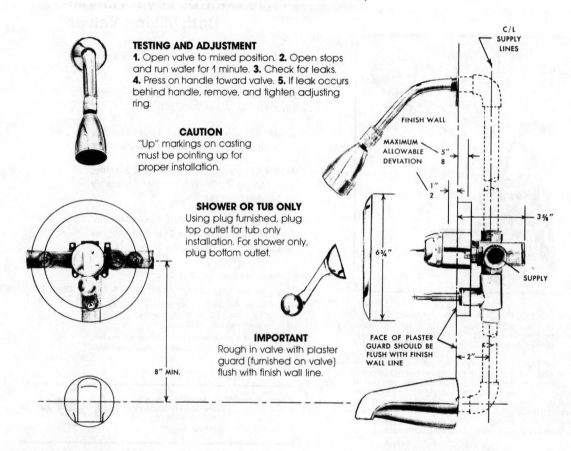

TESTING AND ADJUSTMENT
1. Open valve to mixed position. **2.** Open stops and run water for 1 minute. **3.** Check for leaks. **4.** Press on handle toward valve. **5.** If leak occurs behind handle, remove, and tighten adjusting ring.

CAUTION
"Up" markings on casting must be pointing up for proper installation.

SHOWER OR TUB ONLY
Using plug furnished, plug top outlet for tub only installation. For shower only, plug bottom outlet.

IMPORTANT
Rough in valve with plaster guard (furnished on valve) flush with finish wall line.

C/L SUPPLY LINES

FINISH WALL

MAXIMUM ALLOWABLE DEVIATION — 5" / 8

1" / 2

3⅝"

SUPPLY

6¾"

FACE OF PLASTER GUARD SHOULD BE FLUSH WITH FINISH WALL LINE

2"

8" MIN.

Delta Single Lever
Models 601, 605, 611, 621, 631, 635

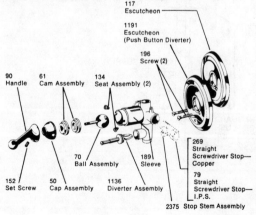

117 Escutcheon
1191 Escutcheon (Push Button Diverter)
196 Screw (2)
90 Handle
61 Cam Assembly
134 Seat Assembly (2)
152 Set Screw
50 Cap Assembly
70 Ball Assembly
1136 Diverter Assembly
189 Sleeve
269 Straight Screwdriver Stop—Copper
79 Straight Screwdriver Stop—I.P.S.
2375 Stop Stem Assembly

Delta Single Lever Pressure Balance
Models 603, 607, 623, 633, 637

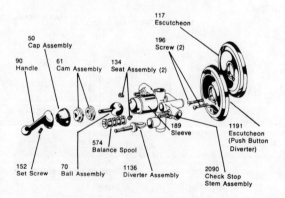

117 Escutcheon
196 Screw (2)
50 Cap Assembly
90 Handle
61 Cam Assembly
134 Seat Assembly (2)
152 Set Screw
70 Ball Assembly
574 Balance Spool
1136 Diverter Assembly
189 Sleeve
1191 Escutcheon (Push Button Diverter)
2090 Check Stop Stem Assembly

Routine Maintenance Instructions

Delta faucets have earned the reputation of superiority in design, engineering, performance and durability in millions of home installations. Routine faucet maintenance to compensate for foreign materials in varying water conditions can restore like new and extend the life and performance of the faucet. And, ease and simplicity of service and maintenance is another advantage of having Delta faucets throughout the home.

A. If you should have leak under handle—tighten adjusting ring by following steps 1 and 9 leaving water supply and faucet turned on. Replace handle as in step 10.

B. If you should have leak from spout—replace seats and springs by following steps 1, 2, 3 and 4. Reassemble following steps 5, 6, 7 and 8. Turn water supply on and follow steps 9 and 10.

Shut off water supply

1. Loosen set screw and lift off handle.

2. Unscrew cap assembly and lift off.

3. Remove cam assembly and ball by lifting up on ball stem.

4. Lift both seats and springs out of sockets in body.

To reassemble faucet

5. Place seat over springs and insert into sockets in body (spring first).

6. Place ball into body over seats.

7. Place cam assembly over stem of ball and engage tab with slot in body. Push down.

8. Partially unscrew adjusting ring and then place cap assembly over ball stem and screw down tight onto body.

Turn on water supply

9. Tighten ring until no water will leak around stem when faucet is on and pressure is exerted on handle to force ball into socket.

10. Replace handle. Tighten handle screw—tight.

PRESSURE BALANCED VALVES

If unable to maintain constant temperature, clean balancing spool as follows:

1. Remove handle and escutcheon.

2. Close stops.

3. Unscrew balancing spool assembly (part No. 574) and remove—slowly— so as not to damage "O" Ring Seals. NOTE: On shower only installations water trapped in shower riser will drain out when balancing spool assembly is removed.

4. Examine valve hole for chips or any other foreign matter.

5. Remove any chips from spool sleeve before attempting to remove spool from inside of sleeve.

6. Remove spool from sleeve and clean all deposits from both sleeve and spool.

7. When spool will slide freely in sleeve, replace assembly in original position.

8. Open stops and check cap of spool assembly and stems of check stops for leaks before reinstalling escutcheon.

Single Control, Duralac Knob Handle Bath Mixing Valves

SUGGESTED SPECIFICATION: Single handle bath mixing valve concealed type for tub and shower. Alcohol resistant Duralac knob handle. Control mechanism shall be of the rotating ball type with replaceable non-metallic seats operating in stainless steel lined sockets. All parts shall be replaceable from the front of the valve. All exposed surfaces shall be chrome plated. Delta Models 602, 612, 622, 632, 642 or equal.

Model 602: Single Control, Duralac Knob Handle, Bath Mixing Valve

Model 612: Single Control, Duralac Knob Handle, Bath Mixing Valve With Tub Spout

Model 622: Single Control, Duralac Knob Handle, Bath Mixing Valve With Shower Assembly

Model 632: Single Control, Duralac Knob Handle Bath Mixing Valve With Shower Assembly and Brass Diverter Spout.

Model 642: Single Control, Duralac Knob Handle Bath Mixing Valve With Tub and Shower Assembly.

Single Control, Duralac Knob Handle Bath Mixing Valves

Model 602: Single Control, Duralac Knob Handle, Bath Mixing Valve

Model 612: Single Control, Duralac Knob Handle, Bath Mixing Valve With Tub Spout

Model 622: Single Control, Duralac Knob Handle, Bath Mixing Valve With Shower Assembly

Model 632: Single Control, Duralac Knob Handle Bath Mixing Valve With Shower Assembly and Brass Diverter Spout.

Model 642: Single Control, Duralac Knob Handle Bath Mixing Valve With Tub and Shower Assembly.   Bright Chrome Only

Special Finishes Available: Satin Chrome, Bright Gold, or Satin Gold, DH Models Only

Meets federal specification WW-P-541/7A Type M Class 2 Oct. 1972

Sweat inlets available on all chrome bath valves. Specify by adding "C" to model number. Not available on special finish models.

Integral screw driver stops available on all models. Specify by adding "With Stops" to model numbers.

Deluxe showerhead featuring cam control provides variable shower patterns. Available on all shower models. Specify by adding "DH" to model number.

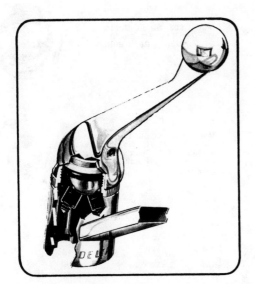

The Rotating Ball is the heart of the single control faucet. The patented design is the most successful and trouble-free single handle faucet on the market today. The ball has ports opening into the center of the ball. Water flow and temperature are controlled by simply lining up the ports in the ball with hot and cold inlet holes in the brass body. Pushing the handle back, controls the amount of flow. Move the handle to the right, you get cold water; move the handle to the left, you get hot water. There's an infinite control of temperature and flow in between. Both operations, flow and temperature, are controlled simultaneously.

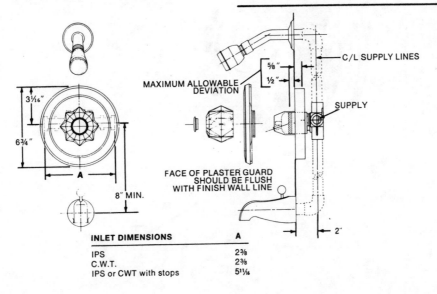

INLET DIMENSIONS	A
IPS	2⅜
C.W.T.	2⅜
IPS or CWT with stops	5¹¹⁄₁₆

Single Control, Duralac Knob Handle Bath Mixing Valves— Push Button Diverters

SUGGESTED SPECIFICATION: Single handle bath mixing valve concealed type for tub and shower. Alcohol resistant Duralac knob handle. Control mechanism shall be of the rotating ball type with replaceable non-metallic seats operating in stainless steel lined sockets. All parts shall be replaceable from the front of the valve. All exposed surfaces shall be chrome plated. Delta Models 606, 636 or equal.

Model 606: Single Control, Duralac Knob Handle, Bath Mixing Valve—Push Button Diverter

Model 636: Single Control, Duralac Knob Handle, Bath Mixing Valve—Push Button Diverter with Tub and Shower Assembly

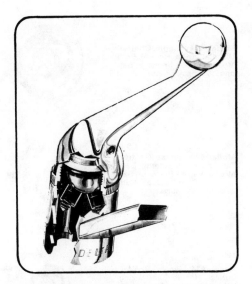

The Rotating Ball is the heart of the single control faucet. The patented design is the most successful and trouble-free single handle faucet on the market today. The ball has ports opening into the center of the ball. Water flow and temperature are controlled by simply lining up the ports in the ball with hot and cold inlet holes in the brass body. Pushing the handle back, controls the amount of flow. Move the handle to the right, you get cold water; move the handle to the left, you get hot water. There's an infinite control of temperature and flow in between. Both operations, flow and temperature, are controlled simultaneously.

Single Control, Duralac Knob Handle Bath Mixing Valves— Push Button Diverters

Model 606: Single Control, Duralac Knob Handle, Bath Mixing Valve—Push Button Diverter

Model 636: Single Control, Duralac Knob Handle, Bath Mixing Valve—Push Button Diverter With Tub and Shower Assembly

Special Finishes Available: Satin Chrome, Bright Gold, or Satin Gold, DH Models Only

Meets federal specification WW-P-541/7A Type M Class 2 Oct. 1972

Sweat inlets available on all chrome bath valves. Specify by adding "C" to model number. Not available on special finish models.

Integral screw driver stops available on all models. Specify by adding "With Stops" to model numbers.

Deluxe showerhead featuring cam control provides variable shower patterns. Available on all shower models. Specify by adding "DH" to model number.

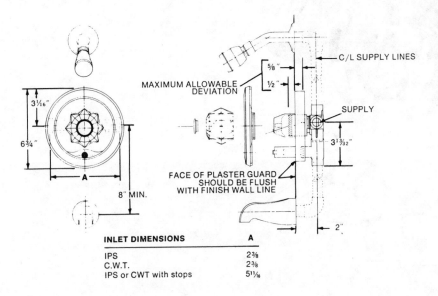

INLET DIMENSIONS	A
IPS	2⅜
C.W.T.	2⅜
IPS or CWT with stops	5¹¹⁄₁₆

SERIES
600
KNOB HANDLE
PRESSURE
BALANCE
CHECK STOPS

Single Control, Duralac Knob Handle Bath Mixing Valves— Pressure Balance & Check Stops

SUGGESTED SPECIFICATION: Single handle bath mixing valve pressure balance concealed type for tub and shower. Alcohol resistant Duralac knob handle. Control mechanism shall be of the rotating ball type with replacable non-metallic seats operating in stainless steel lined sockets. Valves shall be equipped with pressure balance spool and sleeve device of 302 stainless steel which will maintain a pre-set mix of hot and cold water and comply with Federal Specification WW-P-541. All parts shall be replaceable from the front of the valve. All exposed surfaces **shall** be chrome plated. Delta Models 604, 624, 634, 644 or equal.

Model 604: Single Control, Duralac Knob Handle Bath Mixing Valve—Pressure Balance & Check Stops

Model 624: Single Control, Duralac Knob Handle Bath Mixing Valve—Pressure Balance & Check Stops With Shower Assembly

Model 634: Single Control, Duralac Handle Bath Mixing Valve—Pressure Balance & Check Stops With Shower Assembly and Brass Diverter Spout.

Model 644: Single Control, Duralac Knob Handle Bath Mixing Valve—Pressure Balance & Check Stops With Tub and Shower Assembly.

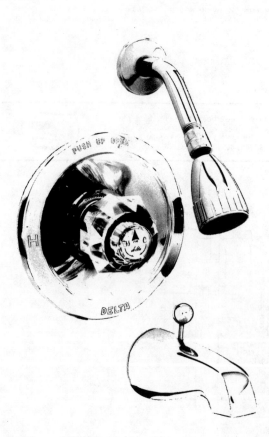

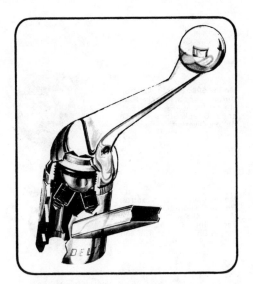

The Rotating Ball is the heart of the single control faucet. The patented design is the most successful and trouble-free single handle faucet on the market today. The ball has ports opening into the center of the ball. Water flow and temperature are controlled by simply lining up the ports in the ball with hot and cold inlet holes in the brass body. Pushing the handle back, controls the amount of flow. Move the handle to the right, you get cold water; move the handle to the left, you get hot water. There's an infinite control of temperature and flow in between. Both operations, flow and temperature, are controlled simultaneously.

Single Control, Duralac Knob Handle Bath Mixing Valves—Pressure Balance & Check Stops

Model 604: Single Control, Duralac Knob Handle Bath Mixing Valve—Pressure Balance and Check Stops

Model 624: Single Control, Duralac Knob Handle Bath Mixing Valve—Pressure Balance and Check Stops With Shower Assembly

Model 634: Single Control, Duralac Knob Handle Bath Mixing Valve—Pressure Balance & Check Stops With Shower Assembly and Brass Diverter Spout.

Model 644: Single Control, Duralac Knob Handle Bath Mixing Valve—Pressure Balance & Check Stops With Tub and Shower Assembly. Bright Chrome Only

Special Finishes Available: Satin Chrome, Bright Gold, or Satin Gold, DH Models Only

Meets federal specification WW-P-541/7A Type P Class 2 Oct. 1972

Sweat inlets available on all chrome bath valves. Specify by adding "C" to model number. Not available on special finish models.

Deluxe showerhead featuring cam control provides variable shower patterns. Available on all shower models. Specify by adding "DH" to model number.

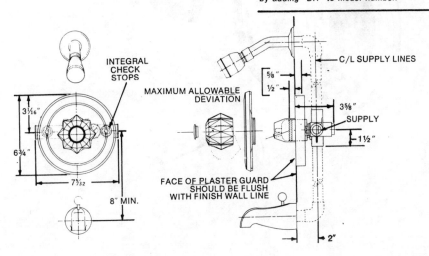

SERIES
600
**KNOB HANDLE
PRESSURE BALANCE
PUSH BUTTON
DIVERTER CHECK
STOPS**

Single Control, Duralac Knob Handle Bath Mixing Valves—Pressure Balance, Check Stops & Push Button Diverters

SUGGESTED SPECIFICATION: Single handle bath mixing valve, pressure balanced concealed pipe for tub and shower. Alcohol resistant Duralac knob handle. Control mechanism shall be of the rotating ball type with replaceable non-metallic seats operating in stainless steel lined sockets. Valves shall be equipped with pressure balance spool and sleeve device of 302 stainless steel which will maintain a pre-set mix of hot and cold water and comply with Federal Specification WW-P-541. All parts shall be replaceable from the front of the valve. All exposed surfaces shall be chrome plated.
Delta Models 608, 638 or equal.

Model 608: Single Control, Duralac Knob Handle Bath Mixing Valve—Pressure Balance, Check Stops & Push Button Diverter

Model 638: Single Control, Duralac Knob Handle Bath Mixing Valve—Pressure Balance, Check Stops & Push Button Diverter With Tub and Shower Assembly

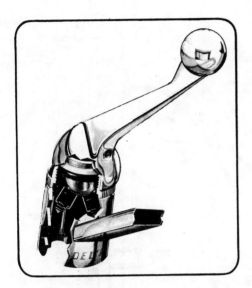

The Rotating Ball is the heart of the single control faucet. The patented design is the most successful and trouble-free single handle faucet on the market today. The ball has ports opening into the center of the ball. Water flow and temperature are controlled by simply lining up the ports in the ball with hot and cold inlet holes in the brass body. Pushing the handle back, controls the amount of flow. Move the handle to the right, you get cold water; move the handle to the left, you get hot water. There's an infinite control of temperature and flow in between. Both operations, flow and temperature, are controlled simultaneously.

Single Control, Duralac Knob Handle Bath Mixing Valves—Pressure Balance, Check Stops & Push Button Diverters

Model 608: Single Control, Duralac Knob Handle, Bath Mixing Valve—Pressure Balance, Check Stops & Push Button Diverter

Model 638: Single Control, Duralac Knob Handle, Bath Mixing Valve—Pressure Balance, Check Stops & Push Button Diverter with Tub and Shower Assembly

Special Finishes Available: Satin Chrome, Bright Gold, or Satin Gold, DH Models Only

Meets federal specification WW-P-541/7A Type P Class 2 Oct. 1972

Sweat inlets available on all chrome bath valves. Specify by adding "C" to model number. Not available on special finish models.

Deluxe showerhead featuring cam control provides variable shower patterns. Available on all shower models. Specify by adding "DH" to model number.

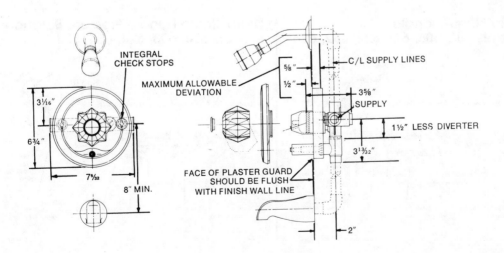

Delta Series

SINGLE HANDLE BATH VALVES

Single handle with or without pressure balance

Installation Guide
(Pressure Balance Model Illustrated)

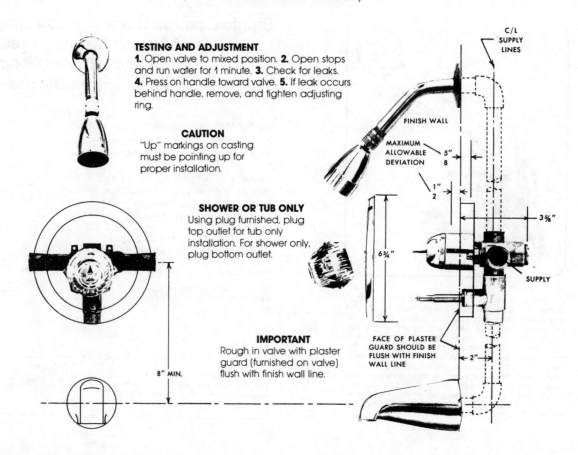

TESTING AND ADJUSTMENT
1. Open valve to mixed position. **2.** Open stops and run water for 1 minute. **3.** Check for leaks. **4.** Press on handle toward valve. **5.** If leak occurs behind handle, remove, and tighten adjusting ring.

CAUTION
"Up" markings on casting must be pointing up for proper installation.

SHOWER OR TUB ONLY
Using plug furnished, plug top outlet for tub only installation. For shower only, plug bottom outlet.

IMPORTANT
Rough in valve with plaster guard (furnished on valve) flush with finish wall line.

C/L SUPPLY LINES

FINISH WALL

MAXIMUM ALLOWABLE DEVIATION

5"/8

1"/2

3 5/8"

SUPPLY

6¾"

FACE OF PLASTER GUARD SHOULD BE FLUSH WITH FINISH WALL LINE

2"

8" MIN.

Delta Single Handle
Models 602, 606, 612, 622, 632, 636

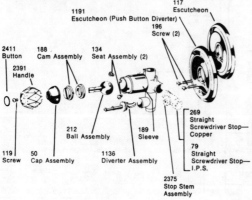

2411 Button
2391 Handle
188 Cam Assembly
134 Seat Assembly (2)
1191 Escutcheon (Push Button Diverter)
196 Screw (2)
117 Escutcheon
119 Screw
50 Cap Assembly
212 Ball Assembly
1136 Diverter Assembly
189 Sleeve
269 Straight Screwdriver Stop—Copper
79 Straight Screwdriver Stop—I.P.S.
2375 Stop Stem Assembly

Delta Single Handle Pressure Balance
Models 604, 608, 624, 634, 638

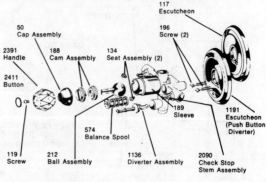

50 Cap Assembly
2391 Handle
2411 Button
188 Cam Assembly
134 Seat Assembly (2)
196 Screw (2)
117 Escutcheon
119 Screw
212 Ball Assembly
574 Balance Spool
1136 Diverter Assembly
189 Sleeve
1191 Escutcheon (Push Button Diverter)
2090 Check Stop Stem Assembly

Routine Maintenance Instructions

Delta faucets have earned the reputation of superiority in design, engineering, performance and durability in millions of home installations. Routine faucet maintenance to compensate for foreign materials in varying water conditions can restore like new and extend the life and performance of the faucet. And, ease and simplicity of service and maintenance is another advantage of having Delta faucets throughout the home.

A. If you should have leak under handle—tighten adjusting ring by following steps 1 and 9 leaving water supply and faucet turned on. Replace handle as in step 10.

B. If you should have leak from spout—replace seats and springs by following steps 1, 2, 3 and 4. Reassemble following steps 5, 6, 7 and 8. Turn water supply on and follow steps 9 and 10.

Shut off water supply

1. Pry off handle button, remove screw and lift off handle.

2. Unscrew cap assembly and lift off.

3. Remove cam assembly and ball by lifting up on ball stem.

4. Lift both seats and springs out of sockets in body.

To reassemble faucet

5. Place seat over springs and insert into sockets in body (spring first).

6. Place ball into body over seats.

7. Place cam assembly over stem of ball and engage tab with slot in body. Push down.

8. Partially unscrew adjusting ring and then place cap assembly over ball stem and screw down tight onto body.

Turn on water supply

9. Tighten ring until no water will leak around stem when faucet is on and pressure is exerted on handle to force ball into socket.

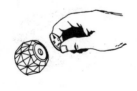

10. Replace handle. Tighten handle screw—tight. Replace handle button with "ON" arrow pointing up.

PRESSURE BALANCED VALVES

If unable to maintain constant temperature, clean balancing spool as follows:

1. Remove handle and escutcheon.

2. Close stops.

3. Unscrew balancing spool assembly (part No. 574) and remove—slowly— so as not to damage "O" Ring Seals. NOTE: On shower only installations water trapped in shower riser will drain out when balancing spool assembly is removed.

4. Examine valve hole for chips or any other foreign matter.

5. Remove any chips from spool sleeve before attempting to remove spool from inside of sleeve.

6. Remove spool from sleeve and clean all deposits from both sleeve and spool.

7. When spool will slide freely in sleeve, replace assembly in original position.

8. Open stops and check cap of spool assembly and stems of check stops for leaks before reinstalling escutcheon.

delex by delta

Scald-Guard™

Single Control, Metal Blade Handle Bath Mixing Valves

SUGGESTED SPECIFICATION: Shower and/or Tub Valves with metal blade handles shall be of mixer type which always open to cold water first and then through warm (mix) to hot. The hot and cold water must shut off separately and the valve shall be constructed so as to prevent any possibility of by-pass through the valve when in closed position. Valve operating mechanisms shall be of the rotating cylinder type and shall be equipped with replaceable non-metallic seats contained in stainless steel lined sockets. (Valves shall be equipped with an adjustable limit stop which may be adjusted to prevent the valve from being opened to full hot water.) All operating parts shall be separately replaceable from outside the wall. Delta/Delex Models 1601, 1611, 1621, 1605, 1635 or equal.

Scald-guard

Model 1601: Single Control, Blade Handle, Bath Mixing Valve

Scald-guard

Model 1611: Single Control, Blade Handle, Bath Mixing Valve With Tub Spout

Scald-guard

Model 1621: Single Control, Blade Handle, Bath Mixing Valve With Shower Assembly

Scald-guard

Model 1605: Single Control, Blade Handle, Bath Mixing Valve—Push Button Diverter

Scald-guard

Model 1635: Single Control, Blade Handle, Bath Mixing Valve—Push Button Diverter With Tub and Shower Assembly

Single Control, Metal Blade Handle Bath Mixing Valves

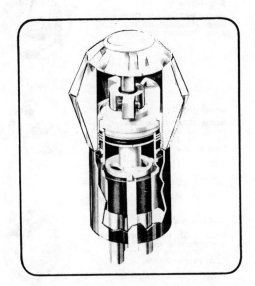

Here's the heart of the revolutionary Delex two handle faucet showing the simplicity of the rotating cylinder valve that ends washer headaches forever. There are no compression valves or washers to crush, tear or replace—the rotating cylinder valve simply shears off the water with no tightening down, no grinding, no force.

Back-to-back installation—Save time and money by eliminating expensive cross piping. A special reverse Scald-Guard valve is available to be used on the back-side of the back-to-back installation. The hot water riser on the right is easily connected to the right side of the "reverse" valve and the cold water riser on the left is easily connected to the left side of the valve, all without cross piping. Cold water will always flow first.

Foolproof maintenance—Once correctly plumbed, the stem unit can be reassembled only one way during maintenance, assuring that cold water will always flow first. This safety feature is not available with most mixing valves. Some can be reassembled incorrectly which is impossible with Scald-Guard.

To specify a reverse Scald-Guard, add "R" to the model number.

Model 1601: Single Control, Blade Handle Bath Mixing Valve

Model 1611: Single Control, Blade Handle Bath Mixing Valve With Tub Spout

Model 1621: Single Control, Blade Handle Bath Mixing Valve With Shower Assembly

Model 1605: Single Control, Blade Handle Bath Mixing Valve — Push Button Diverter

Model 1635: Single Control, Blade Handle Bath Mixing Valve—Push Button Diverter With Tub and Shower Assembly

Bright Chrome Only

Meets federal specification WW-P-541/7A Type M Class 2 Oct. 1972

Sweat inlets available on all chrome bath valves. Specify by adding "C" to model number. Not available on special finish models.

All showerheads equipped with volume control.

Integral screw driver stops available on all models. Specify by adding "With Stops" to model numbers.

Deluxe showerhead featuring cam control provides variable shower patterns. Available on all shower models. Specify by adding "DH" to model number.

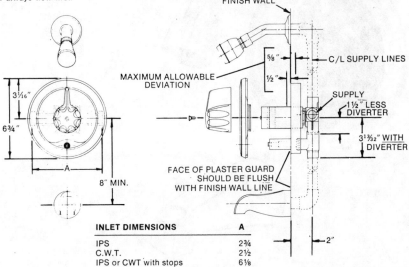

FINISH WALL

MAXIMUM ALLOWABLE DEVIATION

5/8"

1/2"

C/L SUPPLY LINES

SUPPLY

1½" LESS DIVERTER

3¹³⁄₃₂" WITH DIVERTER

FACE OF PLASTER GUARD SHOULD BE FLUSH WITH FINISH WALL LINE

2"

3¹⁄₁₆"

6¾"

A

8" MIN.

INLET DIMENSIONS	A
IPS	2¾
C.W.T.	2½
IPS or CWT with stops	6⅛

delex by delta

Scald-Guard™
Single Control, Blade Handle
Bath Mixing Valves—Pressure Balance

SUGGESTION SPECIFICATION: Shower and/or Tub Valves with metal blade handles shall be of mixer type which always open to cold water first and then through warm (mix) to hot. All operating parts must be separately replaceable from outside the wall. Valve operating mechanisms shall be of the rotating cylinder type and shall be equipped with replaceable non-metallic seats contained in stainless steel lined sockets. Valves shall be equipped with pressure balance spool and sleeve device of 302 stainless steel which will maintain a pre-set mix of hot and cold water and comply with Federal Specification WW-P-541. Valves shall be equipped with check stop (and with an adjustable stop to prevent any possibility of the valve being turned to full hot water.) Delta/Delux Models 1603, 1623, 1607, 1637 or equal.

Scald-guard
Model 1603: Single Control, Blade Handle, Bath Mixing Valve—Pressure Balance

Scald-guard
Model 1623: Single Control, Blade Handle, Bath Mixing Valve—Pressure Balance With Shower Assembly

Scald-guard
Model 1607: Single Control, Blade Handle, Bath Mixing Valve—Pressure Balance and Push Button Diverter

Scald-guard
Model 1637: Single Control, Blade Handle, Bath Mixing Valve—Pressure Balance and Push Button Diverter With Tub and Shower Assembly

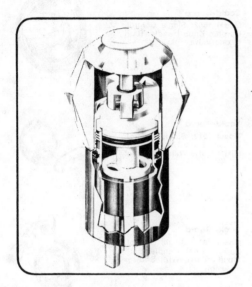

Single Control, Blade Handle Bath Mixing Valves — Pressure Balance

Model 1603: Single Control, Blade Handle Bath Mixing Valve—Pressure Balance

Model 1623: Single Control, Blade Handle Bath Mixing Valve—Pressure Balance With Shower Assembly

Model 1607: Single Control, Blade Handle Bath Mixing Valve—Pressure Balance and Push Button Diverter

Model 1637: Single Control, Blade Handle Bath Mixing Valve—Pressure Balance and Push Button Diverter With Tub and Shower Assembly

Bright Chrome Only

Meets federal specification WW-P-541/7A Type P Class 2 Oct. 1972

Sweat inlets available on all chrome bath valves. Specify by adding "C" to model number. Not available on special finish models.

All showerheads equipped with volume control.

Here's the heart of the revolutionary Delex two handle faucet showing the simplicity of the rotating cylinder valve that ends washer headaches forever. There are no compression valves or washers to crush, tear or replace—the rotating cylinder valve simply shears off the water with no tightening down, no grinding, no force.

Back-to-back installation—Save time and money by eliminating expensive cross piping. A special reverse Scald-Guard valve is available to be used on the backside of the back-to-back installation. The hot water riser on the right is easily connected to the right side of the "reverse" valve and the cold water riser on the left is easily connected to the left side of the valve, all without cross piping. Cold water will always flow first.

Foolproof maintenance—Once correctly plumbed, the stem unit can be reassembled only one way during maintenance, assuring that cold water will always flow first. This safety feature is not available with most mixing valves. Some can be reassembled incorrectly which is impossible with Scald-Guard.

To specify a reverse Scald-Guard, add "R" to the model number.

Deluxe showerhead featuring cam control provides variable shower patterns. Available on all shower models. Specify by adding "DH" to model number.

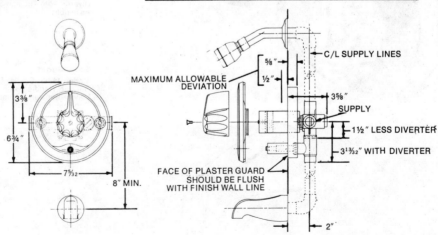

3⅜"

6¾"

7⁵/₃₂"

8" MIN.

MAXIMUM ALLOWABLE DEVIATION

⅝"

½"

C/L SUPPLY LINES

3⅝" SUPPLY

1½" LESS DIVERTER

3¹³/₃₂" WITH DIVERTER

FACE OF PLASTER GUARD SHOULD BE FLUSH WITH FINISH WALL LINE

2"

delex by delta

Scald-Guard™

Single Control, Duralac Knob Handle Bath Mixing Valves

SUGGESTED SPECIFICATION: Shower and/or Tub Valves with alcohol resistant Duralac knob handles shall be of mixer type which always open to cold water first and then through warm (mix) to hot. The hot and cold water must shut off separately and the valve shall be constructed so as to prevent any possibility of by-pass through the valve when in closed position. Valve operating mechanisms shall be of the rotating cylinder type and shall be equipped with replaceable non-metallic seats contained in stainless steel lined sockets. (Valves shall be equipped with an adjustable limit stop which may be adjusted to prevent the valve from being opened to full hot water.) All operating parts shall be separately replaceable from outside the wall. Delta/Delex Models 1602, 1612, 1622, 1606, 1636, 1646 or equal.

Scald-Guard
Model 1602: Single Control, Duralac Knob Handle, Bath Mixing Valve

Scald-Guard
Model 1612: Single Control, Duralac Knob Handle, Bath Mixing Valve With Tub Spout

Scald-Guard
Model 1622: Single Control, Duralac Knob Handle, Bath Mixing Valve With Shower Assembly

Scald-Guard
Model 1606: Single Control, Duralac Knob Handle, Bath Mixing Valve—Push Button Diverter

Scald-Guard
Model 1636: Single Control, Duralac Knob Handle, Bath Mixing Valve—Push Button Diverter With Tub and Shower Assembly

Scald-Guard
Model 1646: Single Control, Duralac Handle, Bath Mixing Valve—Shower Assembly and Tub Spout Diverter

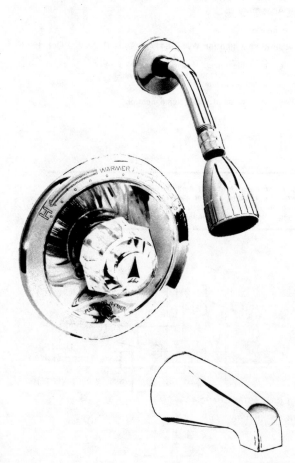

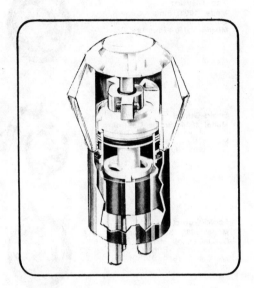

Here's the heart of the revolutionary Delex two handle faucet showing the simplicity of the rotating cylinder valve that ends washer headaches forever. There are no compression valves or washers to crush, tear or replace—the rotating cylinder valve simply shears off the water with no tightening down, no grinding, no force.

Back-to-back installation—Save time and money by eliminating expensive cross piping. A special reverse Scald-Guard valve is available to be used on the backside of the back-to-back installation. The hot water riser on the right is easily connected to the right side of the "reverse" valve and the cold water riser on the left is easily connected to the left side of the valve, all without cross piping. Cold water will always flow first.

Foolproof maintenance— Once correctly plumbed, the stem unit can be reassembled only one way during maintenance, assuring that cold water will always flow first. This safety feature is not available with most mixing valves. Some can be reassembled incorrectly which is impossible with Scald-Guard.

To specify a reverse Scald-Guard, add "R" to the model number.

Single Control, Duralac Knob Handle Bath Mixing Valves

Model 1602: Single Control, Duralac Knob Handle Bath Mixing Valve

Model 1612: Single Control, Duralac Knob Handle Bath Mixing Valve With Tub Spout

Model 1622: Single Control, Duralac Knob Handle Bath Mixing Valve With Shower Assembly

Model 1606: Single Control, Duralac Knob Handle Bath Mixing Valve—Push Button Diverter

Model 1636: Single Control, Duralac Handle Bath Mixing Valve—Push Button Diverter With Tub and Shower Assembly

Model 1646: Single Control, Duralac Handle, Bath Mixing Valve—Shower Assembly and Tub Spout Diverter   Bright Chrome Only

Special Finishes Available: Satin Chrome, Bright Gold, or Satin Gold, DH Models Only

Meets federal specification WW-P-541/7A Type M Class 1 Oct. 1972

Sweat inlets available on all chrome bath valves. Specify by adding "C" to model number. Not available on special finish models.

All showerheads equipped with volume control.

Integral screw driver stops available on all models. Specify by adding "With Stops" to model numbers.

Deluxe showerhead featuring cam control provides variable shower patterns. Available on all shower models. Specify by adding "DH" to model number.

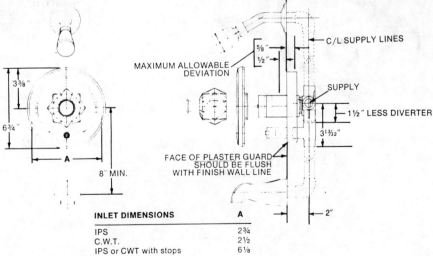

INLET DIMENSIONS	A
IPS	2¾
C.W.T.	2½
IPS or CWT with stops	6⅛

delex by delta

Scald-Guard™

Single Control, Duralac Knob Handle Bath Mixing Valves—Pressure Balance

SUGGESTED SPECIFICATION: Shower and/or Tub Valves with alcohol resistant acrylic knob handles shall be of mixer type which always open to cold water first and then through warm (mix) to hot. All operating parts must be separately replaceable from outside the wall. Valve operating mechanisms shall be of the rotating cylinder type and shall be equipped with replaceable non-metallic seats contained in stainless steel sockets. Valves shall be equipped with pressure balance spool and sleeve device of 302 stainless steel which will maintain a pre-set mix of hot and cold water and comply with Federal Specification WW-P-541. Valves shall be equipped with check stop (and with an adjustable stop to prevent any possibility of the valve being turned to full hot water.) Delta/Delex Models 1604, 1624, 1608, 1638, 1648 or equal.

Scald-Guard
Model 1604: Single Control, Duralac Knob Handle, Bath Mixing Valve—Pressure Balance

Scald-Guard
Model 1624: Single Control, Duralac Knob Handle, Bath Mixing Valve—Pressure Balance With Shower Assembly

Scald-Guard
Model 1608: Single Control, Duralac Knob Handle, Bath Mixing Valve—Pressure Balance and Push Button Diverter

Scald-Guard
Model 1638: Single Control, Duralac Knob Handle—Pressure Balance and Push Button Diverter With Tub and Shower Assembly

Scald-Guard
Model 1648: Single control, Duralac Knob Handle, Pressure Balance Bath Mixing Valve, Shower Assembly and Tub Spout Diverter

Single Control, Duralac Knob Handle Bath Mixing Valves—Pressure Balance

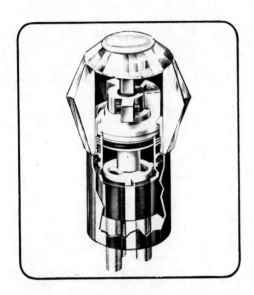

Here's the heart of the revolutionary Delex two handle faucet showing the simplicity of the rotating cylinder valve that ends washer headaches forever. There are no compression valves or washers to crush, tear or replace—the rotating cylinder valve simply shears off the water with no tightening down, no grinding, no force.

Back-to-back installation—Save time and money by eliminating expensive cross piping. A special reverse Scald-Guard valve is available to be used on the backside of the back-to-back installation. The hot water riser on the right is easily connected to the right side of the "reverse" valve and the cold water riser on the left is easily connected to the left side of the valve, all without cross piping. Cold water will always flow first.

Foolproof maintenance— Once correctly plumbed, the stem unit can be reassembled only one way during maintenance, assuring that cold water will always flow first. This safety feature is not available with most mixing valves. Some can be reassembled incorrectly which is impossible with Scald-Guard.

To specify a reverse Scald-Guard, add "R" to the model number.

Model 1604: Single Control, Duralac Knob Handle Bath Mixing Valve—Pressure Balance

Model 1624: Single Control, Duralac Knob Handle Bath Mixing Valve—Pressure Balance With Shower Assembly

Model 1608: Single Control, Duralac Knob Handle Bath Mixing Valve—Pressure Balance and Push Button Diverter

Model 1638: Single Control, Duralac Knob Handle—Pressure Balance and Push Button Diverter With Tub and Shower Assembly

Model 1648: Single control, Duralac Knob Handle, Pressure Balance Bath Mixing Valve, Shower Assembly and Tub Spout Diverter Bright Chrome Only

Special Finishes Available: Satin Chrome, Bright Gold, or Satin Gold, DH Models Only

Meets federal specification WW-P-541/7A Type P Class 2 Oct. 1972

Sweat inlets available on all chrome bath valves. Specify by adding "C" to model number. Not available on special finish models.

All showerheads equipped with volume control.

Deluxe showerhead featuring cam control provides variable shower patterns. Available on all shower models. Specify by adding "DH" to model number.

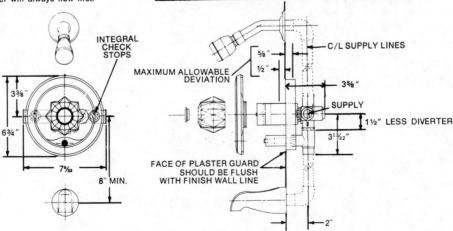

INTEGRAL CHECK STOPS

3⅜"
6¾"
7½₂"
8" MIN.

MAXIMUM ALLOWABLE DEVIATION

FACE OF PLASTER GUARD SHOULD BE FLUSH WITH FINISH WALL LINE

C/L SUPPLY LINES
⅝"
½"
3⅝"
SUPPLY
1½" LESS DIVERTER
3¹³⁄₃₂"
2"

Chapter 4

Repairing Water Closet Tanks

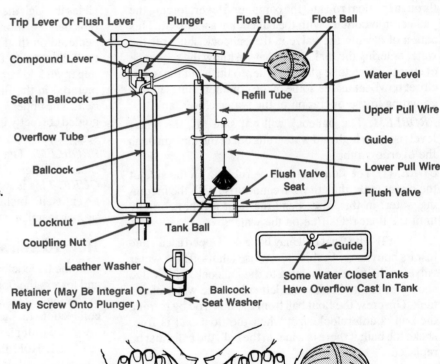

Fig. 1. Most old fashioned tank components.

For years men have had a knack for making simple things complicated. Most of the water closet tanks in use today are as outdated as the Model-T Ford. Everything considered, the maze of wires, floats, rods, levers, tubes, guides, and balls which comprise the working parts of the average water closet tank, work amazingly well. Repair or replacement of some or all of the parts of a water closet tank become necessary when corrosion, mineral build-up, or normal wear cause these parts to malfunction. The replacement of old-fashioned tank parts with

the newer parts shown in this chapter is strongly recommended.

In order to make it easier to repair a water closet tank, let us study the chain of events set in action when the trip lever of the tank is pushed down. Refer to Fig. 1 as we go along. Trip lever (A) pulls up on the upper pull wire (B) which is linked to the lower pull wire (C). The lower pull wire, sliding through the guide (D) is screwed into the top of the tank ball (E). The tank ball, upon being pulled up, becomes free of the water pressure which had been holding it in place on the seat of the flush valve (F) and becomes waterborne. The water then starts to rush out of the tank and into the closet bowl. As the water level in the tank drops, the float ball (G) connected to the ballcock (H) by float rod (I) through compound lever (J) also drops with the water level. As the water level drops, the weight of the float ball and float rod exerts pressure to raise the plunger (K) off of the ballcock washer seat (L). The ballcock, (which is a water valve) is thus turned on and water begins to flow into the tank. The water flowing into the tank is not coming in as fast as the water which is already there is leaving, therefore, the tank ball drops with the receding water and is guided into proper contact with the flush valve seat by the guide. When it is properly seated, the tank ball prevents any more water from leaving the tank. The float ball is now riding the crest of the incoming water and when the water level has reached the water line (O), the float exerts pressure on the plunger through the float rod and the compound lever, forcing the plunger into contact with the ballcock washer seat. This action of closing the valve in the ballcock shuts off the water entering the tank. Now that we know the sequence of events which take place in order to flush out the water closet bowl, let us see some of the common things which go wrong in the operation of the water closet tank.

PROBLEM: The ballcock will not shut off, the water level rises too high and water runs out of the tank through the overflow tube.

CURE: (1) The float ball may be rubbing on the side of the closet tank, thus the ball cannot rise with the incoming water; in this case, bend the float rod horizontally until the float rides free on the water.

(2) The float ball may have developed a leak and lost its bouyancy. In this case, close (shut-off) the water valve on the pipe connecting to the ballcock. This valve should be located below the left side of the water closet tank. Unscrew the float ball from the float rod by turning the ball counterclockwise; when the float ball is free, shake the ball; if there is water in the ball, the ball must be replaced.

(3) Replace the ballcock washer. The ballcock washer is on the bottom of the plunger.

PROBLEM: You have a very high water bill, you suspect a leak in the water closet tank but you cannot find the leak.

CURE: Pour a tablespoon of dark blue or black ink (or food coloring) into the water closet tank. Wait for 3 or 4 minutes then observe the water in the closet bowl. If the water in the bowl has turned dark, there is a leak in the closet tank where the tank ball seats on the flush valve or there is a leak in the overflow tube. Replace the tank ball first, preferably with one of the new 'flapper-type' tank valves, furnished as a complete kit which includes the

flapper type valve, a stainless steel seat, and the necessary cement.

If replacing the tank ball does not stop the leak, the leak is in the overflow tube. If the overflow tube is made of brass, use slip joint pliers to grasp the overflow tube just above the connection to the flush valve. Turn the overflow tube counterclockwise to unscrew the tube from the flush valve. If the overflow tube is made of plastic, the complete flush valve should be replaced.

PROBLEM: The closet tank does not refill after the closet bowl has been flushed.

CURE: (1) The float rod may be bent causing the float ball to drag on the side of the tank. If this happens the float ball cannot drop and the water will not be turned on by the ballcock. Bend the rod horizontally to free the float ball from the side of the tank.

(2) The tank ball does not drop into position to seat properly on the flush valve. The lower pull wire slides through the eye of the guide and is screwed into the tank ball. Position the guide so that the eye of the guide is centered on the flush valve opening. Refer to the closet tank illustration for measurements. Be certain that the upper and lower pull wires are installed as they are pictured in the illustration and that these wires are not bent. The upper and lower pull wires must be straight and guided correctly in order for the tank ball to drop into and seat properly on the flush valve.

PROBLEM: The trip lever must be depressed several times in order to flush the closet bowl.

CURE: This is usually due to too much play in the trip lever itself. Install a new trip lever. Also, check for proper installation of the upper and lower pull wires.

How To Adjust The Water Level In The Tank

Some ballcocks have an adjusting screw which raises and lowers the float rod and thus adjusts the water level in the tank. This screw will be located at the float rod connection to the ballcock. Another way to raise or lower the water level in the tank is to bend the float rod (I) see Fig.1. Hold the float rod firmly with the left hand. Using the right hand, bend the rod up to raise the water level in the tank; bend the rod down to lower the water level. The correct water level is approximately one inch below the top of the overflow tube. Fig. 1 shows the old standard types of ballcock, float ball and float rod, flush valve, upper and lower pull wires and tank ball installed in the water closet tank. As we mentioned earlier, the use of newer type fill valves and "flapper" type tank balls is strongly recommended.

Water Closet Fill Valves

Two types of water closet fill valves which are superior to the old-fashioned ballcock in design and performance are the *Fillmaster* and the *Fluidmaster*.

The *Fillmaster* Fill Valve

The *Fillmaster*, shown in Fig. 2, uses the hydraulic force of the supply water to provide positive opening and closing of the fill valve. Using no float rod or float ball, the *Fillmaster* measures the water level in the tank by means of a diaphragm which responds to the head (or height) of water in the closet tank. The water level in the tank can be adjusted to the desired level by turning the knob marked "ADJ". Each full turn of the knob changes the water level by one inch. The *Fillmaster* fill valve modulates the flow of water into the tank. Water seepage through the flush valve is replaced with no cycling noise. The valve is constructed of plastic, rubber and stainless steel. An antisiphon version of the *Fillmaster* is available for use in cities where plumbing codes require back-siphonage prevention. A flapper-type tank ball is included in a *Fillmaster* kit to completely modernize the major control devices of a water closet tank all at one time. See Figs. 2 through 9 for installation and operation sequence of the *Fillmaster*.

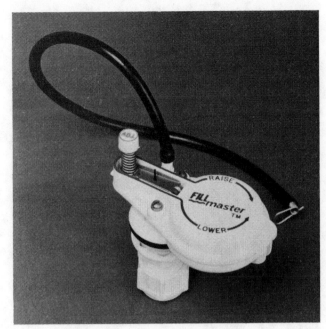

Courtesy J H Industries, Inc.

Fig. 2. The *Fillmaster* valve unit.

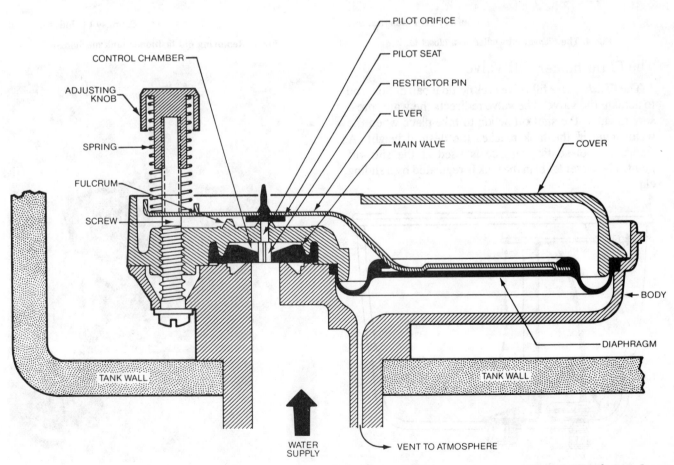

Courtesy J H Industries, Inc.

Fig. 3. Cutaway view of the *Fillmaster* valve.

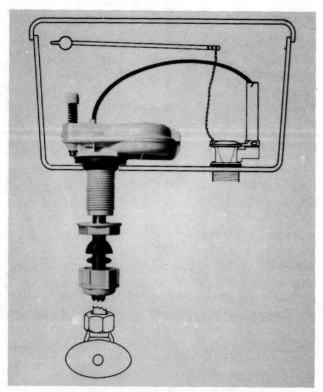

Fig. 4. The *Fillmaster* installed in a closet tank.

The *Fluidmaster* Fill Valve

The *Fluidmaster* fill valve uses no float ball or float rod to actuate the valve. The valve redirects the water pressure to cause the shut-off action to take place when the water level in the tank reaches the desired height. A stainless steel seating surface is used at the shut-off point. The water level in the tank is regulated by a sliding clip.

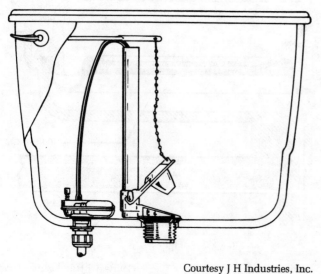

Fig. 5. Typical *Fillmaster* installation.

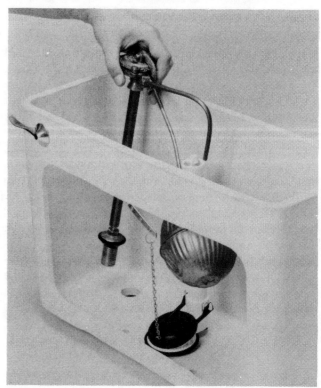

Fig. 6. Removing old fashioned tank mechanism.

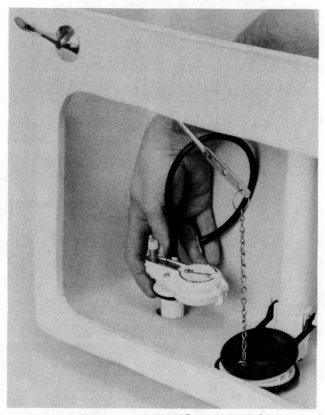

Fig. 7. Inserting *Fillmaster* valve in tank.

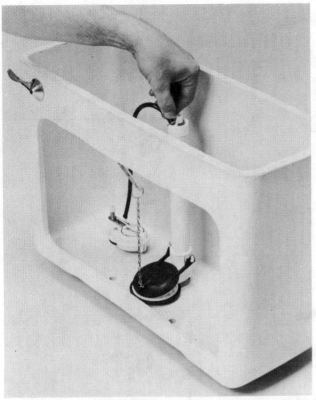

Fig. 8. Clipping rubber refill tube to overflow pipe.

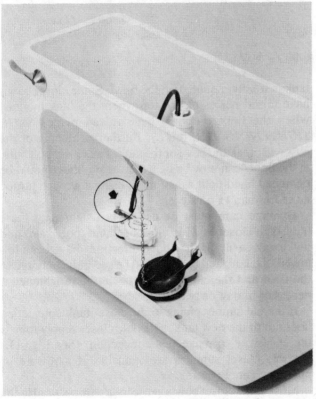

Fig. 9. Water level adjustment knob.

The *Fluidmaster* operates in a full-on or full-off position, there is no modulation in its action. Water leakage at the tank ball, through the flush valve, is signalled by the sound of the valve refilling the tank, when the tank has not been flushed.

In addition to the Model 100 and Model 200, the *Fluidmaster* is also available in an antisiphon Model 400 for use in cities where the plumbing code requires back-siphonage protection. See Fig. 10. The *Fluidmaster* "Flusher-Fixer" kit, (a flapper type tank ball) with a stainless steel seating surface, can be installed to modernize the working parts of the water closet tank. See Fig. 11.

How To Install Or Reset A Closet Combination

The following instructions relate to installing or resetting a close-coupled (tank mounted on the bowl) closet combination. The bowl is secured to the floor by bolts which are inserted under a closet flange. The closet flange is part of the drainage piping. Fig. 12 shows how the bolts are inserted under the flange. Closet bolts are available in ¼" and 5/16" sizes—5/16" closet bolts, being heavier, are much the best.

Insert the closet bolts under the closet flange. Set the closet bowl on end, and slide the wax ring over the horn on the bowl. Pick the closet bowl up, holding it around the rim and set it carefully over the closet flange. When the bowl is centered over the closet flange, the closet bolts will protrude through the holes in the bottom of the bowl; push down on the bowl to "set" the wax ring. Drop the chrome-plated closet bolt washers on to the bolts and start the open closet nuts on the bolts. Screw the nuts down hand-tight. If you are simply resetting an old water closet combination, the bolts securing the closet tank to the bowl will be reassembled just as they were originally assembled. If you are installing a new closet combination, follow the installation and assembly instructions furnished with the new combination.

After setting the closet bowl, the tank can be installed the easiest while sitting on the bowl. Holding the tank, sit down on the bowl facing the wall in back of the toilet. Set the tank down on top of the bowl, line up the holes (either two or three) in the tank with the holes in the bowl and insert the bolts, with the rubber washers under the head, through the tank holes and through the holes in the bowl. While holding down on the bolt in the tank with one hand, slide the rubber washer, flat steel washer and lock washer over the bolts and start the nuts on the bolts with the other hand. A small adjustable wrench can be used to tighten the nuts on the bolts. A long screwdriver is necessary to hold the bolts in the tank while tightening the nuts at the bottom.

Tighten the nuts evenly, going back and forth — turning each nut approximately 2 turns, at a time, until the

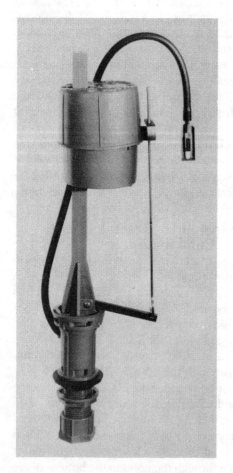

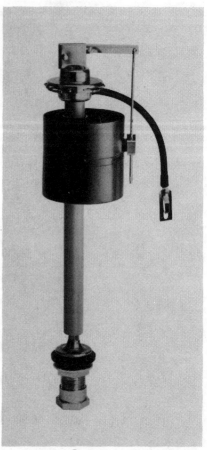

Courtesy J H Industries, Inc.

(A) Model 100.   (B) Model 200.   (C) Model 400.

Fig. 10. *Fluidmaster* fill valve.

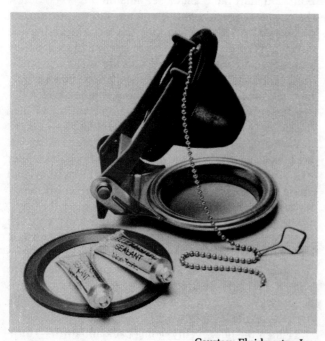

Courtesy Fluidmaster, Inc.

Fig. 11. *Fluidmaster* flapper-type tank ball.

tank is solidly mounted on the bowl. Some closet bowls have raised places on the top where the tank mounts, when the tank touches these raised places *STOP TIGHTENING THE NUTS*. Further tightening of the nuts will break the bowl or the tank. The closet bolt nuts, at the floor, can now be tightened again, they should be tightened hand-tight again, then, using a wrench turned no more than ½ turn each.

Extreme care must be used when tightening these nuts. The closet bowl is vitreous china (porcelain) and will break easily. If the above directions have been followed, the tank and bowl should be securely mounted to the floor and the closet bolt nuts can be retightened if necessary a day or two later.

If the old supply piping to the closet tank connection does not fit the new tank, a flexible closet supply tube is the easy way to make this connection. (See Fig. 13.) Flexible closet supplies are available at hardware or plumbing supply stores.

Most of the closet bowls in use today are secured to the floor or closet flange by closet bolts only. If the bowl you are setting has two extra holes in the base, towards the

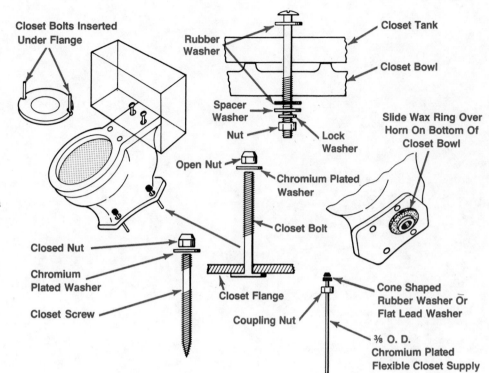

Fig. 12. Bowl and tank installation instructions.

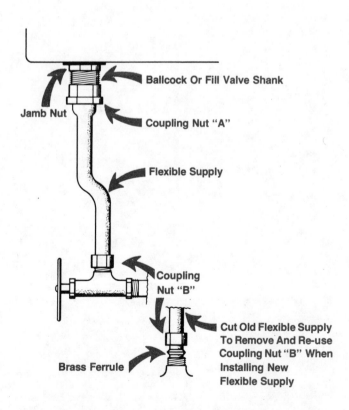

Fig. 13. Installation procedures for installing a new supply line.

front, it will require the use of closet screws. Before installing the wax ring on the closet bowl, set the bowl in place on the flange, insert a short pencil in the front holes in the bowl and mark the floor under the front bowl holes. Remove the bowl and install the closet screws in the center of the marked holes. The top of the screw should protrude above the floor approximately 1⅝'' or about ⅜'' above the base of the bowl when it is set in place.

Note that the closet screw has a closed top nut, the closet bolt has an open-top nut. When the wax ring has been mounted on the bowl, set the bowl in place, tighten the closet bolts first, then the closet screws. Extreme care must be taken not to over-tighten either the screws or the bolts as the bowl is porcelain and will break easily. If you are in doubt, leave the bolts and the screws slightly loose; after the bowl has been in place for a day it will have settled into place and the bolts and screws can be tightened if necessary.

Chapter 5

What To Do About Stopped-Up Drains

Grease, hair and soap build-up over a period of time in drain piping are the most common causes of drain stoppages inside the home. Roots and broken drain tile are responsible for most stoppages in the sewer from the home to the main sewer in the street or alley. If all the drains in a home run slow or are completely stopped-up, it is safe to assume that the sewer from the home to the main sewer is blocked. This is not a job for the average householder to attempt to do. The equipment necessary to do this job is cumbersome, heavy and requires experience in its use. The operator has to 'feel' his way as he works the cutter through the sewer. He knows from the way the equipment handles when he encounters tree roots, or a broken tile. If all the drains in your home are stopped-up, leave this job to the experts and call your local plumber or a company specializing in drain cleaning. After the roots have been removed, a copper sulphate solution should be poured into the sewer periodically, preferably through a clean-out opening to slow down or prevent re-growth of roots. Copper sulphate solutions are sold by plumbing shops and hardware stores for this purpose.

Industrial types of chemical drain cleaners used by plumbers is very effective, but they usually contain very strong acids and should *only* be used by skilled professionals. Use of the types of chemical drain cleaners found on grocery shelves, etc., is *NOT* recommended.

Kitchen Sink Drains

A plunger is often very effective in unstopping sink drains. When a plunger is pushed down over the drain, a shock wave is transmitted through the water in the drain, and steady use of the plunger, causing a series of shock waves, will often break a stoppage loose. If you have a two compartment sink, use one plunger to seal the opening on one side of the sink, while using the plunger on the

other side. If the plunger does not open the drain, it will be necessary to run a cable through the drain piping to get to the stoppage. A cleanout plug may be installed in the drain piping under the sink for this purpose, if there is no cleanout plug, disconnect the sink trap and insert the cable into the drain piping at this point. Refer to Fig. 1.

If there is water in the sink, above the cleanout plug or above the trap, dip or sponge out the water in the sink or place a large pan or bucket under the trap before removing the trap to catch any water in the sink or in the piping. Insert the cable into the drain piping, rotating the cable (spinning the 'top') will help to work the cable around turns or fittings in the piping. Working the cable forward and back will help to break through the obstruction. After reaching and breaking through the obstruction, remove the cable, screw the cleanout plug into the cleanout or reconnect the trap, and run water into the drain to make certain the drain is open. When the drain is open, turn the hot water faucet on for three or four minutes. The hot water will melt the congealed grease in the drain piping.

Lavatory Drains

The procedure for unstopping a lavatory drain is basically the same as unstopping a kitchen sink drain. The most common cause for a stopped-up lavatory drain is an accumulation of hair. If the lavatory has a pop-up drain, lift out the stopper in the drain. Some stoppers have a lock-in device, i.e. they are secured to the lever which is connected to the lift rod. Other stoppers twist and lift out. Hair will often catch on the bottom of the stopper and cause the lavatory drain to stop up. Removing the stopper and cleaning out the accumulation of hair will open the drain if this is the problem.

The plunger is a tool which also is very useful for unstopping lavatory drains. The lavatory has a drain

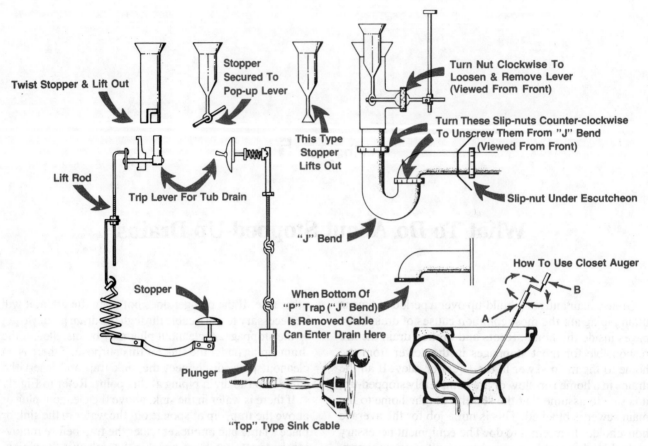

Fig. 1. A typical hand plunger.

connection, also an overflow connection. The overflow opening may be on either the front or the back side of the bowl, near the top. Block the overflow connection with a wet cloth (or a small plunger) when plunging the drain connection. The plunger, used as described earlier to unstop kitchen sinks, is usually very effective in unstopping lavatory drains. If the drain piping is stopped up, use a smooth jaw monkey wrench or *Ridgid* offset hex wrench to turn the slip-nuts on the trap counterclockwise to loosen. When the slip-nuts are loose, slide the 'J' bend or bottom of the 'p' trap down off of the drain fitting. Slide the escutcheon forward, away from the wall, to expose the trap connection at the wall. If the trap has a slip-nut connection at the wall, unscrew the slip-nut and pull the balance of the trap out of the drain piping.

If the trap is soldered into a lead pipe at the wall or has a threaded connection, (other than a slip-nut) it would be best to leave the rest of the trap connected at the wall and thread a small sink cable (5/6 or ⅜) into the drain piping through the trap connection as illustrated. Spin the cable as it works its way through the drain piping, this will help to work the cable around fittings in the piping and the spinning action of the cable will also help clean the drain

line. The top-type sink cable shown in Fig. 2 is very useful for unstopping lavatory drains.

Bathtub Drains

The cause of most stopped-up bath drains is also hair. Quite often hair will collect around the pop-up drain parts, a spring, a plunger or a stopper. If an accumulation of hair has caused the stopped-up drain, taking the pop-up parts out and cleaning them will usually open the drain. To disassemble a typical pop-up drain, remove the screws securing the overflow plate. Pull the overflow plate forward and upward and work the lift rod out. Remove any hair, soap or grease from the spring or plunger. The stopper, in the drain outlet of the tub, can be removed by pulling up on the stopper and working it out of the drain opening. Remove any accumulation of hair, soap or grease from the stopper. If after cleaning the plunger, spring, or stopper, the bath drain is still stopped, stuff a wet rag into the overflow pipe to block it and use a plunger on the drain opening. This plunging action will usually open the drain.

If using the plunger on the drain fails to open the drain, then a small cable, such as the top-type cable should be used. Feed the cable into the drain at the overflow connection, rotating the drum will help the cable work its way around the fittings in the bath trap. When the drain is open, reassemble the drain connections.

Water Closet Bowls

The most common cause for a stopped-up toilet bowl is the sanitary napkin carelessly disposed in the toilet bowl. Next common troublemakers are combs, toys, or pencils. A large size plunger, especially made for plunging toilet bowls, will usually force sanitary napkins and accumulated wastes on through the bowl. Toys, especially soft plastic toys which bend, can usually be forced through a closet bowl with a good plunger. Straight objects, pencils and combs will often be washed half way through the trap but will catch or jam into the final turn of the trap. The only practical way to dislodge a pencil, comb, or similar straight object is to use a closet auger. If a plunger fails to unstop a closet bowl, see Fig. 1 use a closet auger in the following manner:

Grasp the closet auger at point (A) with one hand and the handle (B) with the other hand. Pull the handle up and out, until the spring end stops against the end of the guide tube. Insert the spring end of the auger into the entrance to the trap in the closet bowl. Push down on the handle of the auger, forcing the spring into the closet bowl and turn the handle in a clockwise direction while forcing the spring into the bowl. When the handle has come to rest against the guide tube, turn the handle clockwise three or four times, this action will help to break up or catch the offending article in the spring end of the auger. Pull the handle up and out and repeat the process again if necessary. A word of caution, closet bowls are made of china and are easily breakable, turning the auger handle while inserting the spring end into the closet bowl will minimize the chances of breaking the closet bowl.

Chapter 6

Replacing and Repairing Automatic Water Heaters

Water heater manufacturers guarantee their products for 5, 10, or 15 years when used in the home. In most cases, this is a prorated guarantee, based on the theory that the product user pays only for the amount of use he received from the product. In other words, if a water heater is guaranteed for 10 years, and starts to leak in the sixth year, then the owner received 6/10 of the service he paid for and the heater will be replaced at a cost of list price less 40%. This is not really a bargain because the owner has to pay the full price for installation. The guarantee is for the product only and does not cover installation costs. Here again, high labor costs are forcing many homeowners to install their own replacement water heaters.

This chapter will cover the repair or replacement of gas fired water heaters only. The water piping connections to electric water heaters are basically the same as for gas fired heaters. The heating elements on electric water heaters are made for different voltages, such as 208, 220, 230 volts, and although a ten percent (10%) variation in voltage is usually considered acceptable, the conditions in any given locality may vary considerably. Also, while I believe almost all of the people who read this book can repair or replace the items covered here, I do not believe most homeowners or "do-it-yourselfers" are familiar enough with electricity to wire their water heaters. If you have an electric water heater to replace, call a qualified electrician to disconnect your old heater. Install your new heater — make the piping connections, and call the electrician to come back and connect the new heater. Of course, if you are a qualified electrician, you can take care of the complete project.

Repairing Gas Fired Water Heaters

If the tank develops a leak, replace the heater. A leak usually shows up as just a damp spot on the floor near or under the heater. It may leak for a day or two rust up and stop leaking and then start leaking again. It is not practical to try to repair a leaking tank.

The relief valve may also start leaking. This may be due to an increase in water pressure in the water main or may simply be a defective or worn-out relief valve. If the water pressure in your area has been increased, install a new temperature and pressure relief valve. The water utility in your area should be able to recommend the proper relief valve.

Automatic gas water heaters have a safety feature called a safety pilot. The safety pilot action depends on the thermocouple. The thermocouple element, when heated by a flame, generates an electric current. This electric current in turn holds a gas valve in open position and permits gas to flow to the main burner. If for any reason the gas pilot light on the water heater goes out, the thermocouple cools, and the main burner is shut off automatically. To relight the heater, the control must be set to the pilot position and the pilot control button, or knob, must be held down (depressed) while the pilot light is relighted. The knob, or button, must be held in open position (to permit gas flow to the pilot light) until the thermocouple has generated sufficient voltage to hold the pilot control in open position automatically. This generally takes about one minute. Release the button or knob, (if the pilot stays on) and turn the control knob to "on" position. Turn the thermostat, or temperature control knob to the desired water temperature and if the water in the tank is below the set temperature, the main burner will light.

If the pilot light goes out when the pilot knob or button is released, check the thermocouple coupling nut connection to the control. (See Fig. 1.) The thermocouple coupling nut must be tight. If the thermocouple coupling nut is tight, follow the relighting instructions on your

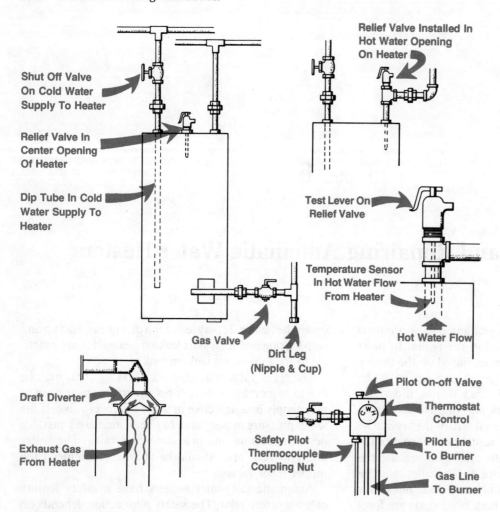

Fig. 1. Typical water and gas connections for a gas fired water heater.

water heater and observe the pilot flame. The tip of the thermocouple element must be in the pilot flame and the tip of the element should be heated to a cherry red. If the element is in the flame properly and the pilot light goes out when the safety pilot control is released, the thermocouple is defective. Replacement thermocouples are readily available at plumbing shops or hardware stores. They are available in various lengths to fit any water heater, and are called universal type thermocouples because the kit includes several types of adapters to fit any burner.

To replace the defective thermocouple, turn off the main gas valve to the heater. Use a small adjustable wrench, turn the thermocouple coupling nut clockwise to loosen and remove the thermocouple tubing from the control. Disconnect the thermocouple from the mounting on the main burner. Replace the defective thermocouple with a universal type, 30 millivolt replacement thermocouple, of the same length as the original one if possible. The replacement thermocouple should be coiled when you receive it. Uncoil it carefully, do not kink or damage the tubing, and if the replacement is longer than the original, uncoil only enough to reach

from the burner to the control mounting point. Connect the replacement thermocouple to the water heater control. The coupling nut has a very fine thread; be sure this nut is not crossthreaded when connection is made. If the nut is started straight, it should turn about four turns clockwise by hand before tightening. Finish tightening with an adjustable wrench. Mount pilot end of thermocouple to the pilot burner. Turn on main gas valve and light the pilot. Observe the thermocouple tip for correct mounting in pilot flame.

If the thermostat control on the water heater becomes defective, I advise replacing the water heater. In my opinion, the cost of the new control, installed on an old water heater, is prohibitive and unwarranted.

Installing a New or Replacement Gas-Fired Water Heater

Fig. 1 shows the piping to a typical gas-fired water heater. The gate or globe valve is in the cold water supply to the water heater. The cold water supply to a top-connected water heater must enter the tank through a dip tube. The dip tube forces the cold water to the bottom of the tank, thus preventing the cold water from tempering or cooling off the hot water in the top of the tank. If the cold water enters the tank at or near the bottom of the tank, no dip tube is needed. Some manufacturers make their water heaters with the dip tube removable, i.e. the dip tube can be lifted out of one opening and inserted in another opening. This is a definite advantage at times when replacing a water heater, in order to avoid changing existing piping. If you find that the dip tube will not fit out, then the opening in which the dip tube is inserted must be used for the cold water connection to the tank.

When installing a new or replacement water heater, it is advisable to also install a new temperature and pressure-type relief valve on the heater. Temperature and pressure-type relief valves are obtainable at pressure ratings of from 75 to 175 P.S.I. and temperature rating of 210 degrees. Your local water utility should be able to recommend the proper pressure rating of the relief valve if you are in doubt on this point.

Some water heaters are manufactured with three (3) connections in the top, one is the cold water inlet, one is the hot water outlet and the center connection is for the relief valve. If your water heater has only two openings, one for cold water inlet and one for hot water outlet, then the temperature and pressure-relief valve should be installed as shown in Fig. 1 in the hot water outlet, with the temperature sensor of the relief valve in the flow line of the hot water leaving the tank. A dirt leg, as illustrated, should be installed on the gas connection to the water heater. The dirt leg will trap any moisture, flakes of rust, etc., that may be present in the gas piping to the heater. The draft diverter which is furnished with a water heater, must be installed on the top of the heater, or in the vent pipe from the water heater to the chimney or flue. The draft diverter permits the burned gases to escape up the vent pipe, while at the same time the diverter prevents a backdraft from blowing down the vent pipe and blowing out either the pilot flame or the main burner flame. Use pipe thread lubricant on all the piping connections. Test all gas piping connections with soap suds solution as follows:

Use one teaspoon of household detergent, such as *Ivory Snow* liquid, in one half-cup of water, shake or stir to make suds. Turn gas valve on, but do not light heater. Use small paint brush and cover all gas fittings and threads with soap suds. If there are any leaks in the piping, or sand holes, cracks etc., in the fittings, the leaks will show as bubbles when the suds are applied. If the gas piping shows no sign of leakage, follow the instructions furnished with the heater to light the pilot flame. In any event, do not light the pilot flame until the tank is filled with water.

Chapter 7

How To Work With Copper Tubing

Joining two or more pieces of copper tubing, or repairing a leak in copper tubing, is not a difficult operation. There is a very simple way to make repairs or add pipe and fittings to an existing piping system, which I will describe later in this chapter.

The most common and least expensive method to make repairs to, or changes in, a copper piping system is to solder (or "sweat") the joints. There are a few simple rules for soldering copper joints. If you will learn them and follow these rules, you can make perfect solder joints every time.

1. The male end of the piping and the female end of the fitting to be joined together must be clean and bright.
2. Heat must be applied at the right place on the fitting. Capillary action will then pull the solder into the joint.
3. Use plain 50/50 wire solder and a noncorrosive solder paste. **DO NOT USE** acid-core solder or rosin-core solder.
4. The piping being joined *must* be dry. It is impossible to solder a joint properly with water in the pipe. There may be times when you find that a valve will not shut off completely and will allow a trickle of water to flow through the pipe. You can stop this water flow long enough to enable you to solder a joint in the pipe in the following manner: Obtain a slice of bread and break it into small pieces. Insert these pieces of bread into the pipe and pack it tightly, using the eraser end of a pencil as a tamping tool, this will stop the flow of water and enable you to solder the joint. When the solder joint has been made and the water pressure is turned on in the pipe, the "bread plug" will literally disintegrate and can be washed out of the pipe through an opened valve or faucet. If your faucets have aerators on the spouts, remove the aerators while flushing out the piping.
5. If you are working on a closed system of piping, open a valve or turn on a faucet as shown in Fig. 1.

A closed system of water piping is a system which is connected to a water supply line at one end and to valves, fixtures, or appliances at the other end. The heat which is applied to the pipe and the fitting during the soldering operation causes a build-up of air pressure in the pipe system.

The built-up pressure, if it is not relieved by opening a faucet on the piping system, will cause the solder to be blown out of the soldered joint, leaving a hair-line crack which will then leak when the water is turned on.

Clean the male ends of the copper pipe and the female ends of the copper fitting with sandcloth or sandpaper. When clean, they will be bright and shiny. Then coat the cleaned ends with solder paste. Join the pieces together and apply heat from your torch. Unroll about six inches of wire solder from the spool and apply the end of the solder to the joint. As soon as the solder starts to melt, apply the heat to the center of the fitting and all the way around it. Follow the flame around with the solder. As soon as the solder has melted into the joint all the way around, remove the heat and allow the joint to cool. When all of the joints have been soldered, close the valves or faucets which were opened and turn the water back on. **Notice:** Do not put tip of flame directly on joint to be soldered—the flame will contaminate the joint and the solder will not flow into the joint.

As I stated earlier, there is an easy way to make repairs or add fittings or piping to an existing piping system which requires no soldering. Ferrule-type compression fittings are used. These fittings are readily available at most plumbing shops and hardware stores. When using this type of fitting, insert tubing into fitting recess and

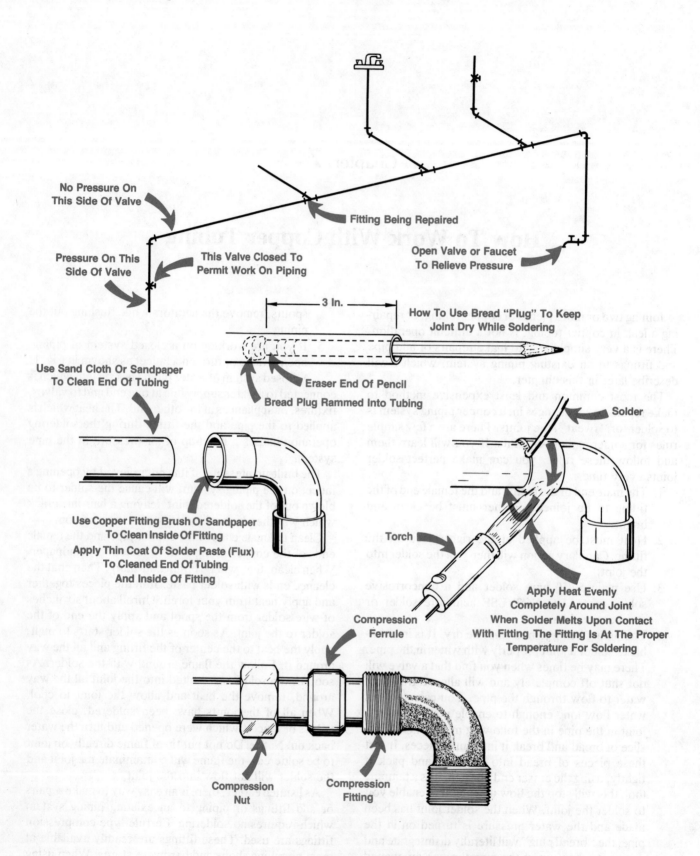

Fig. 1. Making water tight connections with solder or compression joints.

tighten compression nut onto fitting. Tightening the compression nut will lock the ferrule onto the tubing and if the compression nut is properly tightened, the joint will be water tight.

The ferrule-type compression fitting requires no solder, or solder paste, nor does it require cleaning of the tubing or the fittings. There is still another way of installing fittings in copper piping. This last method requires the use of a flaring tool, or a flaring block. "Hard" copper can not be flared unless it is annealed, or softened. Therefore, flare type fittings are designed for use on soft copper tubing, such as water service piping from the water main in the street into a building, or fuel oil piping from the oil tank to the furnace or boiler.

Pictured in Fig. 2 are some of the common solder-type copper fittings. Street fittings are any fittings with one male end and one or more female end. A street tee would have one male end, two female ends. Copper street fittings have no restriction at the male ends and work very well in copper pipe installations. Tees and ells are available both as straight pipe size fittings and as reducing ells or tees. A tee is read (or described) in the following manner and as pictured; end 1, end 2, side 3. A straight ½" tee is ½ by ½ by ½ (½ x ½ x ½). A reducing tee, such as ¾ x ½ x ¾" tee would be ¾" one end, ½" other end, ¾" on the side. The rule for describing a reducing tee is always (1) end size, (2) end size, (3) side size. Male and female adapters are used when connecting copper tubing to iron pipe. Copper tubing for plumbing use is measured by I.D. (Inside Diameter).

Copper tubing is measured and cut using the same rules as for steel piping. The most accurate method of measuring is the end to center measurement. The end to center measurement of a copper tubing fitting is very

90° Ell

45° Ell

90° Street Ell

45° Street Ell

Tee

Wing (Or Drop) Ell

Copper To F.I.P. Ell

Union

Coupling

Reducing Coupling

Copper To M.I.P. Adapter
M.I.P.=Male Iron Pipe

Copper To F.I.P. Adapter
F.I.P.=Female Iron Pipe

Reducing Tee

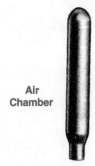

Air Chamber

Courtesy Nibco, Inc.

Fig. 2. Various solder connection fittings.

easy to figure because the fitting, whether it is a ferrule type fitting or a solder-type fitting, is recessed for the piping. The distance from the end of the piping recess to the center of the fitting is the end to center measurement of the fitting. The formula for figuring 45° breaks in the piping is the same as shown in the chapter on how to measure and cut steel piping (90° break x 1.41).

Copper plated hangers, straps or pipe hooks should always be used for hanging or securing copper piping. Copper plated hangers prevent electrolysis or corrosion between dissimilar metals.

Chapter **8**

Installing A Garbage Disposer

When replacing a garbage disposer, first disconnect the electric power to that particular circuit. Then disconnect the wiring to the old disposer. Disconnect the drain connection from the old unit. If a dishwasher drain is connected to the old disposer, disconnect the dishwasher drain at the disposer. If the old disposer is held in place on the mounting flange by a clamp, (similar to a large hose clamp) loosen the clamp and the disposer can then be removed. The method of mounting disposers in a sink varies from one manufacturer to another. Most disposers are mounted on a flange which is secured to the sink by means of three screws pressing against a pressure plate. The pressure plate, in turn, is secured by a snap ring which fits into a groove in the flange as shown in Fig. 1 A.

To remove the old mounting flange, turn the hex nuts on the screws between the two plates counterclockwise to loosen, then turn the screws counterclockwise until the pressure plates are loose. Use a screwdriver blade to pry the snap ring out of the groove in the flange, the steel plates will drop free and the flange can be pushed up and lifted out of the sink.

If the new disposer replaces a basket strainer, use a 14-inch pipe wrench and turn the slip-nut on the bottom of the basket strainer counterclockwise. Unscrew the slip-nut from the basket strainer. Use a 14-inch pipe wrench to loosen the locknut on the bottom of the basket strainer. The basket strainer can then be lifted up and out of the sink. The installation instructions furnished with the new disposer will explain how to mount the disposer in the sink and how to connect the electrical wiring. If the new disposer replaces an old disposer, the drain connections may not have to be changed to receive the drain from the disposer. The new disposer can be rotated, after mounting, to any desired position, to make the connection to the present drain piping. If the new disposer replaces a basket strainer it may be necessary to use a longer tailpiece and set the tee in the connected waste closer to the trap. See Fig. 1 B. If the disposer is mounted in a single compartment sink, it may only be necessary to shorten the trap connection.

Repairing Garbage Disposers

Garbage grinders or disposers (after a period of use) often become very noisy in operation. In most cases this is because the metal lugs mounted on the revolving plate in the disposer become worn and loose. The lugs are secured to the plate by bolts or rivets. When the disposer is in operation, the centrifugal force of the spinning plate throws the lugs outward. The lugs are designed to clear the inside of the disposer but as the disposer is used, the lugs, bolts, or rivets, become worn at the pivot mounting point and hit the inside of the disposer when it is operating. Usually the rivet or bolt is sheared off, leaving the lug loose in the grinding chamber. When one lug is off of the revolving plate, the plate is thrown out of balance resulting in wear and damage to the disposer bearings.

It is not feasible to repair a disposer which has reached this stage of breakdown. The labor and parts charges will usually exceed the cost of a new disposer. A defective capacitor in the motor section will often cause a disposer not to operate. If the disposer is in good mechanical condition, replace the capacitor. The yellow pages of the phone book under ''Garbage Disposals'' should have listings of dealers handling repair parts. To remove the capacitor and have it checked and replaced, first turn off the electrical circuit to the disposer.

If you are not certain which circuit controls the disposer, pull the main switch and disconnect all electric power to the home while working on the disposer. Remove the cover on the bottom of the disposer and locate the capacitor. It is usually approximately one inch in

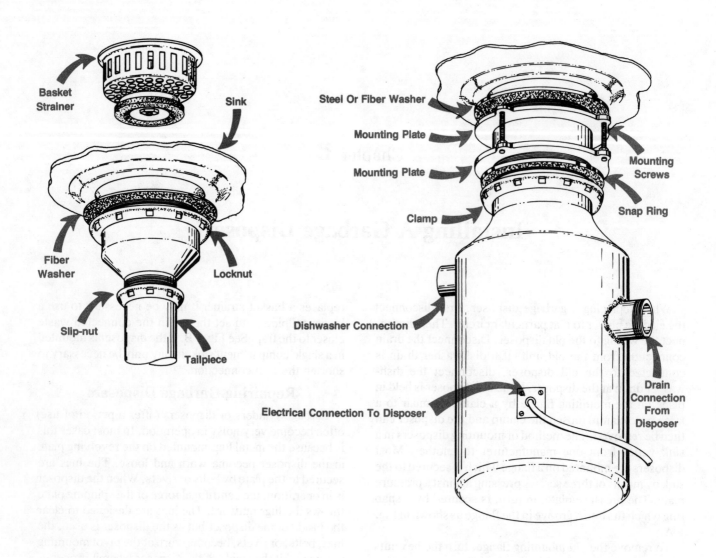

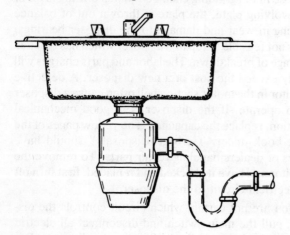

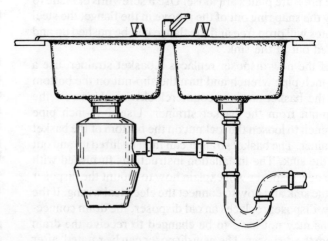

Fig. 1. Various garbage disposer installations.

diameter and two inches long, with two wires connecting it to the motor terminals. Observe the polarity markings on the capacitor before removing it. If it is necessary to remove any other wires in the process, tag the wires so they can be reconnected to the proper points. Pull the bare ends of any disconnected wiring apart and wrap the bare ends with several layers of plastic electrical tape. If it is possible to leave the electrical circuit which furnishes power to the disposer off until the repairs are completed, it is best to do so.

Quite often, usually after a lengthy period of usage, water will be seen leaking out of the bottom of the disposer, through the motor section. Inspect the mounting flange at the bottom of the sink and the drain connection on the disposer. If there is no leak at these points, the leak is probably internal — through the bearings — and the disposer should be replaced immediately.

Electrical Wiring for Garbage Disposer

Electric power for the garbage disposer can sometimes be picked up from a nearby receptacle. In the newer homes, these kitchen circuits are on a 20 amp. circuit. If your home does not have a separate 20 amp. circuit for the kitchen area, I recommend a new circuit be installed from your service box into the sink area. If a receptacle is

available, your first step would be to be certain that power is disconnected to this receptacle. Remove the receptacle from the wall by loosening the mounting screws on top and bottom of the receptacle. You will find on the receptacle 4 screws, two of which are brass or dark, and two of which are silver or light in color.

Your second step is to select a switch box that is suitable for installation in a finished wall. There are several different types on the market, but the one we will deal with is the type that is manufactured by the RACO Corporation. It may be purchased in most electrical supply stores. It has two expansion type clips which expand as they are being screwed down and thus holds the box tightly in the wall. This switch box is furnished with a tracing template of the size needed for the box. Place the template on the wall where the switch is desired; take a pencil and trace around the template and cut the hole for the switch box. Fish the wire from the receptacle into the switch box and another wire from the switch box to the garbage disposal.

The easiest method is to use a nonmetallic sheath cable, commonly known as Romex. The wire should normally be 12 gauge three conductors — a black wire; a white wire; and a bare wire or green-covered wire. Connect the black wire in the Romex cable to the brass screw

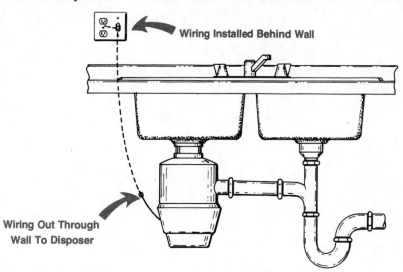

Fig. 2. Wiring circuit for disposer.

Wiring Installed Behind Wall

Wiring Out Through Wall To Disposer

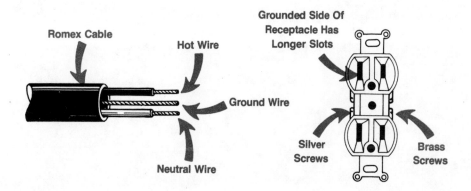

Romex Cable

Hot Wire

Ground Wire

Neutral Wire

Grounded Side Of Receptacle Has Longer Slots

Silver Screws

Brass Screws

on the receptacle; the white wire to the silver screw on the receptacle; the bare wire to the bare wire connection on the switch box. In the switch box you connect the black wire on one side of the switch and the ground wire to the grounding screw of the box; the white wire is connected with a wirenut to the white wire in the piece of Romex going to the garbage disposal. The black wire going to the garbage disposer is attached to the other screw on the switch; the bare wire is also attached to the grounding screw in the switch box.

In the garbage disposer, you will find a black and a white wire protruding from the lower portion of the garbage disposer and also a green screw for grounding Attach the black wire to the black wire; attach the white wire to the white wire; attach the bare wire to the green grounding screw.

Install the box in the wall and the switch into the box; cover with a switch plate and restore the power. Wiring connections must comply with local codes.

Chapter 9

Installing Dishwashers

Under the counter-type automatic dishwashers are becoming very popular. They are easily installed in the cabinetwork of a kitchen, since most cabinets are build in increments of 24 inches. Automatic dishwashers require a hot water connection and a drain connection. The best location for an automatic dishwasher is next to the sink cabinet. Most manufacturers of electric dishwashers specify that the hot water supply pipe to the dishwasher must not be less than ⅜" I.D. Soft copper tubing is the easiest to install for this purpose. The pipe connection at the dishwasher inlet valve should be ⅜" I.P.S. (iron pipe-size). An adapter fitting either elbow type or straight type, depending on the make of dishwasher, should be used at this point. The adapter should be ⅜" M.I.P. (male iron pipe) on one side, ⅜" I.D. tubing, ferrule type on the other side.

Install a tee between the valve and the sink faucet in the hot water pipe to the sink, as shown in Fig. 1. There should be a union connection; either a union, a coupling nut, or a flexible connection where the pipe connects to the sink faucet. If the piping to the sink is galvanized pipe, add a tee with a ½" side opening and reconnect the sink faucet. Use a ½" M.I.P. X ⅜ I.D. ferrule-type adapter to connect the ⅜" I.D. tubing from the tee to the dishwasher inlet. If the piping to the hot water side of the sink is ½" I.D. copper (⅝" O.D.), install a ½ x ½ x ⅜ ferrule-type tee, (½ I.D. x ½ I.D. x ⅜ I.D.) If the piping is ⅜ I.D. copper (½" O.D.) use a ⅜ x ⅜ x ⅜ (I.D.) ferrule-type tee. Use of a ferrule type fitting will eliminate the need for soldering when installing dishwashers. Use teflon tape or pipe joint cement on the male I.P.S. threads. Use two wrenches (one for a back-up) when tightening ferrule connections.

Instructions for the electrical connections to the dishwasher will be found in the installation instructions furnished with the dishwasher.

If the dishwasher drain pipe is connected to the drain pipe from the kitchen sink, instead of into a garbage disposer, the connection must be made on the inlet side of the trap on the sink drain piping. A special fitting, called a dishwasher tailpiece is made for this purpose and can be purchased at most plumbing shops. Hot water heater hose, the same kind as used on automobiles, is available at auto-supply stores or service stations, in ½", ⅝", or ¾" I.D. sizes and makes an excellent piping material for drain connections on dishwashers. Use copper adapters — ¾" M.I.P. (male iron pipe) x ½" I.D. copper; or ½" M.I.P. x ½" O.D. copper solder-type adapters and hose clamps. The exact size of the adapters will vary, depending on the make of your dishwasher.

Electric Circuit To Dishwasher
(See also chapter on garbage disposers)

Electric power for a dishwasher can be obtained from a receptacle which is close to the dishwasher. Before removing the receptacle from the wall, be certain the electric power is disconnected to the receptacle. Loosen the screw holding the receptacle plate and remove the plate. Loosen and remove the two screws holding the receptacle and pull the receptacle out of the box. Loosen the clamp holding the Romex wire which feeds the receptacle and drive out the unused knockout plug behind it. Fish a piece of Romex 12-2 with ground wire down the wall and bring it out through the wall in the sink cabinet space. Connect the ground (bare) wire in the Romex going to the garbage disposer to the ground wire in the receptacle box. Connect the black wire to the unused brass screw on the receptacle. Connect the white wire to the unused silver screw on the receptacle. Tighten the clamp holding the two pieces of Romex in the box. Push the receptacle back into the box, replace the two screws holding the receptacle and replace the cover plate. Connect the black wire to the black wire on the dishwasher, the white wire to the white wire on the dishwasher and the bare wire to the green ground screw on the dishwasher. Reconnect the power to the receptacle circuit.

If it is planned to install and connect both a dishwasher and a garbage disposer on the same circuit, it is recommended that a new circuit be installed from the electric service cabinet to the sink area to avoid an overload condition. All wiring must comply with local codes.

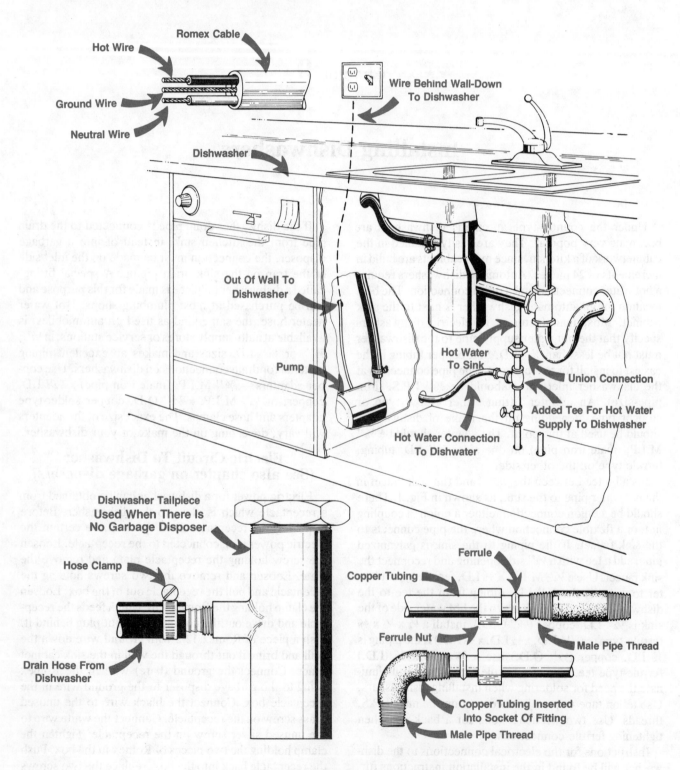

Fig. 1. Installation procedure for automatic dishwasher.

Chapter **10**

How to Measure and Cut Pipe

There are many times in the course of a remodeling project, or when adding new plumbing fixtures or repairing a leak, that knowing how to measure the needed pipe can save time and work. Example (A) in Fig. 1 shows tees added into existing piping and new piping added, (as connections for an automatic washer). In this case, we will assume the pipes are on the same level against the floor joists on the ceiling of a basement. Because the hot and cold piping is on the same level, it will be necessary to use fittings to enable the new pipe to pass under, or over, the existing pipe.

Examples (B), (C), and (D) show different ways in which this can be done. In examples (B) and (C), the new piping is on the same level. In example (D), the new hot-water pipe would be above the cold-water pipe. In examples (B) and (C), the two pieces of veritcal piping would be the same length. In example (D), the new vertical hot-water pipe would be longer than the new cold-water pipe. If you have your own pipe dies and vise, the necessary cuts and threads can be made on the job. A small job, such as illustrated, would not justify the cost of renting these tools. Most hardware stores which stock pipe and fittings will also thread pipe.

The correct and most accurate way to measure out lengths of pipe is to make an "end-to-center" measurement. Fig. 1 A shows a section of piping with a tee, nipple, and union installed between points (3) and (4). "End-to- center" means from the end of the pipe, with the fitting "made up," or tightened onto the pipe, to the center of that fitting. In order to make an end-to-center pipe measurement, it is necessary to know how to measure the end-to-center distance of a fitting. Fig. 2 A illustrates a full size drawing of a ½″ elbow—X is the distance the thread makes up into the fitting, (½″). The distance Y is from the end of the thread of the pipe to the centerline of the fitting.

In Fig. 1 E we show a 3-inch nipple "made up" into a tee. The nipple length (3″) plus the end-to-center measurement of the fitting (⅝″) gives an end-to-center pipe measurement of 3⅝ inches. Suppose you want to cut a piece of pipe to measure 24¾ inches end-to-center of a tee; the actual length of the pipe would be 24⅛ inches. The 24⅛ inches plus the ⅝ inch (end-to-center of the fitting) would equal 24¾ inches end of the pipe to center of the tee, or the desired length. Also shown in Fig. 2 is a full size ½-inch tee, ½-inch 45° elbow, and a ½-inch union.

Referring again to Fig. 1 F the view is an isometric drawing of the piping as shown in (A).

From the front side, facing any fixture, the hot-water pipe should always be on the left side. The tees shown in (A) should be placed in such a manner to bring the piping to the desired location, with the hot-water pipe on the left. When a fitting is added to piping between other fittings (3) and (4), and (3) and (4) cannot be turned, a union must be used to join the new pipe. The distance between (5) and (6) is the end-to-end measurement of the present piping. After a tee, a nipple, and a union are added, or cut into the present piping, this distance must remain the same, therefore, some pipe must be cut out to make room for the tee, union, and nipple.

Fig. 1 A explains how the actual cut lengths of pipe are determined. For the purpose of this explanation, the original piping length between (5) and (6) is 66 inches. If the center of the tee in the cold-water pipe is to be 24 inches from the makeup point (5), then the actual cut pipe length (7) would be 19⅝ inches (19⅝″ + 4⅜″ = 24″). The other cut piece (B) would then be 66 inches minus 19⅝ + 5″ (or 24 inches end-to-center of tee plus ⅝ inches to make-up of tee) or 41⅜ inches end-to-end.

To determine the end-to-center measurement of (9) in Fig. 1 F and B and Fig. 3A, and using the type of pipe

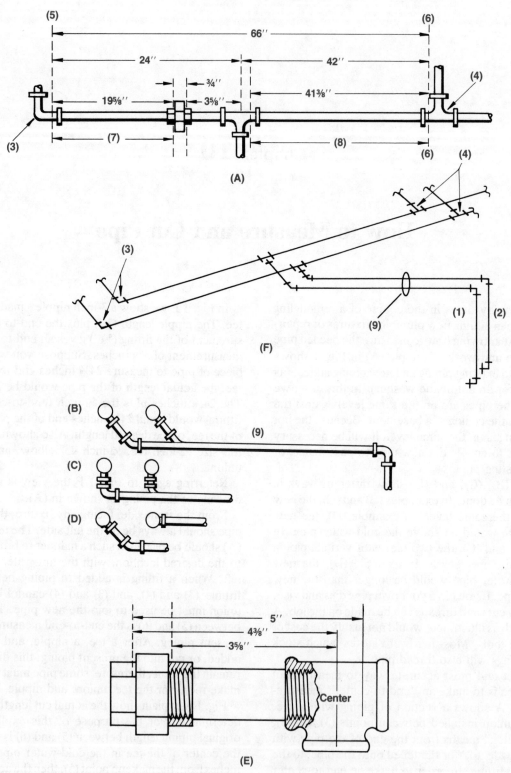

Fig. 1. Various ways of measuring pipe.

support shown, the distance from the wall line to the center of the pipe support, when the support is mounted on the wall, is ¾ inches. If a rule is held against the wall as shown and the end of the pipe (where the pipe makes up into the 45° elbow) to the wall measurement is 34 inches, then the end-to-center measurement of this pipe would be 34 inches minus ¾ inches (to center of pipe support) or 33¼ inches end-to-center of the elbow. The drop pieces of pipe (1) and (2) can be cut to any desired lengths. All of the end-to-center measurements given in example are for ½-inch pipe; end-to-center measurements for other sizes of pipe and fittings are arrived at in exactly the same way. The tables in Fig. 3 B show these measurements for other common pipe sizes.

Quite often it is necessary to offset around a light fixture, a post, or other obstructions when installing new runs of pipe. There is a formula for figuring the lengths of pipe when offsets are necessary. A 45° fitting should be used when making offsets because there is less friction loss (and therefore less pressure drop) through a 45°

fitting than there is through a 90° fitting. For the purpose of this illustration, we will assume it is necessary to offset both the hot and cold water pipe 14 inches in order to go around a light fixture. The formula for this offset is the square break, 14″ x 1.41. The result of this is the center-to-center distance between two 45° ells. Subtract the end-to-center measurements of the fittings, (2-45° elbows) and the answer is the length of the pipe. Fig. 2 C shows how to figure the offset. In this illustration, Y is the distance of the square break; X is the center-to-center measurement of the 45° break; therefore, X = 14″ x 1.41 or 19.74. We round off the .74 to .75 and the

Fig. 2. The proper way to measure fittings.

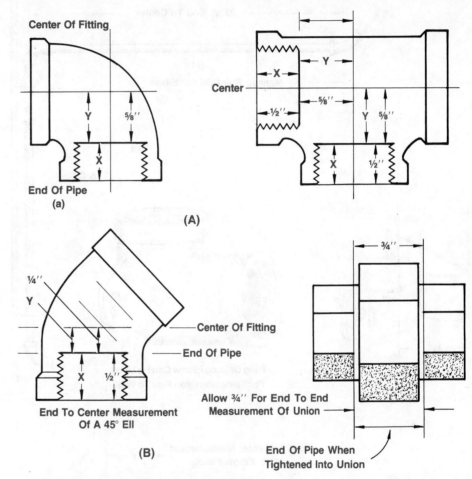

The Measurements Above Are For ½″ Fittings

Other Fittings Are Measured In The Same Way

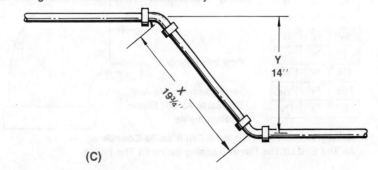

answer 19.75 (19¾'') is the center-to-center measurement of the 45° break. Referring to Fig. 2 B , the end-to-center measurement of a ½-inch 45° elbow is ¼ inches. There is a 45° elbow at each end of the 45° break, so we deduct ¼'' x 2 = ½''. Subtract ½ inch from the center-to-center measurement and the answer; 19¼ inches is the length of the pipe in the 45° break.

When pipe cutting and threading is being done, care must be taken to ream the burr from the cut ends. This can be done with a file if a regular type pipe reamer is not available. The new pipe thread being cut will be the right length when the new thread is flush with the outside face of the pipe die. A thread that is too long will be loose and prone to leak; a thread that is too short will be hard to start in a fitting.

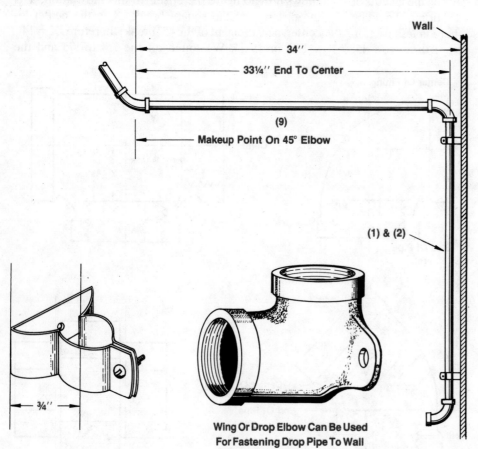

Fig. 3. Makeup measurements.

Wing Or Drop Elbow Can Be Used
For Fastening Drop Pipe To Wall

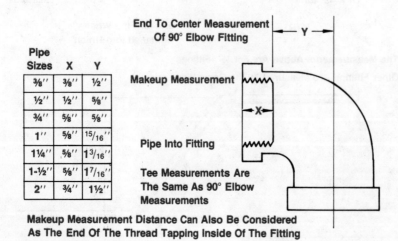

End To Center Measurement
Of 90° Elbow Fitting

Makeup Measurement

Pipe Into Fitting

Tee Measurements Are
The Same As 90° Elbow
Measurements

Makeup Measurement Distance Can Also Be Considered
As The End Of The Thread Tapping Inside Of The Fitting

Pipe Sizes	X	Y
⅜''	⅜''	½''
½''	½''	⅝''
¾''	⅝''	⅝''
1''	⅝''	15/16''
1¼''	⅝''	1 3/16''
1-½''	⅝''	1 7/16''
2''	¾''	1½''

Chapter 11

Repairing or Replacing Sink and Lavatory Traps

Many times a small repair job turns into a big job when it need not be. One of the most common repair jobs in the house is replacing a trap on a sink or lavatory. Tube-type brass traps, usually chromium plated, are most often used on sinks and lavatories. Quite often the lower part of the trap, called a J-bend, will split open. This is usually due to stress caused when the trap was extruded. As a temporary repair, the trap can be soldered by using a small propane torch, acid flux or solder paste and solder. It will be necessary to file the chrome plating off all along the split part to be soldered. The bright clean brass surface exposed after the chrome plating is filed off will take solder very well.

To make this temporary repair, loosen the slip-nuts on the J-bend and pull down to remove it from the tailpiece on the sink or lavatory. If the rubber slip-nut washers on the other half of the trap and on the sink or lavatory tailpiece stick to the J-bend, remove them and save them for use when reassembling the trap. File the chrome plating off ¼ inch on each side of the split until the brass surface of the trap is exposed. Coat the exposed brass with solder paste or acid flux and heat the area to be soldered with the propane torch. Unroll 6 or 7 inches of wire solder from the spool and touch the end of the solder to the trap. When the solder starts to run, play the heat from the torch over the area to be soldered until the whole area that was cleaned has been coated. Remove the heat and allow the J-bend to cool and then reassemble the trap.

J-bends are available at most hardware stores for replacement. There are several methods of connecting traps to the drain pipe. The common practice in many cities in years gone by was to use a solder bushing at the wall, if the trap was a P-trap, or at the floor if the trap was an S-trap. To use a solder bushing, the chromium plating

was filed off the end of the trap and the solder bushing was then slipped over the end of the trap. The solder bushing was heated and soldered on the end of the trap. The trap was then threaded into the pipe fitting at, or behind, the wall line and connected to the fixture it served.

Another way in which traps were connected to the drain pipe was to slide the trap into a nipple in the drain pipe at the wall line and connect the trap to the nipple by tightening a slip-nut to the nipple.

Still another way was to solder the trap to a piece of lead drain pipe which extended to the wall line, or in the case of an S-trap, to a lead drain pipe at the floor. Fortunately, this method is seldom used today.

The most widely used method of installing traps on lavatories and sinks is to use a compression to iron pipe size adapter to make the connection between the trap and the drain pipe. This fitting is shown in Fig. 1. This type of fitting is made for use with galvanized steel pipe, DWV copper drain pipe and DWV PVC plastic drain pipe.

Another way in which traps have been connected to drain pipe in homes or buildings in which DWV copper drainage pipe is used, is to slip the end of the trap into the DWV copper tubing and solder the joint between the trap and the DWV tubing. Now that we know the most common ways in which traps are connected to drains, we will go back and see how to replace the present trap.

If the trap is connected to the drain by a solder bushing, the solder bushing will probably be badly corroded and cannot be unscrewed from the drain fitting. Loosen the slip-nuts on the J-bend and remove. Insert the handle of a screwdriver into the trap and turn the trap counterclockwise. If the solder bushing is not badly corroded, it may then unscrew from the drain fitting. If the solder

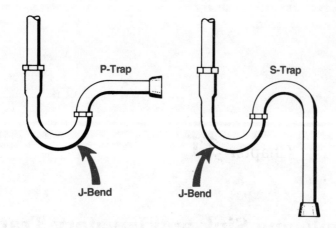

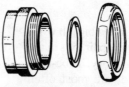

Male Adapters With Slip-joint Connection

Female Adapters With Slip-joint Connection

Fig. 1. Various connections used in drain traps.

Face Solder Bushing

Hex Solder Bushing

Slip Nut & Washer

bushing is corroded, the trap may either twist off or pull out of the solder bushing. In either case it will be necessary to use a hammer and a small cold chisel to split the solder bushing and remove it.

Care must be taken not to damage the threads in the drainage fitting while removing the solder bushing. Once the solder bushing has been removed, compression to iron pipe adapter can be threaded into the drainage fitting, the new trap inserted into the compression adapter and the compression nut on the fitting can be tightened. After this type adapter has been installed, the trap can be easily removed in the future for repair or replacement.

If the trap is connected to the drain pipe by a slip-nut loosen the slip-nut by turning it counterclockwise, remove the old trap, slide the new escutcheon on the trap, then the new slip-nut and washer and tighten.

Now we come to the difficult job. If the trap is soldered to a lead drain pipe in the wall, a soldering iron must be used to melt the old solder from the lead joint, the old trap removed and the new one soldered into the lead pipe. The services of a skilled plumber are required in this case. This is also true if the trap is an S-trap and is soldered to a lead pipe at the floor line. This is a job for an expert. If the trap is connected to the drain by a compression to iron pipe, or a compression to DWV copper, or plastic adapter, loosen the slip nut; remove the old trap; slide a new escutcheon on the trap; slide the slip-nut and washer on the trap; insert the trap into the adapter and tighten the slip nut on the adapter. When the trap is connected to the sink or lavatory tail piece, the repair is completed.

S-traps are prohibited by most plumbing codes. Slip-joints are permitted only in trap seals or on the inlet side of the trap. Plumbing codes and interpretations vary from state to state or within a state. Repairs as outlined in this chapter must comply with local codes or ordinances.

Chapter **12**

Installing a Homemade Humidifier

A good effective humidifier is a necessity in today's homes. The coldest weather in winter is produced by high pressure areas sweeping in from Arctic regions. This Arctic air is very cold and extremely dry. In homes heated by hot-air furnaces, the cold dry air is heated and circulated through the home. Furniture with glued joints comes apart. Static electricity is formed when walking across certain types of carpet and a shock is felt when metal is touched. Medical doctors advise the installation of humidifiers to alleviate the ill-effects of dry air on the human body.

A great many homes today are built on a concrete slab and the plumbing pipes and the heating air supply ducts are located under the concrete slab. The hot-air furnace used with this type construction is called a counter-flow furnace. When a counter-flow furnace is used, the hot air supply ducts are located under the floor and the return or cold air ducts are located above the ceiling. When the home is of this type construction, a homemade trouble-free humidifier can be installed in the hot air supply ducts for a fraction of the cost of conventional humidifiers. The humidifier can be manually or electrically operated to keep the humidity within the desired range. In order to know how to make the necessary electrical connections, it is first necessary to understand the sequence of action in the operation of a furnace.

For the purpose of this explanation we will assume the furnace is equipped with a gas burner. The actual operation is essentially the same if the furnace is equipped with an oil-fired burner. When the room thermostat senses a drop in temerature, below the set level, a switch in the thermostat closes. (When we say a switch in the thermostat closes we mean that an electrical circuit is completed, and current flows in the circuit.) An open switch interrupts the flow of current in that circuit. When the thermostat switch is closed, electrical current flows in the circuit to the gas valve in the furnace, the gas valve is energized and the gas valve is turned on. The main burner then lights and heat builds up in the heat exchanger. Many furnaces have a combination fan-limit switch to control the operation of the fan and the bonnet/heat exchanger temperature of the furnace. When the temperature of the air in the heat exchanger has reached the set level, the fan switch closes and the fan is turned on. The cooler air is pulled back from the rooms through the cold air return ducts and pushes the heated air back into the rooms through the hot air supply ducts.

When the temperature in the heat exchanger reaches the setting of the limit switch, the limit switch is opened and the electrical current to the gas valve is cut off, thus shutting off the gas burner. The fan continues to operate, circulating the hot air back to the rooms in the house until the desired setting of the thermostat is reached. Most thermostats have a heat-anticipating feature built into them. This feature regulates the point at which the fan will be turned off. The operation of the humidifier described in Fig. 1 requires only two electrical connections. The two wires from the solenoid valve must be connected to the two wires going to the fan motor. Thus, when the fan motor is energized the solenoid valve is also energized; the solenoid valve is opened and water will spray out of the tubing into the hot air ducts under the furnace. The hot dry air from the furnace will pick up the moisture from the water spray and raise the humidity of the air to the desired level. When the humidity is at the desired level, the manually operated switch shown in the illustration can be turned off. This interrupts the circuit to the solenoid valve, shutting the solenoid valve off but leaving the fan in operation. When the humidity falls below the desired level, the switch can be turned back on, putting the humidifier back into action.

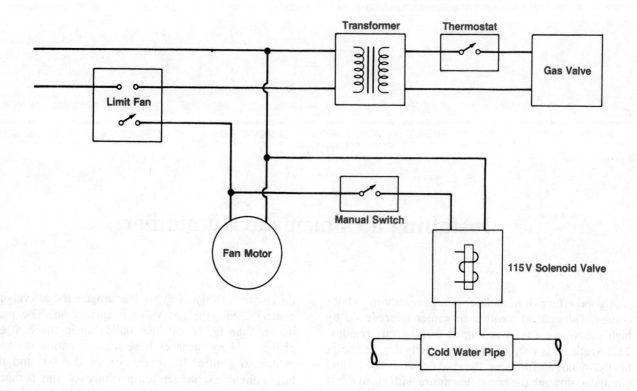

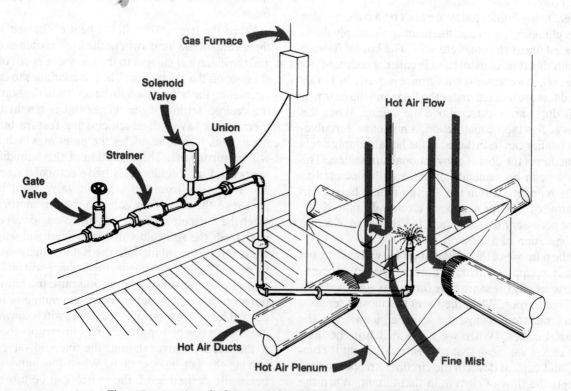

Fig. 1. Construction of a homemade humidifier and the electrical connections.

If a more sophisticated system is desired, a humidistat can be installed to control the operation automatically. Wall or duct type humidistats are available for this purpose. The humidistat will sense the moisture content of the air, and when the moisture content falls below the desired level, a switch in the humidistat will close (completing an electrical circuit to the solenoid valve) turning on the solenoid valve. When the humidity reaches the desired level, the switch in the humidistat opens, the circuit to the solenoid valve is also opened and the solenoid valve is shut off.

The fan voltage is normally 120 volts so the solenoid valve must also be rated for 120 volt operation. If a line voltage (120 volts) humidistat is used, the proper size wire must be installed from the solenoid valve to the humidistat. Local electrical codes must be followed when installing this wiring. The author recommends that a qualified electrician install necessary wiring and that he also makes the required connections. For either of these installations the solenoid valve must be a normally "closed" type, that is, when there is no electrical current to the coil on the valve, the valve is shut off. When the coil is energized, the valve opens. To prevent any scale or other foreign matter from lodging in the solenoid valve, a strainer should be installed in the pipe as shown in the illustration. Solenoid valves and strainers are available at refrigeration supply stores.

Installation of the pipe from the solenoid valve to the plenum chamber under the furnace will of necessity depend on the manner in which the furnace is installed. It is here that the ingenuity of the home-owner will be used. The pipe from the solenoid valve should extend into one of the heating ducts leading from the plenum box. The pipe must be removable, the tip end will occasionally develop a formation of lime or scale and will have to be cleaned. A larger pipe could serve as a sleeve through which the humidifier pipe would enter the plenum. It would then be a simple matter to loosen the union in the humidifier pipe and remove the pipe for inspection. The pipe is assembled as shown in Fig. 1. The end of the soft copper tubing is pinched shut with pliers. Pinching the tubing shut will not keep water from coming out under pressure, it will force the water out as a fine spray or mist. The hot dry air from the furnace will absorb the moisture from the spray and raise the humidity level in the home.

Galvanized steel piping with galvanized malleable iron fittings could of course be used instead of copper tubing as shown in the illustration. The pipe inserted through the sleeve and into a duct opening should be ⅛ inch I.D. (¼ inch O.D.) soft copper tubing. If galvanized pipe and fittings are used up to the gate valve, strainer and solenoid valve use a ½ inch male iron pipe to ¼ inch O.D. ferrule type compression adapter, screwed into the solenoid valve to adapt from iron to copper pipe. If this fitting is used, since it is in effect a union, no other union is necessary at this point.

Chapter 13

How to Install A Sump Pit and Sump Pump

Wet basements, due to ground water seeping in through floors and walls, are a problem in many areas. This situation can be helped by installing a sump pump in the basement.

The pit for the pump should be located near a drain opening in the sanitary drain pipe, as close as possible to an outside wall (unless existing conditions make this impossible.) The concrete floor where the pit is to be located must be cut out, the hole in the concrete floor should be 8 inches larger than the tile pipe used for the pit. An 18 inch diameter x 24 inch long concrete or vitrified tile pipe is ideal for this purpose. Dig the hole for the tile approximately 2 inches deeper than the length of the tile and 6 inches wider than the outside diameter of the tile; put 2 inches of sand in the hole to bed the tile in.

Before setting the tile pipe in the hole, drill six ½ inch holes in the tile, eight inches down from the top. The holes will be approximately 11 inches apart. Use an electric drill with a carbide tipped drill bit, a hammer type drill motor is best for this purpose. After drilling the holes, set the tile down in the pit hole with the top of the tile level with the concrete floor.

A concrete base, 3 inches thick, for the sump pump should be poured in the bottom of the tile and concrete will also be needed to patch the floor around the outside of the tile. Dry concrete mixes in bags are available at most hardware stores.

If any water collects in the bottom of the tile before the sump pump base is poured, dip out as much water as possible and then pour the base. Do not be concerned if water collects in the tile after the base is poured, the

concrete will set up under water in 8 to 12 hours. Fill the space around the outside of the tile to within 4 inches of the top of the concrete floor with gravel. (The gravel should be ¾ inch to 1 inch in diameter.)

After the gravel is placed, mix the concrete and patch the floor around the tile. When the concrete in the bottom of the tile has set up, the pump can be placed and the pipe installed. A check valve should be placed in the discharge piping as shown in Fig. 1.

A check valve will hold water in the piping and prevent sewer gas from entering the pit if the pump is connected directly to a sanitary drain pipe. The check valve will also prevent the water which has been pumped up and out of the pit from running back into the pit when the pump shuts off.

Either a vertical type check valve or a swing-type check valve may be used; the vertical type valve is preferable since it is less noisy in operation. The valve must be installed correctly with the disc in the closed position when the pump is not in operation. A union should be installed in the pipe below the check valve to permit the pump to be removed for servicing. The pump discharge can be piped to a suitable opening in the sanitary drain pipe, such as a plugged cleanout opening. The plug can be removed and a tee or wye installed. The pump discharge can then be piped into the side opening of the tee or wye and the end of the tee plugged for use as a cleanout opening. The pump discharge can also be piped into an open receptacle, such as a laundry tray; or it can be discharged through the basement wall (above ground level) and into the downspout or rain leader.

171

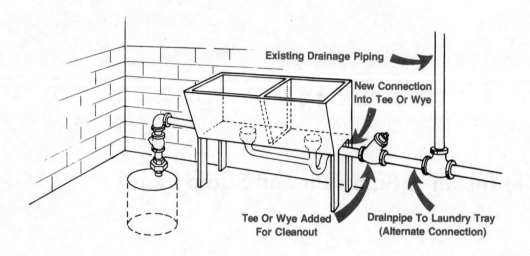

Existing Drainage Piping

New Connection
Into Tee Or Wye

Tee Or Wye Added
For Cleanout

Drainpipe To Laundry Tray
(Alternate Connection)

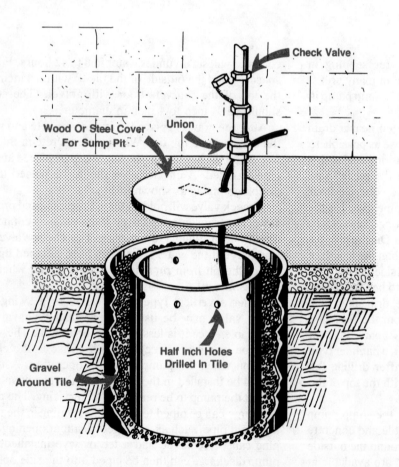

Check Valve

Wood Or Steel Cover
For Sump Pit

Union

Half Inch Holes
Drilled In Tile

Gravel
Around Tile

Fig. 1. Installing instructions for a sump pump.

Chapter 14

Drainage, Waste And Vent Pipe

Remodeling projects or changes of fixtures of fixture locations may require a change in drainage, waste and vent piping. Drainage and waste piping will work satisfactorily if installed with a minimum "fall" or grade of ⅛ inches to the foot. Generally speaking, a grade of ¼″ to the foot is the proper grade. Vent piping is a very important part of plumbing work. A vent pipe provides a flow of air to or from a drainage system. When water is released through a sink or when a toilet is flushed, the flow of water through a trap and into the drainage pipe tends to produce a vacuum. The vent pipe is an extension of the drainage pipe. The air present in the vent piping prevents a vacuum from being formed.

If there is no vent pipe on a fixture, a vacuum could be formed; this in turn could cause the water in the trap to be siphoned out. The purpose of a trap on a plumbing fixture is to prevent sewer gas from entering the house. If the water seal in a trap is broken by siphonage or evaporation, sewer gas can enter the home or building. Sewer gas can also enter the home or building if a back pressure (pressure built up in the sewer system) should back up into the building drainage system. Back pressure could blow the seals on fixture traps and sewer gas could then enter the building. Proper venting of a building drainage system can prevent back pressures by relieving any build-up of pressure through the main vent. Sewer gas is not only foul smelling and dangerous to breathe, it can also be very explosive.

PVC DWV plastic pipe and fittings are being extensively used for drainage, waste, and vent piping. Many plumbing codes which formerly permitted only lead, cast iron, copper, or steel piping for drainage waste and vent piping, have been updated to permit the use of plastic pipe for these purposes. Plastic pipe and fittings are not subject to corrosion or rust and require no special tools for installation. A hand saw can be used to cut the pipe, although if a saw is used, a mitre box is recommended in order to obtain square cuts. A special tubing cutter, Fig. 1, is excellent for use in cutting plastic pipe; the cutter has a built-in reamer for removing the burr left in the pipe when the pipe is cut. The burr can also be removed with a half round file or a pocket knife.

No. 205-P Tubing Cutter
(For Cutting Plastic)
Capacity ¼″ Through 2⅜″ O.D.

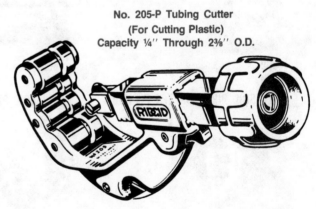

Courtesy The Ridge Tool Co.

Fig. 1. Plastic tubing cutter.

PVC DWV pipe and fittings are easily joined by the use of solvent and PVC cement. It is important to use only the right type solvent and cement when joining PVC pipe and fittings; use only PVC solvent and cement for PVC pipe. The solvent is used for cleaning the pipe and fittings and the cement is then applied when joining the pipe and fittings together. The fittings must be positioned exactly right, at the right angle and in the right direction when pipe and fittings are joined. When properly cleaned and assembled, the cement "sets up" immediately upon contact. Full instructions for use will be furnished with the solvent and cement and the instructions must be follwed for a satisfactory installation.

4807
5807
DWV 90° ELL P x P

4806
5806
DWV 45° ELL P x P

4807-2
5807-2
DWV 90° FTG ELL Ftg x P

4806-2
5806-2
DWV 45° FTG ELL Ftg x P

4810-13
5810-13
DWV 45° Y P x F x P

4812
5812
DWV LONG TURN TY P x P x P

4811
5811
DWV TEE P x P x P

4851
5851
DWV CLOSET FLANGE P

4800
5800
DWV SOIL PIPE ADAPTER P x Hub

4805
5805
DWV SOIL PIPE ADAPTER P x Spigot

4801-2-F
5801-2-F
DWV FTG FLUSH BUSHING Ftg x P

4801-7
5801-7
DWV TRAP ADAPTER P x SJ

4880
5880
DWV P-TRAP W/CLEANOUT
P x SJ x Cleanout W/Plug

4816
5816
DWV FTG CLEANOUT

Courtesy Nibco, Inc.

Fig. 2. Various PVC waste and vent fittings.

The fittings shown in Fig. 2 are only a few of the wide line of PVC (poly-vinyl-chloride) DWV waste and vent fittings now available.

DWV (drainage, waste & vent) type copper tubing and DWV copper fittings are used extensively for drainage waste and vent piping. The fittings and pipe are joined by soldering them together. DWV copper pipe and fittings are light in weight, quickly and easily assembled and not subject to rust and corrosion. DWV fittings are recessed

drainage type fittings and thus provide a smooth, unbroken passageway for water or waste drainage.

Drainage fittings are made with a recess at the threaded ends so that when nipples or pipe are installed into the fittings there is a smooth passageway for drainage water or wastes. Tees, 90° elbows, long turn TY's and wyes are made in both straight pipe sizes and reducing sizes, as a 1½ x 1½ x 1¼ sanitary tee, or a 1½ x 1½ x 1¼ wye — drainage tees and wyes are "read" or described — end/end/side, or as pictured A-B-C-1½ x 1½ x 1¼. Street 90° elbows and street 45° elbows are not recommended for use due to the restriction in the male threaded end and also because they are hard to gasp with a pipe wrench.

CPVC (Chlorinated Poly vinyl chloride) pipe and fittings are now being used for hot and cold water piping. The pipe and fittings are rated at a working pressure of 100 P.S.I. (pounds per square inch) at 180°F. The piping can be cut using a plastic tubing cutter (Fig. 1) or a saw. Joints between the pipe and fittings are solvent welded, using CPVC primer and CPVC solvent cement. Metal transition fittings are used for connecting plastic pipe to metal piping or for connecting plastic pipe to valves. CPVC plastic adapters are also made for connecting plastic pipe and fittings to metal piping. CPVC ells and tees are made in straight sizes and in reducing sizes. (¾ x ½ ells) (¾ x ½ x ½ tees) etc.

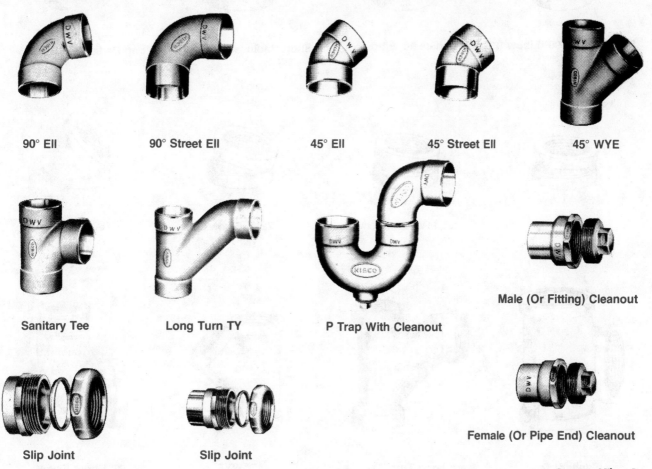

90° Ell	90° Street Ell	45° Ell	45° Street Ell	45° WYE

Sanitary Tee	Long Turn TY	P Trap With Cleanout	Male (Or Fitting) Cleanout

Slip Joint	Slip Joint	Female (Or Pipe End) Cleanout

Courtesy Nibco, Inc.

Fig. 3. Copper DWV fittings.

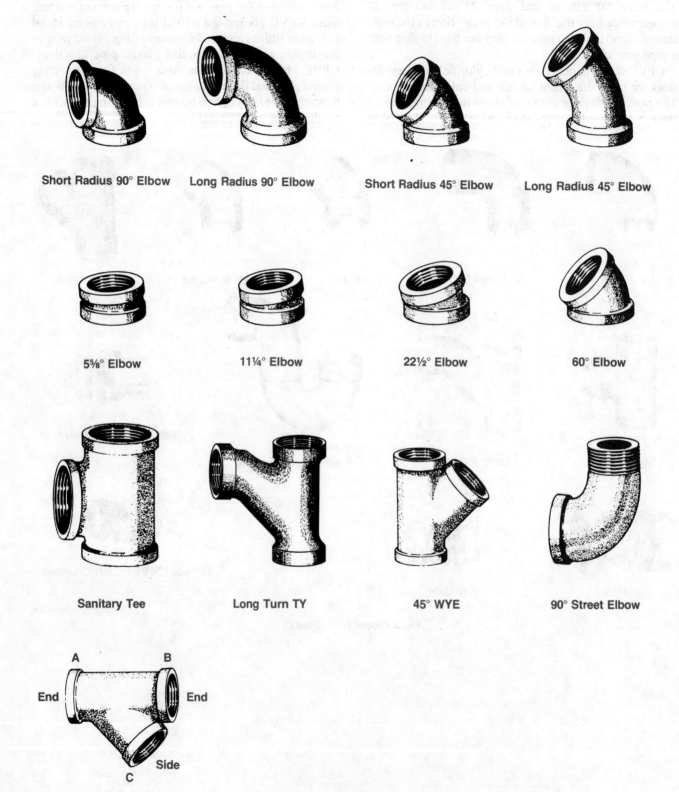

Short Radius 90° Elbow   Long Radius 90° Elbow   Short Radius 45° Elbow   Long Radius 45° Elbow

5⅝° Elbow   11¼° Elbow   22½° Elbow   60° Elbow

Sanitary Tee   Long Turn TY   45° WYE   90° Street Elbow

A   B
End   End
Side
C

Fig. 4. Cast iron drainage fittings.

| 90° Ell | 45° Ell | Coupling | Tee | Adapter |

| 90° Street Ell | Transition Adapter | CPVC Solvent And Primer |

Courtesy Nibco, Inc.

Fig. 5. Plastic hot and cold water pipe fittings.

Chapter 15

Working With Cast Iron Soil Pipe

Cast-iron soil pipe is the most durable material available for drainage waste and vent piping. Special tools are needed when working with soil pipe. Cold chisels or soil pipe cutters may be used to cut cast-iron soil pipe to the desired length. A yarning iron, a packing iron, and inside and outside caulking irons are needed to make the joint between two pieces of soil pipe. A lead melting furnace is used to heat the lead pot. When the lead is hot, a ladle is used to pour lead into the joint. When horizontal joints are made, a joint runner is used to hold the molten lead in the joint until the lead has solidified. Molten lead is dangerous to work with but some simple precautions will minimize the danger.

Never add lead which is wet or even slightly damp into a pot of melted lead. *Never* put a wet or even slightly damp ladle into a pot of melted lead. Hold the ladle above the pot of the melted lead until the ladle is warm and dry. When lead has to be added to the pot, lay a cake of lead in the ladle and hold the ladle and cake of lead over the pot until all moisture has evaporated from the ladle and the lead. If a damp or wet ladle, or a damp or wet cake of lead is inserted into the molten lead in the pot, some of the moisture will be carried under the surface of the molten lead. The moisture will be instantly converted into steam. The steam then literally explodes, throwing hundreds or thousands of particles of molten lead in all directions. Severe burns and loss of eye sight are just two of the serious consequences which could occur in such an accident.

Never pour melted lead into a wet joint or a joint in which the oakum is wet or damp. Again, water under the hot lead will convert into steam and an explosion could result. Use only oiled or tarred oakum in soil pipe joints. Oiled or tarred oakum resists moisture.

Cast-iron soil pipe is made in standard lengths. S.H. (single hub) is made in 5 feet and 10 feet lenghts. Measurements for cut lengths should be end-to-center measurements. The spigot end is inserted into the hub, the end measurement is taken from the point where the spigot end meets the bottom of the hub.

In Fig. 1, a piece of soil pipe must be 42 inches from end-(of pipe)to-center of ¼ bend. The 4 inch ¼ bend (Y) measures 8 inches from end-to-center. Therefore, 42 inches (total end-to-center length) minus 8 inches (end-to-center of the ¼ bend) equals 34 inches. The piece of pipe must be cut 34 inches long. An end-to-center measurement of a wye (Y) or a bend (1/8 - 1/6 - 1/5 - 1/16″) is made as shown in Fig. 2.

Cast-iron soil pipe is used extensively for drainage, waste and vent pipe. Cast-iron has the characteristic of rusting on the surface but protecting itself against the continued rusting which would in time destroy the metal. The initial coat of rust adheres strongly to the metal and prevents further rusting action. Cast-iron pipe and fittings can be joined in several different ways but the conventional way is to use lead joints. When two pieces are to be joined, the spigot end of one piece is set into the hub of the other piece.

Oakum is generally sold in individual boxes of 5 lbs. each; the strands of oakum are usually approximately 2 feet in length. Several strands are twisted together; the oakum should be untwisted and used one strand at a time. Place one end of the strand at the top of the hub and push the oakum down into the hub with the yarning iron. Follow the strand of oakum around the hub, ramming the oakum into place in the hub, using the yarning iron. Oakum should be rammed into the hub, leaving a 1 inch space between the top of the oakum and the top of the

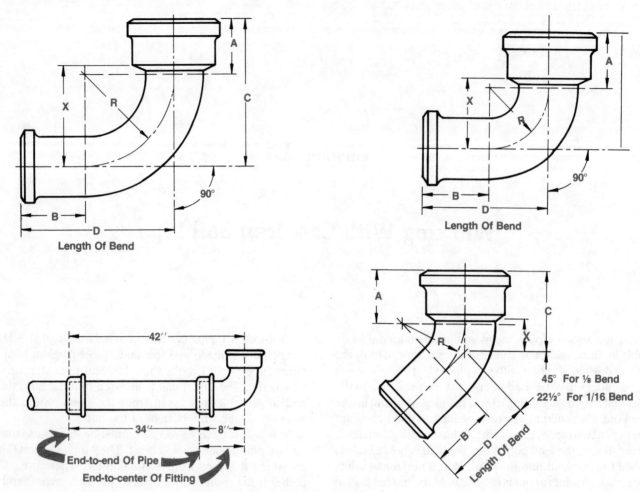

Fig. 1. How to measure bends.

hub. The oakum should be packed tight in the hub using a hammer to drive the packing iron for final packing. The lead is then poured into the joint; the joint should be filled to the top of the hub in one pour. The molten lead will solidify in a very few seconds. When the lead has solidified, use a hammer and inside caulking iron and go around the inside of the joint. When the inside of the joint has been caulked, use the outside caulking iron and go all around the outside edge of the joint.

There are two important points to remember when caulking a joint: (1). Use a light-weight hammer, not over 12 ozs., and do not hit the caulking iron too hard. Lead is soft and when too much force is used on the caulking iron, the lead will be expanded against the outside edge of the joint. The expansion of the lead can crack the cast iron hub. (2) The oakum, not the lead, is the key to a water-tight joint. the old joke among plumbers "do a good job, use lots of oakum" (lead was much more expensive than oakum) was not really a joke after all. If the oakum is tightly packed in the joint, the joint will be waterproof, and the lead only serves to hold the oakum in place. See Fig. 3.

TABLE 1. ¼ Bend

Size (inches)	Dimensions in inches					
	A	B	C	D	R	X
2	2¾	3	5¾	6	3	3¼
3	3¼	3½	6¾	7	3½	4
4	3½	4	7½	8	4	4½
5	3½	4	8	8½	4½	5
6	3½	4	8½	9	5	5½
8	4⅛	5½	10⅛	11½	6	6⅝
10	4⅛	5½	11⅛	12½	7	7⅝
12	5	7	13	15	8	8¾
15	5	7	14½	16½	9½	10¼

TABLE 2. ⅛ Bend

Size (inches)	Dimensions in inches					
	A	B	C	D	R	X
2	2¾	3	4	4¼	3	1½
3	3¼	3½	4-11/16	4-15/16	3½	1-15/16
4	3½	4	5-3/16	5-11/16	4	2-3/16
5	3½	4	5⅜	5⅞	4½	2⅜
6	3½	4	5-9/16	6-1/16	5	2-9/16
8	4⅛	5½	6⅝	8	6	3⅛

⅛ Bend

¼ Bend

1/16 Bend

1/16 Bend

Long Sweep ¼ Bend

Base Ell With Front
Cleanout

Sanitary Tee

Sanitary Tee With
Center Inlet

Plain Closet Bend

Hub Closet Bend

Wye(Y)

Combination Wye
And ⅛ Bend

45° Offset

P Trap

Plain Closet
Flange

Slotted Closet
Flange

Plain Roof
Increaser

Tapped Roof
Increaser

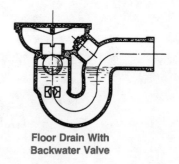

Floor Drain With
Backwater Valve

Fig. 2. Various bends and fittings.

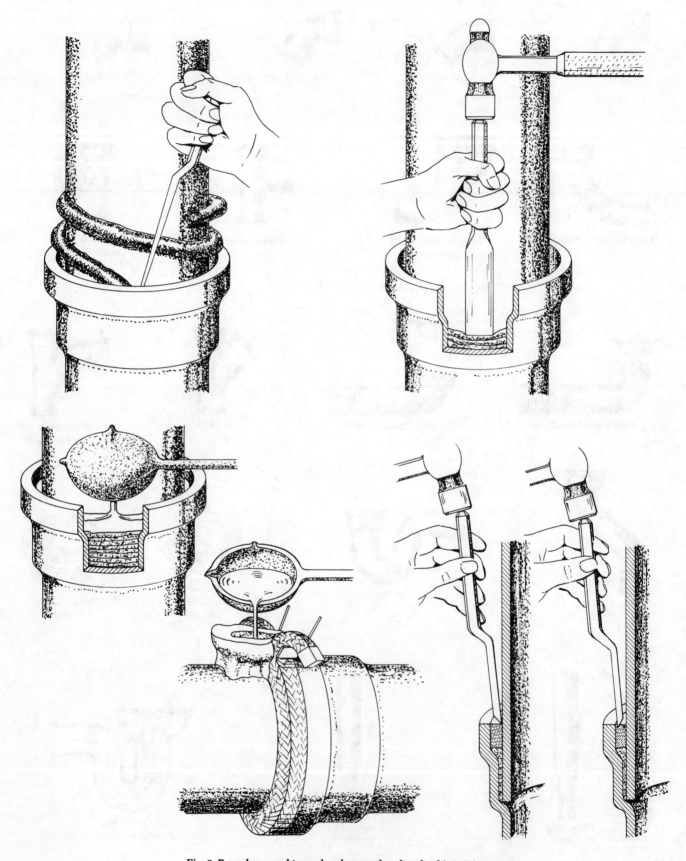

Fig. 3. Procedure used to pack oakum and molten lead into joints.

TABLE 2. ⅛ Bend

Size (inches)	Dimensions in inches					
	A	B	C	D	R	X
10...............	4⅛	5½	7	8⅜	7	3½
12...............	5	7	8-5/16	10-5/16	8	4-1/16
15...............	5	7	8-15/16	10-15/16	9½	4-11/16

TABLE 3. Long sweep

Size (inches)	Dimensions in inches					
	A	B	C	D	R	X
2..................	2¾	3	10¾	11	8	8¼
3..................	3¼	3½	11¾	12	8½	9
4..................	3½	4	12½	13	9	9½
5..................	3½	4	13	13½	9½	10
6..................	3½	4	13½	14	10	10½
8..................	4⅛	5½	15⅛	16½	11	11⅝
10...............	4⅛	5½	16⅛	17½	12	12⅝
12...............	5	7	18	20	13	13¾
15...............	5	7	19½	21½	14½	15¼

See Fig. 4.

(A) Yarning iron — used to insert oakum into the space between the spigot end and the hub.

(B) Packing iron — used to pack the oakum tight in the joint.

(C) Inside caulking iron — used to tighten the lead joint against the pipe.

(D) Outside caulking iron — used to tighten the lead joint against the hub.

(E) Gate chisel — used to cut off the 'gate' formed when pouring a horizontal joint.

(F) Joint runner — a braided asbestos rope which is placed around a horizontal joint to hold the molten lead until the lead has solidified. The joint runner is held in place by a clamp. A new joint runner should be soaked in light oil until the runner is soft and pliable.

(G) Propane tank, burner, hood and lead pot — used to melt lead.

(H) Cold chisel and hammer — can be used to cut soil pipe. Use soapstone or chalk to mark on pipe where the cut is to be made. Place a 2 x 4 under the mark for the cut, use the chisel and hammer, rolling the pipe at the same time and cut a groove around the pipe, on the cut mark. Tap the chisel lightly the first time around the pipe; the second and third times around the hammer blows can be harder, the pipe should break on the cut mark the third time around.

(I) Chain type soil pipe cutter — the chain setting is adjustable for various sizes of soil pipe and the ratchet action makes cutting of soil pipe a quick and easy job.

(J) Level—Use to establish the proper slope or grade of a pipe run. The grade, or fall, should be not less than ⅛ inch per foot; ¼ inch per foot is the best rate of fall. When using a level, the bubble always goes to the high side.

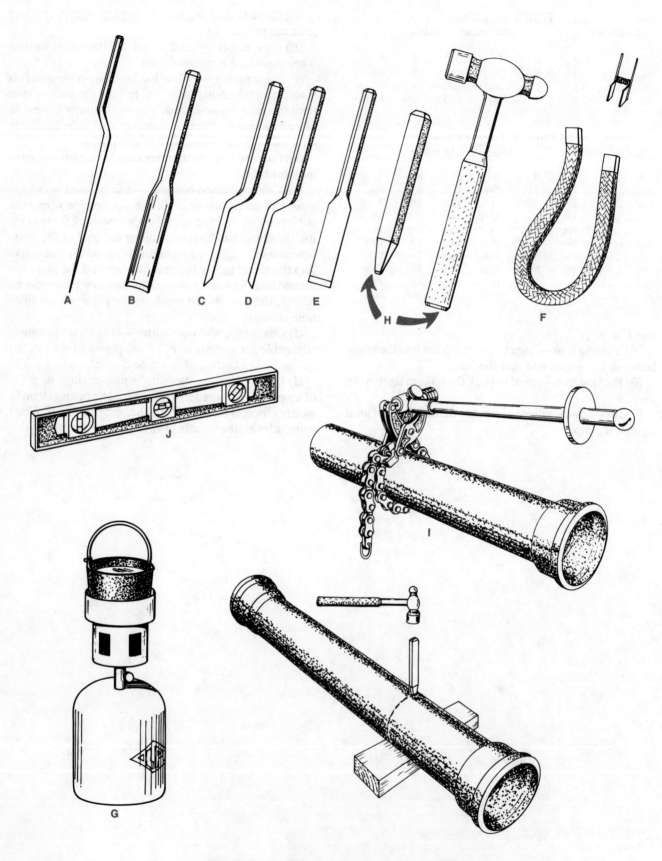

Fig. 4. Tools used when working with cast-iron sail pipe.

Chapter **16**

Remodeling A Kitchen

It may be desirable when remodeling a kitchen to install new wall and base cabinets and a new kitchen sink. Replacement of an existing garbage disposer (or installation of a new garbage disposer) may be included in the project. The addition of an automatic dishwasher unit may also be included. These additions or replacements of items usually mean some changes or additions to the present water and drainage piping in the kitchen area.

Laminated plastic is the material which is most popular for counter tops. The counter top can be cut out to receive the new sink. An electric sabre saw is the right tool to use for this purpose. Sinks may be self-rimming or may require the use of a special rim. In either case, the sink is clamped to the counter top by use of special clamps. The clamps and instructions for their use will be furnished with the sink or the special rim.

A great many of the older homes have wall-type sink faucets and the water piping to the faucet will be concealed in the wall. If the new sink is made with openings for a deck-type faucet, the water piping should be changed to run up through the floor, inside the sink cabinet. It will depend on the way in which the piping was installed originally as to how this change can best be made. It may be possible to cut the present piping below the floor and extend new piping up through the floor into the new cabinet. See Fig. 1 for suggestions on how the water piping may be changed. Fig. 2 shows some typical drainage connections. The chapters on dishwasher in-

stallation and garbage disposer installation will be helpful if these appliances are to be installed.

If the present sink faucet is wall-mounted and if it is desirable to reuse it, it will be necessary to remove the faucet. Refer to Fig. 3 (Chapter 3) where a common type at this time. It is good plumbing practice to install valves on the hot and cold water piping to each fixture. The valves for the hot and cold water piping to the kitchen sink should be underneath the sink, in the sink cabinet. These valves must be turned off before removing the sink faucet. Refer to Fig. 3 (chapter 3) where a common type of wall-mounted sink faucet is shown. With this type of faucet, turn the coupling nuts counterclockwise to loosen. When the coupling nuts are loosened, the body of the faucet will be free from the faucet shanks. The shanks must be removed in order to tile behind them. The seat wrench is the tool to use for this purpose. Insert the seat wrench into the opening in the shank, use an adjustable wrench to turn the seat wrench counterclockwise to remove the faucet shanks. Ceramic tile, applied with mastic, will add approximately ⅜ inches to the present wall. If the faucet shanks have female threads, it will be necessary to remove the present pipe nipples in the wall and use ½ inch longer pipe nipples when reinstalling the faucet. If the shanks have male pipe threads, ½ inch I.P.S. extension couplings should be used to allow for the added thickness of the tile. Refer to Fig. 3 (Chapter 3) to check which threads your faucet has now, in order to make the correct change.

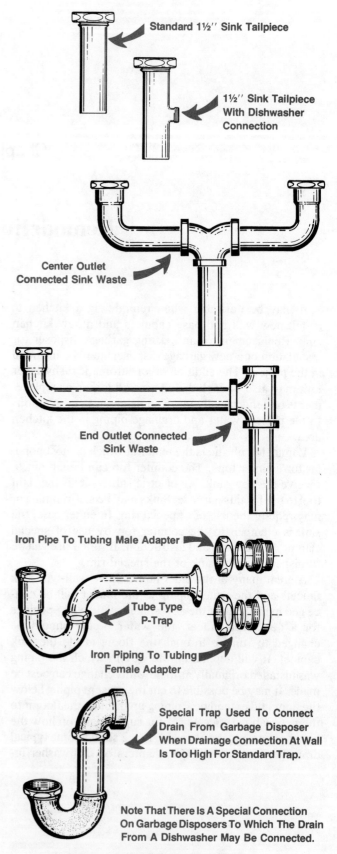

Standard 1½'' Sink Tailpiece

1½'' Sink Tailpiece With Dishwasher Connection

Center Outlet Connected Sink Waste

End Outlet Connected Sink Waste

Iron Pipe To Tubing Male Adapter

Tube Type P-Trap

Iron Piping To Tubing Female Adapter

Special Trap Used To Connect Drain From Garbage Disposer When Drainage Connection At Wall Is Too High For Standard Trap.

Note That There Is A Special Connection On Garbage Disposers To Which The Drain From A Dishwasher May Be Connected.

Fig. 2. Typical drain connections.

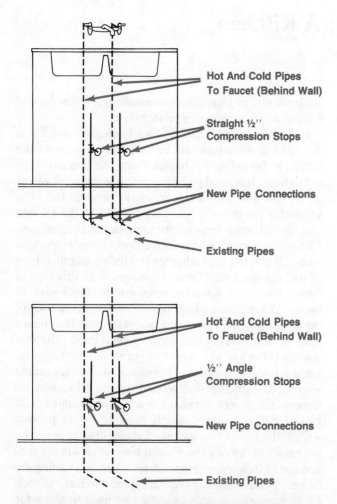

Hot And Cold Pipes To Faucet (Behind Wall)

Straight ½'' Compression Stops

New Pipe Connections

Existing Pipes

Hot And Cold Pipes To Faucet (Behind Wall)

½'' Angle Compression Stops

New Pipe Connections

Existing Pipes

If The Pipes Which Supplies Water To The Faucet Extend Up To Other Fixtures, The Openings Which Supplied The Faucet Must Be Plugged Or Capped. If The Pipes Serve No Other Fixtures, Make The New Connections To The Exisiting Pipes With 90° Elbows And Abandon The Old Pipes In The Wall

Existing Pipes Are Shown As Dotted Lines.

New Pipes Are Shown As Heavy Solid Lines.

Fig. 1. Revamping old water pipes for new faucets.

Chapter 17

Remodeling Bathroom

Replacement of old style plumbing fixtures is usually the reason for a bathroom remodeling project. The trend is to vanity-type lavatories and replacement of old style water closet combinations with more compact types. Ceramic tile can be installed on the walls and floors for a bright, easy-to-clean bathroom. It may be necessary to raise the old closet flange, install a new flange if ceramic floor tile is installed. the standard method of setting water closet bowls is to use a wax ring between the bowl and the closet flange. Wax rings are now made with a new plastic extension which will extend down into the closet flange.

If ceramic tile is used on the walls, the tile will add approximately ⅜ inches to the present wall. Measure the distance from the new wall line to the center of the closet flange. Most free-standing (tank mounted on the bowl) toilets are made for a 12 inch rough in. "Rough in" means the distance from the center of the flange to the finished wall line. If the distance from the center of the flange to the finished wall is less than 12 inches, it will be necessary to buy a closet combination made for a 10 inch rough in. See the chapter on installing or resetting closet combinations for installation instruction.

Flexible closet supply lines make it very easy to re-connect the water supply to the water closet. Flexible closet supply lines are made of chrome plated soft copper and can be bent and shaped to fit the new installation. A very useful and inexpensive tool for bending flexible lavatory supplies is a bending spring. Flexible supply lines are made in ⅜ inch O.D. and ½ inch O.D. sizes. The ⅜ inch size is the most common. The spring will cost less than a flexible supply tube and you would be wise in using the tool if bending the tube is required. (See Fig. 1.) When the spring is slipped over the flexible tube and bent and shaped, the bending spring will prevent any kinking of the tubing. If the existing shutoff valve on the water supply to the closet combination has a flexible tubing connection and it is necessary to use a new flexible supply line, the old supply tubing can be cut and the compression nut can be used with the new supply. It will be necessary to use a new brass or neoprene ferrule on the new supply tube. When brass ferrules have been used they are tightened and sealed to the flexible supply tube. It is not practical to try to remove a brass ferrule for reuse.

When a new lavatory is installed, some change will probably be necessary in the water supplies and the waste connection. The standard pipe size for a lavatory drain is 1¼ inches, the trap should be a 1¼ inch tube type P-trap. If the tailpiece or drain connection on the new lavatory is higher than the old one, an extension tailpiece may be necessary. An extension tailpiece with a slip nut is the best type to use for this purpose. If the extension tailpiece has to be cut to fit, use the large tubing cutter, and ream the tubing after it has been cut. The tubing must be reamed to remove the burr caused by cutting. The cutter has a reamer built on it.

Some types of lavatory faucets have a ⅜ inch O.D. copper tubing water connections. Instead of using flexible lavatory supplies, ⅜ inch O.D. copper tubing can be used from the compression type stop valve and the connection between the tubing on the faucet and the tubing from the stop valve can be made with a ⅜ inch O.D. ferrule type compression union. The ferrule type compression union requires no flaring or soldering.

It is often necessary or desirable to replace an existing bathtub. In an effort to lower prices, many builders in recent years have installed enameled steel bathtubs. Steel tubs are lower in initial cost, light-weight and easy to set in place. Steel tubs are similar in appearance to cast-iron tubs but the steel tub has the disadvantage of being easily chipped by a falling object. The advantage

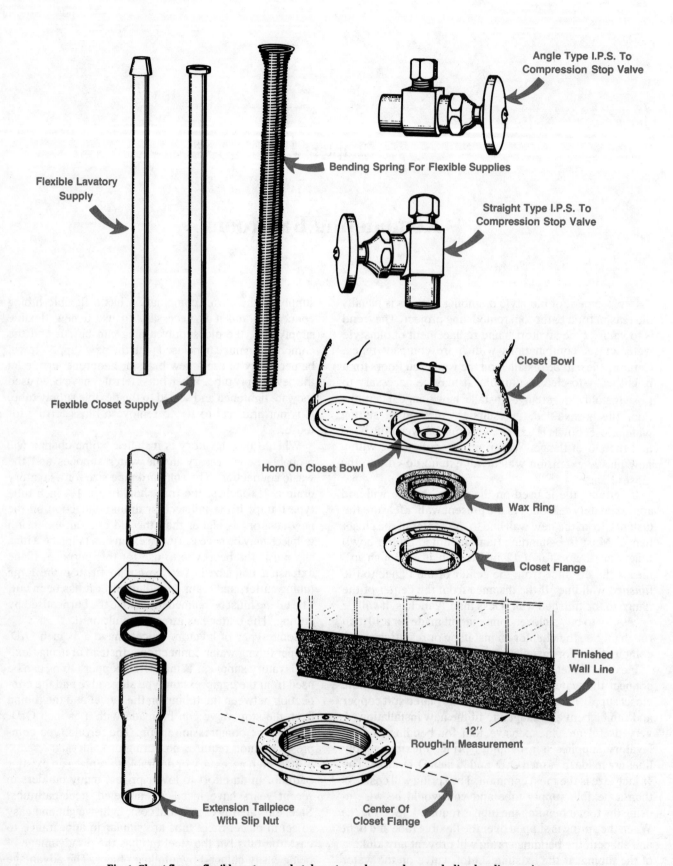

Flexible Lavatory Supply

Flexible Closet Supply

Angle Type I.P.S. To Compression Stop Valve

Bending Spring For Flexible Supplies

Straight Type I.P.S. To Compression Stop Valve

Closet Bowl

Horn On Closet Bowl

Wax Ring

Closet Flange

Finished Wall Line

Extension Tailpiece With Slip Nut

Center Of Closet Flange

12″ **Rough-In Measurement**

Fig. 1. Closet flange, flexible water supply tubes, stop valves, and a supply line bending spring.

offered by a cast-iron, acid-resisting (AR), enameled bathtub is that it is not easily chipped. See Fig. 1.

Corner tubs are identified as LH (left hand) or RH (Right hand) corner tubs. When facing the front (or long) apron, if the finished corner with apron is on the left, the tub is a left hand (LH) corner tub. If the finished corner with apron is on the right, the tub is a right hand (RH) corner tub.

Recessed tubes are identified as LH (left hand) or RH (right hand) tubs. When facing the apron (or front) of the tub, if the waste and overflow connections are at the left end of the tub, it is a LH tub. If the water and overflow connections are at the right end of the tub, it is a RH tub. Various types of shower and bath fittings (over-rim fillers) are available to fit any bathtub. In the left hand tub, the valves or faucets on the K-7004 *Dalton* fitting/turn the water on or off, the water flows into the bathtub or is diverted into the shower head by the diverter spout. In the right hand tub, the K-7014 *Dalney* fitting turns the water on or off, but the flow is directed either into the tub or up and through the shower head by the center valve of the fitting. The diverter type spout is not used with this type of fitting. Recessed tubs are made in several different sizes but the most popular and widely used size is the 5 foot (60 inches) tub. Bathtubs are also made 14 inches and 16 inches in depth. If an existing 14 inch tub is to be replaced with a 16 inch deep tub, the wall will have to be cut two inches higher to accomodate the new, larger tub. In order to remove an existing recessed tub, the wall will have to be cut out approximately 2 inches above the tub rim. The amount of wall which will have to be removed will depend on the layout of the bathroom. A typical problem which may be encountered is shown in Fig. 3.

In this illustration the bathtub is chipped or damaged and replacement is necessary. The water closet must be removed in order to remove the tub. Step number one is to turn off all of the water in the house at the main cold water shut-off valve. The cold water supply to the water closet tank is connected to the ballcock in the tank by the coupling nut at the bottom left side of the closet tank. Turn the nut clockwise to loosen and remove it from the ballcock shank. Flush the closet tank and sponge out the remaining water. Loosen and remove the nuts securing the closet tank to the bowl and lift off the tank. Loosen and remove the nuts securing the closet bowl to the floor. If the bowl seems difficult to move, 'bump' it gently with the heel of your hand at front and sides to loosen it. When it is loosened, lift it up and off of the closet flange and bolts. Remove the closet bolts. The cold water supply piping and valve to the water closet tank must be removed before the old tub can be moved out. Turn the pipe nipple and valve counterclockwise to loosen and remove. The drywall or plaster will have to be cut out as

shown in Fig. 3 in order to slide the tub forward. When the drywall or plaster has been removed, a pipe plug can be screwed into the fitting and the water can be turned on again to the rest of the house. The last remaining obstacle to the removal of the tub is the drain connection. There should be a slip-nut at the inlet connection to the trap as shown in Fig. 4. Turn this slip-nut counterclockwise to free it from the trap. When the slip-nut is free, lift or pry the tub up to lift the tailpiece out of the trap. It should now be possible to slide the old tub out and remove it.

Installing a Replacement Tub

The replacement tub should be uncrated outside and carried into the bathroom. Boards or heavy cardboard should be placed on the floor and over the closet flange to prevent damage to the tub apron when sliding the tub into place. When sliding the tub in place, do not pry against the apron. Lift at the ends and back to work the tub into place. When the tub is in place lay a level on the top of the tub at the front, (over the apron) and level the tub. Place the level on top of one end and level the end. (Wooden wedges cut from a 2 x 4 are excellent for leveling shims.) When the tub is level both ways, front and end, reach under the tub at each rear corner and mark the stud at the bottom of the tub flange. Slide the tub out approximately 18 inches and measure from the pencil mark on the stud to the floor. Cut two pieces of 2 x 4 this measured length and nail one to each rear corner stud. This will provide good support for the tub at the rear corners. Steel tub hangers can be purchased with the tub and used instead of 2 x 4's if desired. (Fig. 4.) When the 2 x 4's or the hangers are in place, the tub can be set back into place.

The waste and overflow piping should be installed next. The exact instructions for assembling these parts will depend on the brand and type of waste and overflow piping that has been purchased. Installation instructions will be furnished with the parts.

The wall in back of the water closet can now be repaired. The water must be turned off again at the main valve and the plug in the water closet supply fitting must be removed. If drywall is used, cut the hole for the supply nipple and valve, nail the dry wall to the studs and insert the nipple and valve into the fitting; tighten the nipple and the valve. When the nipple and the valve are tightened, turn the supply valve off (turn the handle clockwise) and the main valve can be turned back on.

Before any work is done on repairing the wall around the tub, the tub should be protected. Tub covers are sold for this purpose, or the tub can be papered. Several layers of paste and paper should be applied to the exposed enamel surface of the bathtub, including the apron front. (Fig. 4.) After the paper has dried, usually overnight, the wall can be repaired around the tub. When the bathroom walls are finished, the toilet can be reset. See the chapter on resetting closet combinations.

Replacing Wall-hung Lavatories

Lavatories may be either wall- mounted or mounted in a counter top. To install a wall-mounted lavatory, a bracket (or pair of brackets) must be securely fastened to the wall. The bracket(s) serve as hangers for the lavatory. When an old lavatory is replaced with a new one, the old brackets will generally not fit the new lavatory and the height of the bracket may also be different. When a house is under construction, the plumber knows where the various fixtures will be placed. If a wall-hung lavatory is to be used, he will nail a 'backing board' to the studs in the wall at the lavatory location. (See Fig. 5.)

The screws holding the bracket to the wall penetrate through the finished wall and into the backing board. The bracket is then fastened securely to the wall. When the lavatory is placed on the brackets it will also be solidly mounted. When a new lavatory is purchased, the supplier can furnish the buyer with a 'rough-in drawing' showing where the brackets should be mounted. Fig. 5A shows a "rough-in" drawing for a popular *Kohler Co.* lavatory. The height and location of the mounting brackets is clearly shown. There are provisions for using long wood screws to further secure the lavatory, (E) and (F). The woodscrews used at (E) and (F) should have round heads and be chrome, nickel or cadmium plated. Round plated washers should also be used with these screws.

The first step when replacing an existing lavatory is to make certain that a backing board is in the right place.

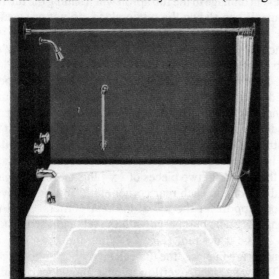

LEFT HAND TUB

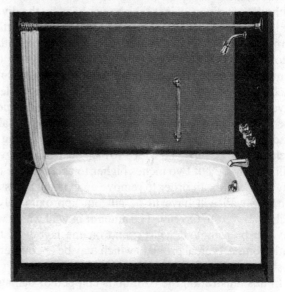

RIGHT HAND TUB

K-515-SA	DYNAMETRIC 16 inches high		K-516-SE
Bath Complete	K-515-SA with Safeguard bottom, Triton II trim (Illustrated)	K-516-SE with Safeguard bottom, Triton II trim (Illustrated)	
	K-516-SA with Safeguard bottom, Triton II trim	K-515-SE with Safeguard bottom, Triton II trim	
Bath Only	K-515-S with Safeguard bottom, outlet at left	K-516-S with Safeguard bottom, outlet at right	
	K-516-S with Safeguard bottom, outlet at right	K-515-S with Safeguard bottom, outlet at left	
Shower and Bath Fitting	K-7004-T Dalton fitting with Valvet units, diverter spout, K-7370 shower head	K-7014-T Dalney diverter fitting, with Valvet units, over rim spout, K-7370 shower head	
Drain	K-7150-R 1½" pop-up	K-7150-R 1½" pop-up	
Wall Grip Rail	K-9613 18"	K-9613 18"	
Curtain Rod	K-9710	K-9710	
Curtain Hook and Chain	K-9720 — (Curtain and pins are not furnished.)	K-9720 — (Curtain and pins are not furnished.)	
Size	4½', 5', 5½'	4½', 5', 5½'	
Height	16"	16"	

Fig. 2. Typical corner

The bracket(s) should be placed against the wall at the correct height. The bracket should be centered over the waste opening.

Fig. 5A shows the top of the holes in the bracket being 33¼ inches above the finished floor. Set a level across the top of the bracket as shown in Fig. 5B. When the bubble is centered, the bracket is level. While the bracket is held level, mark the wall at the bracket holes. A small (8p or 10p) nail can be driven into the wall at these points. If there is a backing board behind the wall, the nail will be securely held; if there is no backing board, the nail will be loose. With this type bracket it would only be necessary to make two test holes to locate the backing board, one at the hole in the extreme left end of the bracket and one at the hole in the right end.

If the nail encounters resistance at both these points, the bracket can be securely mounted. If there is no backing board at these points, the wall should be cut as shown in Fig. 5C and a backing board should be securely nailed to the studs. If the cut out area is in a plastered wall, plaster board can be nailed on to the studs and the surface repaired with patching plaster; if the wall is dry wall construction, a piece of dry wall board can be nailed to the studs and the joints between the original dry wall and the patch can be filled with dry wall spackle and sanded smooth when the spackle is dry. The repaired area can be finished as desired with ceramic tile, paint, etc.

If the wall is to be tiled, the tile should be applied and when the tile is grouted, the bracket should once again be

RIGHT HAND CORNER TUB

LEFT HAND CORNER TUB

K-521-SA	DYNAMETRIC 16 INCHES HIGH	K-524-SE
Bath Complete	**K-521-SA** with Safeguard bottom, Triton II trim (Illustrated) **K-524-SA** with Safeguard bottom, Triton II trim	**K-524-SE** with Safeguard bottom, Triton II trim (Illustrated) **K-521-SE** with Safeguard bottom, Triton II trim
Bath Only	**K-521** with Safeguard bottom — left corner, left outlet **K-524** with Safeguard bottom — right corner, right outlet	**K-524** with Safeguard bottom — right corner, right outlet **K-521** with Safeguard bottom — left corner, left outlet
Shower and Bath Fitting	**K-7004-T** Dalton fitting with Valvet units, diverter spout, K-7370 shower head	**K-7014-T** Dalney diverter fitting with Valvet units, over rim spout, K-7370 shower head
Drain	**K-7150-R** 1½″ pop-up	**K-7150-R** 1½″ pop-up
Wall Grip Rail	**K-9613** 18″	**K-9613** 18″
Curtain Rod	**K-9700**	**K-9700**
Curtain Hook and Chain	**K-9720** — (Curtain and pins are not furnished.)	**K-9720** — (Curtain and pins are not furnished.)
Size	5′	5′
Height	16″	16″

Courtesy Kohler Co.

and recessed bathtubs.

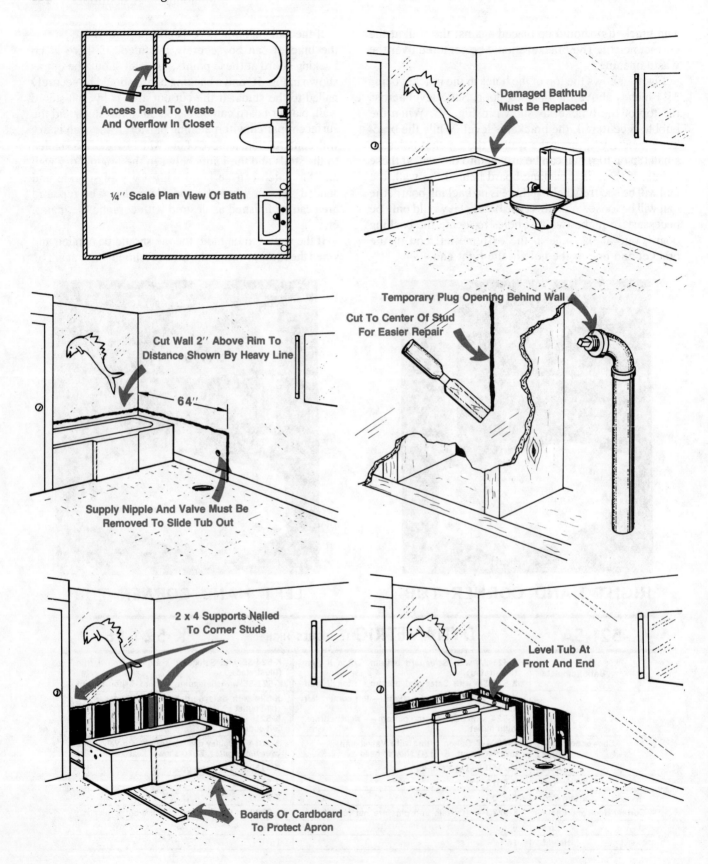

Access Panel To Waste And Overflow In Closet

¼″ Scale Plan View Of Bath

Damaged Bathtub Must Be Replaced

Cut Wall 2″ Above Rim To Distance Shown By Heavy Line

64″

Supply Nipple And Valve Must Be Removed To Slide Tub Out

Temporary Plug Opening Behind Wall

Cut To Center Of Stud For Easier Repair

2 x 4 Supports Nailed To Corner Studs

Boards Or Cardboard To Protect Apron

Level Tub At Front And End

Fig. 3. Tub installation procedures.

K-7014 built-in diverter, shower and bath supply.
K-7150-R pop-up drain.

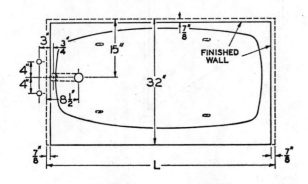

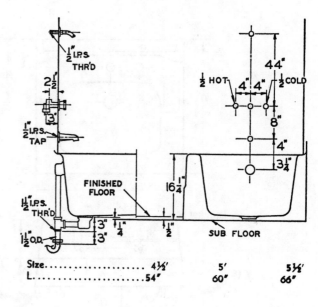

Size............................ 4½'   5'   5½'
L................................ 54"   60"   66"

Measurements may vary ⅛".

Ancora Bath Hangers

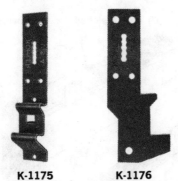

K-1175   K-1176

Kohler Ancora bath hangers are made of steel with an upper and lower shelf. They are practical for use on studding, furring strips and flat walls.

Provide a firm anchorage holding the bath securely against wall.

Four Ancora hangers are used to anchor a recess bath; three for corner baths.

Available in two types —
K-1175 For mounting on furring strips and flat walls.
K-1176 For mounting on face of studdings.

Protek provides a safe and inexpensive medium for the protection of the lustrous enameled surface of the bath while the bathroom is being built.

It comes in powder form, and when mixed with water forms a paste, which, when applied with several thicknesses of ordinary newspapers, produces, when dry, a tough and strong covering.

Protek will not etch or injure the high glaze of Kohler enamel. To remove, soak the paper with water. It peels off easily.

K-1180

Courtesy Kohler Co.

Fig. 4. Typical tub installation dimensions and drain connections.

placed at the correct height, leveled, and the wall marked for the bracket screws. The ceramic tile can be drilled for the screws using a ¼ inch masonry bit. The bracket should then be fastened to the wall, using ¼ inch diameter woodscrews, 2½ to 3 inches long, and leveled before the screws are completely tightened.

Place the lavatory onto the bracket and push down on the rear corners of the lavatory to be certain it is firmly seated on the bracket. If the lavatory is not level when seated on the bracket, lift it off, loosen the screws on the low side, tap the bracket up or down slightly, retighten the screws and replace the lavatory on the bracket. When the lavatory is leveled and firmly seated, the wall can be marked at the holes (E) and (F). If the wall has been tiled, the lavatory should be removed from the brackets to drill the tile. If the lavatory is mounted on

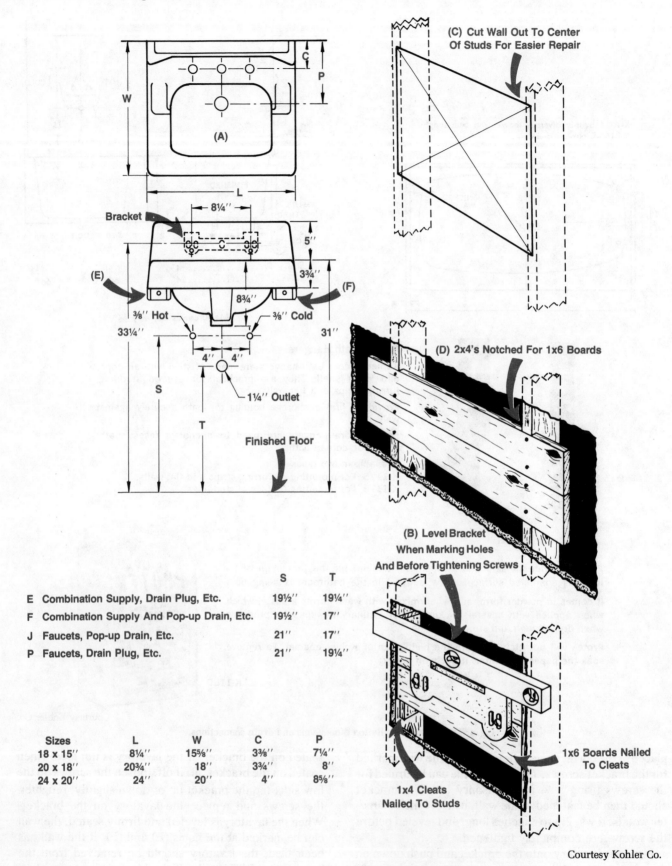

E	Combination Supply, Drain Plug, Etc.	S	T
		19½″	19¼″
F	Combination Supply And Pop-up Drain, Etc.	19½″	17″
J	Faucets, Pop-up Drain, Etc.	21″	17″
P	Faucets, Drain Plug, Etc.	21″	19¼″

Sizes	L	W	C	P
18 x 15″	8¼″	15⅝″	3⅜″	7¼″
20 x 18″	20¾″	18″	3¾″	8″
24 x 20″	24″	20″	4″	8⅝″

Courtesy Kohler Co.

Fig. 5. Various steps in mounting a bathroom lavatory.

plaster or drywall walls, it will not be necessary to remove the lavatory. The holes for the screws at (E) and (F) can be started with an 8p nail. A word of caution here, vitreous china is glass; the screws at (E) and (F) should not be over-tightened. When the washer under the screw head is snug against the porcelain enamel stop tightening. If the screw is over-tightened, the lavatory will break. A ⅛ inch toggle bolt or small molly anchors can be used instead of woodscrews for securing the lavatory at points (E) and (F). Backing boards can also be nailed to the studs as shown in Fig. 5D.

Chapter 18

Installing Ceramic Tile

For a bright, cheerful, easy-to-clean kitchen, install ceramic tile on the wall. The initial expense may be more than paint, wallpaper or other methods of wall covering but once ceramic tile is installed your problems over the years for maintenance are cut down considerably. Keeping the tile washed clean is all that is needed. The tile, mastic, and grout are readily available at tile stores, discount houses, and mail-order department stores. The tile cutters and nibbling pliers are the only special tools needed and these can usually be rented from the company where the tile is purchased. The purchase of these tools is advised rather than rental because the rental charge over a period of days will amost equal the total purchase price. Once these tools are purchased they can be used on the next project, bathroom walls and floors are good areas for ceramic tile.

The savings effected by doing these tile jobs yourself will pay for the needed tools many times over. The other tools needed for this job are a level, (at least 24 inches long) a notched trowel for spreading mastic and a few miscellaneous hand tools.

If you are planning to remodel your kitchen and want to install ceramic tile while you are remodeling, this chapter will take you step-by-step through ceramic tile installation.

Basic Rules for Installing Ceramic Wall Tile

Start at the top! The only exception to this rule is if the tile is to be installed from floor to ceiling. Ceramic tile is usually installed in bathrooms to a height of 54 to 56 inches above the floor and the wall above the tiled area can be painted or papered.

Ceramic tile in a kitchen is usually extended from the floor to the bottom of the wall cabinets. The exposed walls above the tiled area are also painted or papered.

Starting at the top row of tile is important because the top row must be level, or the whole effect will be ruined. No room is perfectly square, no floor is perfectly level. It will be necessary to cut the tile in one vertical row and one horizontal row. The vertical row of cut tile should be at a corner if possible; the horizontal row of cut tile should be the first row above the cove base tile at the floor. There is an exception to this rule also: when tiling a kitchen, the cut row of horizontal tile should be at the top of the counter top or top of the backsplash on the counter top.

Preparation: Measure first then buy the ceramic tile, mastic and grout. Tools needed:

Tile cutters,
Nibbling pliers,
Level (at least 24 inches long),
Notched trowel,
A few hand tools, screwdriver etc,
Clean rags.

Ceramic tile should not be installed over wallpaper. If the kitchen walls are papered in the area to be tiled, use a steamer or scraper and remove all paper down to the plaster or drywall.

Ceramic tile can be applied over painted walls. It will be necessary to wash the walls with a good household cleaner to remove any grease film that may be present. After washing, wipe the walls dry and score the surface with a screwdriver blade or some other similar sharp tool. Make diagonal strokes so as to cut diamond shaped scoring. The lines of the scoring should be 2 to 3 inches apart. If a quarter-round or other moulding strip is used at the top of the backsplash of the cabinets, remove this molding strip. The finished tile should extend down to the top of the backsplash.

Where wall cabinets are used in the kitchen area, the bottom edge of the cabinets will usually be between 54 and 55 inches above the floor. The standard counter top area height in a kitchen is 36 inches above the floor. The counter top area will rarely be perfectly level, therefore, the cut pieces of tile should be used in the first row of horizontal tile above the counter top, as shown in Fig. 1. A good carpenter's level will be needed when laying off the first row of tile. Start at one corner or end, put the level against the bottom of the cabinets and draw a pencil mark at the bottom edge of the level. Continue around or along the wall to be tiled, moving the level and marking the wall. A string stretched along the wall using a line level can also be used. You have now established a level line to work from. Wall tile usually is sized at 4 or 4¼ inches square and is made with raised edges as shown in Fig. 3, to insure that the spacing is correct between the tiles. Correct spacing is necessary in order to properly grout the tile. Use a rule and measure the distance from the pencil mark to the bottom of the cabinets. If this distance is the same all along the bottom of the cabinets then the bottom of the cabinets are level and the first horizontal row of tile can be started with the top of the tile against the bottom of the cabinets. Most often, the bottoms of the cabinets will not be level. If this is true, measure up from the top of the counter to a point approximately 46 inches above the floor, place the top of the level with one end on the mark, hold the level so that the bubble is centered between the level lines and draw a

pencil mark along the top of the level. The level should be at least 24 inches long. Move the level along so that one end touches the line you have drawn; continue as before until you have drawn a level line on the wall in the area to be tiled. This is the line of the top of the first row of tile to be installed. The next step is to establish the center of the area to be tiled. Generally speaking, this should be the center line of the kitchen sink, especially if the sink is centered on a window opening.

The wall type mastic will have the application directions on the can. Apply the mastic as directed but do not cover the level line or the center mark. Place the first tile so that the top of the tile is at the horizontal level line and one corner is at the center mark. Continue placing the tiles along the horizontal line, taking care that the first tile placed is not pushed sideways. Note that the back side of the tiles have lines or grooves. Place the tiles so that the lines or grooves are horizontal; this will keep the tiles from sliding down the wall. If the tile ends at a corner, and a cut piece of tile must be used, make the cut at the corner (or the end tile in the row). If it is desirable to stop the tiling at a given point along the wall, not at a corner, mark this point, and place the level in a vertical position at this point, center the plumb line bubble and draw a vertical line.

A cap tile should be used with the rounded edge on the plumb line. A corner cap should be used at the top

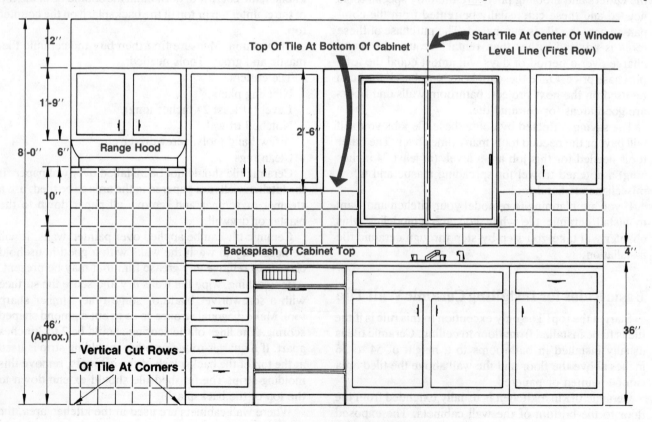

Fig. 1. Measuring for tile installation.

corner. If it is necessary to use a cut piece of tile, the cut piece should be next to the cap tile at the end. For a professional appearing installation when finished, the cap tile should not be cut. When the first row of tile has been set to the end, or the corner, go back to the center mark and work away from it, to the other end or corner. When the first horizontal row has been set, place the top of the level against the bottom of the first row of tile and check to make certain that the tile has not slipped down. If the tile has slipped down, pushing up very lightly on the level will bring the tile back up to the level point; when the first horizontal row has been set, work up and down from this row. If the counter top is 36 inches above the floor and the level line was marked at 46 inches as suggested, and if the backsplash on the cabinet top is approximately 4 inches high, the first row of tile above the cabinet top backsplash will be made up of cut pieces of tile. Measure each space before cutting tile since each cut piece may vary slightly from the one next to it. The top row of tile, if it is placed against the bottom edge of the wall cabinets will probably also be made up of cut pieces. Again, measure each space before cutting the tile. Each cut piece may be slightly different therefore the measuring at this point of tiling is very important.

Adjustments to Electrical Fixtures

Electric switches and plug receptacles will undoubtedly be in the area to be tiled. It is necessary to loosen the screws securing the switches and receptacles. The fuses and/or circuit breakers controlling the electric switches and receptacles will be located in the main electric panel. Remove the fuses or turn the circuit breakers to the off position before and during the work on or around the electric switches and receptacles. Plug the cord of an appliance into the receptacle(s) to make certain that the circuit is not energized before removing the cover plate.

Double check by turning light switches on before removing the fuses or turning the circuit breakers off; when the proper fuse is removed or the proper circuit breaker is turned off, the lights will be out. There may be more than one fuse or circuit breaker controlling the kitchen area, therefore, check to make certain that all lights and all receptacles in the kitchen area are not energized. Remove the screws holding the cover plates on the switches and receptacles and remove the cover plates. Remove the screws securing the switches and receptacles to the wall boxes and pull the switches and receptacles out of the wall boxes approximately 1 inch. Use the tile cutter and/or nibbling pliers to fit the tile to the edge of the wall boxes. When the tile has been cut and set around the wall boxes, push the switches and receptacles back into the wall boxes. The mounting ears of the switches and wall boxes will rest on the outside face of the tile. The receptacles and switches are mounted to the

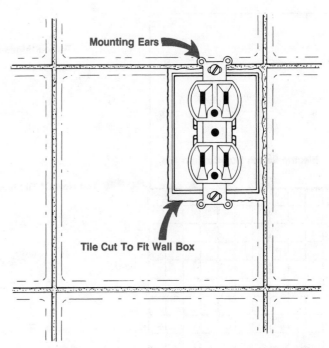

Fig. 2. Mounting ears of switches and receptacles on outside of tile.

wall boxes with 6-32 flat head screws. The old 6-32 screws may be too short when the switches and receptacles are mounted against the new wall surface. Longer screws are readily available at good hardware stores if replacement of screws is necessary. The areas to be tiled will vary according to the individual kitchen arrangement.

If there is a range hood installed above the kitchen stove, it is advisable to tile the wall up to the bottom of the range hood. In the other areas of the kitchen, the top of the cap row of tile should be level with the bottom of the wall cabinets, or approximately 54 to 55 inches above the floor. Directions for mixing and application of the tile grout will accompany the grout and should be closely followed. The ceramic tile should be allowed to set 18 to 24 hours before the grouting is done. Remove any mastic from the face of tile with recommended solvent, such as mineral spirits, before applying grout.

How To Measure The Desired Areas For Tile

The bottom row of tile may be set to the top of the baseboard and quarter round may be removed and cove-base tile can be used for the bottom row. Cove-base tiles are also made in corner pieces. Order inside or outside corners when tile is purchased. Measure the length and the height of the area to be tiled. The length (in feet) times the height (in feet) will equal the square footage area — 4½ feet (54") high x 8 feet length = 36 square feet. Allow for breakage or for bad cuts of tile when figuring tile amount needed. Ceramic tile is generally packed 10 square feet to the box. See Fig. 3.

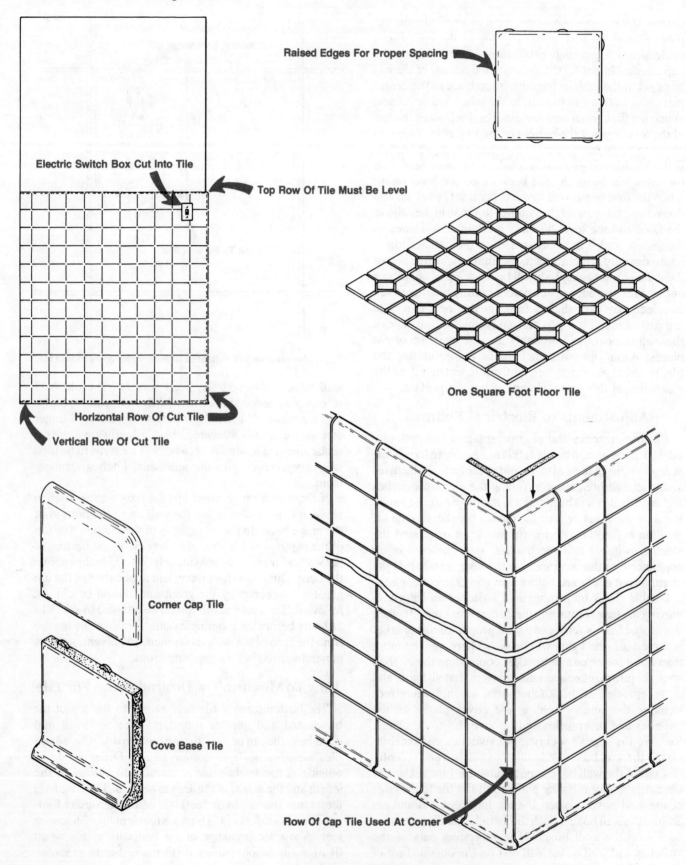

Raised Edges For Proper Spacing

Electric Switch Box Cut Into Tile

Top Row Of Tile Must Be Level

Horizontal Row Of Cut Tile

Vertical Row Of Cut Tile

One Square Foot Floor Tile

Corner Cap Tile

Cove Base Tile

Row Of Cap Tile Used At Corner

Fig. 3. Various stages of tile installation.

Installing Ceramic Floor Tile

Installation of a ceramic floor is often desirable when remodeling a bathroom. A ceramic tile floor can be installed over a concrete floor, such as a house built on a slab. The bathtub need not be taken out; the new tile floor can extend up to the bathtub apron. The old vinyl asbestos or vinyl rubber floor should be removed down to the concrete and the concrete should be scraped clean. Floor tile can be purchased in one foot square sheets, with paper backing. A special floor type mastic should be used and the paper backing is left on the tile. Start in the middle of the bathroom and work to the outside edges.

The water closet bowl should be taken up and the tile should extend under the bowl area, to the edge of the closet flange. When the water closet bowl is reset, use a special wax ring made for this purpose and the closet flange may not have to be raised to make up for the added floor thickness of the tile. When this ring is used, turn the plastic end of the wax ring away from the closet bowl. The plastic end will extend into the closet flange. The addition of the concrete floor as detailed in the following paragraphs adds considerable weight which must be carried by the floor joists.

Before attempting this remodeling project, the owner should seek advice from a qualified builder as to whether it is feasible to do this work. It may be necessary to add extra floor joists or otherwise beef up the floor joists before starting this project.

When ceramic tile is to be installed in a bathroom with a wood or hardboard sub-floor, the floor must be prepared for a tile installation. The water closet must be removed, exposing the sub-flooring, Fig. 4A. The sub-flooring must then be taken up, exposing the floor joist Fig. 4B. The floor joists must be cut as shown in Fig. 4C and 1 x 4 cleat boards must be nailed to the sides of the floor joist as shown in Fig. 4D. Sub-flooring boards are then nailed to the cleats and a minimum of 2 inches of concrete grout is then poured over the sub-floor and troweled smooth. When the concrete has thoroughly dried, usually about 24 hours, the ceramic tile floor can be installed.

Note that in Fig. 4D, the bridging (the cross braces between the joist) has been replaced. The bridging must be nailed back before the sub-flooring is nailed in place since it will be impossible to nail it afterwards. Depending on the placement of the floor joist, a bulkhead may have to be built to carry the concrete floor to the apron of the bathtub.

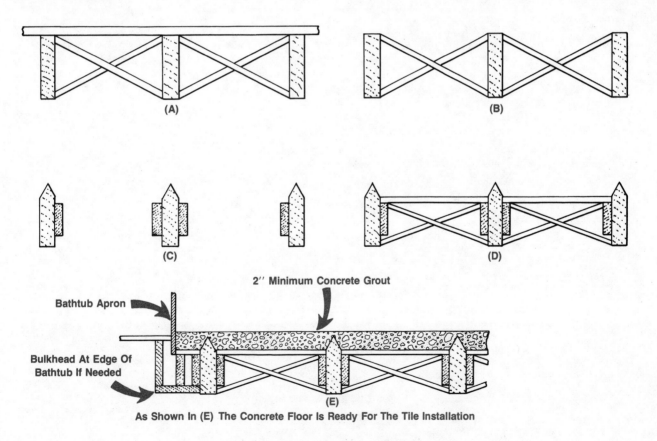

(A)

(B)

(C)

(D)

2″ Minimum Concrete Grout

Bathtub Apron

Bulkhead At Edge Of Bathtub If Needed

(E)

As Shown In (E) The Concrete Floor Is Ready For The Tile Installation

Fig. 4. Preparing the floor for ceramic tile.

Chapter **19**

Septic Tanks and Disposal Fields

Septic tanks and disposal fields are used by many residents of suburban or rural areas which have no sewage treatment facilities. Precast concrete septic tanks, properly sized and installed correctly (and built to comply with local codes and health dept. regulations) will give many years of satisfactory service. *Percolation Tests:* In order for a septic tank system to work properly, the effluent (discharge) from the tank must be disposed of. The best way to do this is to install a disposal field. The rate of absorption of water into the ground at a given point is the basis for determining the size of the disposal field. Percolation tests establish the rate of absorption.

The tests are made by digging three holes in the area where the disposal field is to be located. These holes should be from 2 ½ to 3 feet in depth. This is the correct depth for a disposal field. Fill the holes with water. When the water has been absorbed into the ground, begin the test. (This saturates the ground and simulates the actual working conditions of the soil when the disposal field is installed.)

Water should be poured into the holes to a depth of ten (10) inches from the bottom of the hole — do not add any water during the test. Lay a board across the top of the hole and measure the distance from the bottom of the board to the top of the water. Record this distance and the time. (Local Health authorities usually have forms for this purpose.) After one hour, again measure the distance from the board to the top of the water. The difference between the first measurement and the second one is the inches per hour absorption rate. Continue the test by recording the distance down to the water level until the test has been conducted for five consecutive hours or until the water has been absorbed. If the water is absorbed before the five hour test period, the test will be ended when the water has been absorbed. Generally speaking, the last hourly rate of absorption during the test is the figure to use in computing the size of the disposal field. If the rate of absorption is less than 1 inch in 60 minutes, a disposal field will probably not perform satisfactorily in that area.

Septic Tanks

Local regulations will probably govern the minimum size of the septic tank. One rule commonly used is to figure that the storage capacity of the tank should equal the number of gallons of sewage entering the tank in a 24 hour period. At the rate of 100 gallons per person per day, a septic tank for a four person household (or a two bedroom home) should have a minimum capacity of 400 gallons storage. If a garbage disposer is used, the capacity should be increased 50 percent. (From 400 to 600 gallons for a four person household.) In actual practice, the minimum size of any septic tank should be 1000 gallons.

Septic tanks function by a combination of bacterial action and gases. Solids entering the tank drop to the bottom. Bacteria and gases cause decomposition to take place, breaking down the solids into liquids and in the course of this process the indissoluble solids, or sludge, settles to the bottom of the tank. Decomposition in an active tank takes about 24 hours. The sludge builds up on the bottom of the tank so that periodically (perhaps once a year or only once in ten years, depending on usage) the tank must be cleaned or pumped out. In the cleaning or pumping process, only the sludge on the bottom of the tank should be removed. The crust, formed on the water level at the top of the tank, should not be disturbed. The only exception to this rule is if the crust has become coated with grease and the bacterial action of the tank thus destroyed. If this should happen, the crust on top will have to be removed.

203

Bacterial action will begin again when the top is placed on the tank and the tank sealed. Special compounds can be purchased to hasten the resumption of bacterial action but these compounds are rarely, if ever, really needed.

The top of the septic tank should be located at a minimum depth of 12 inches below ground level. The actual depth will probably be somewhat greater due to the depth of the sewer entering the septic tank. The septic tank must also be located at least 75 feet away from any well and downhill so that all drainage is away from the well. Local regulations must be followed as to placement of the septic tank.

The Disposal Field

The disposal field must be sized using the rate of absorption established in the percolation tests. If the absorption rate is one inch in 60 minutes, then the factor 2.35 times the number of gallons of sewage entering the septic tank per day will determine the number of square feet of trench bottom area needed. Using the recommended figure of 100 gallons per day per person, if the absorption rate is 1 inch in 60 minutes and there are four persons in the household, the formula 2.35 times 400 = square feet of trench bottom needed for the disposal field. If the disposal field has five fingers (trenches), each trench would need 188 square feet of trench bottom. If the trench is 2½ feet (30 inches) wide, then each trench would be 75 feet long.

If the absorption rate established in the tests was one inch in 10 minutes, then the factor .558 times 100 gallons per person per day would be used. Again using the base of four persons in the household, then .558 times 400 = square feet of trench bottom needed. With five fingers or trenches installed in the disposal field, each trench should have 45 square feet of trench bottom. If the trench is 2½ feet wide then each trench should be 18 feet long. The figures used in these examples are from the Indiana State Board of Health recommendation. Local regulations in your area may vary somewhat. The rules and regulations governing sizes of disposal fields in other areas should be followed.

Disposal fields serve two purposes. The fingers of a disposal field provide storage for the discharge of a septic tank until the discharge can be absorbed by the earth. The fingers also serve as a further step in the purification of the discharge through the action of bacteria in the earth. Liquids discharged into the disposal field are disposed of in two ways. One is by evaporation into the air. Sunlight, heat and capillary action draw the subsurface moisture to the surface in dry weather.

In periods of wet or extremely cold weather, the liquids must be absorbed into the earth. The discharge from the septic tank should go into a siphon chamber and be siphoned into a distribution box. The distribution box

should direct the discharges so that approximately the same amount of liquid enters each trench.

A siphon (also called a "dosing" siphon) is desirable for this reason:

The siphon does not operate until the water level in the siphon chamber reaches a predetermined point. When this level is reached, the siphonage action starts and a given number of gallons is discharged into the disposal field. The sudden rush of this liquid into the distribution box and then into the separate fingers of the disposal system, insures that each finger will receive an equal share of the discharge. The discharge thus received by the fingers will have time to be absorbed before another siphon action occurs. A siphon is not essential to the operation of a septic tank and disposal field, however, a siphon will improve the efficiency of the system. The lateral distance between trenches in a disposal field will be governed by ground conditions and local regulations.

The construction of the disposal field will vary, due to soil conditions; basically, a typical trench will be as shown in Fig. l. A layer of gravel is placed in the bottom of the trench, field tile or perforated drain tile is then laid on the gravel. If field tile (farm tile) is used, the tiles should be spaced ¼ inch apart and should slope away from the distribution box at the rate of 4 inches fall per 100 feet. The space between the tile should be covered by a strip of heavy asphalt-coated building paper or by a strip of asphalt roofing shingle.

The trench should then be filled to within 6 inches of ground level with gravel. Topsoil should then be added to bring the finished trench to ground level. Since some settling of the topsoil will occur, it is wise to mound the earth slightly over the trenches to allow for settling.

Grease Trap

Grease, which is present in dishwashing water, can destroy the action of a septic tank and if allowed to get into the disposal field, will coat the surface of the earth in the disposal field and prevent absorption. Grease is present in sewage; the septic tank can dispose of this; if the grease from the kitchen can be trapped before it reaches the septic tank, the life of the septic tank and the disposal field will be greatly prolonged.

A grease trap must be large enough so that the incoming hot water will be cooled off as soon as it reaches the grease trap. A 400 gallon septic tank is ideal for use as a grease trap. The water present in this size tank will always be relatively cool, and the grease, present in the dishwashing water, will congeal and rise to the top of the tank. The inlet side and the discharge side of the tank have baffles, or fittings turned down.

Thus, the water entering and leaving the tank is trapped. The grease will congeal and float to the top of the tank and relatively clear water will flow out of the

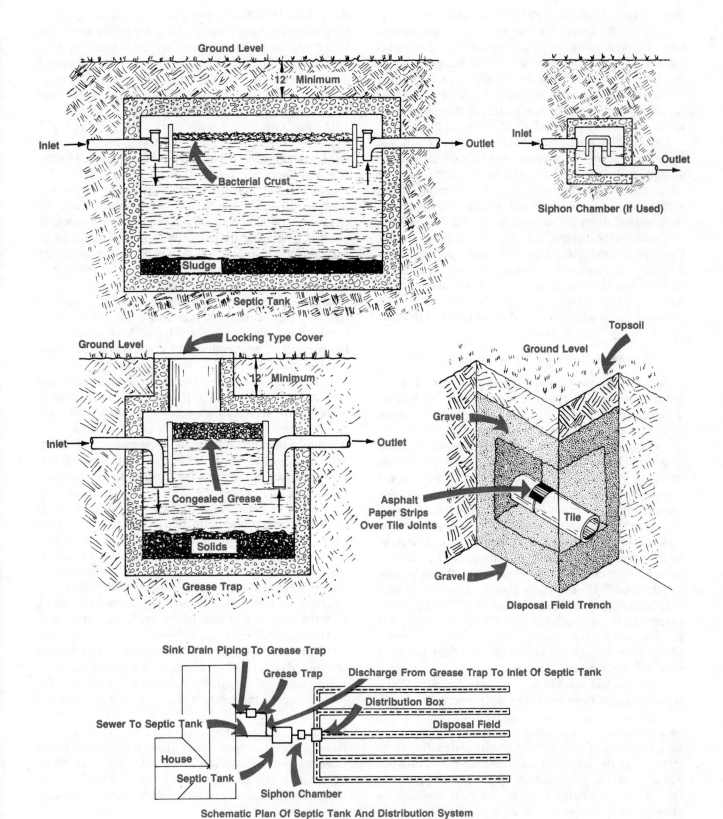

Fig. 1. The construction of a typical septic tank.

grease trap and into the septic tank. Drain cleaning compounds which contain lye, (sodium hydroxide) should never be used with a septic tank installation. The drainage piping from the kitchen sink will not be connected to the drainage system of the house when a grease trap is used. The kitchen sink drainage should be piped out separately and the grease trap located as near as possible to the point where the sink drain line exits the house.

The top of the grease trap should be within 12 inches of the ground level. A manhole can be extended up to ground level, with a light-weight, locking-type manhole cover; this will provide easy access for skimming off the grease. A periodic check should be made and the grease skimmed off when it reaches a depth of 3 or 4 inches.

The tile in the disposal field will vary in depth according to local conditions. The septic tank supplier, if unfamiliar with a siphon and siphon chamber, can get this information. If a siphon is not used, the outlet of the septic tank is piped directly to the distribution box. The siphon outlet, if a siphon is used, must be lower than the inlet to the siphon chamber.

Aerated Sewage Treatment Plants

As we mentioned earlier, septic tanks and disposal fields have, until recently, been the only method by which residents of areas not served by central sewage systems could dispose of their sewage. The central sewage systems use a combination of bacterial action and aeration to break down sewage. If the plant is operating properly, the discharge or effluent is a clear, odorless liquid which can be piped to any running stream, discharged into a storm sewer or routed through a disposal field where, by a combination of absorption and evaporation, it is finally disposed of. A septic tank and disposal field, or finger system, will give satisfactory service provided that one essential condition exists at the septic tank and disposal field or finger system location. The ground must be able to absorb water at the disposal field or finger system depth, at a rate equal to 100 gallons per person per day, and it must be able to do this the year around. When this condition exists, the disposal field will work very well for a period of time.

When the soil conditions in an area are such that a septic tank installation will not work properly, the installation of an aeration type sewage treatment plant should be strongly considered. Small, one-household size aeration type sewage treatment plants using the same methods as the large central plants are now available. These plants are very efficient; when operating properly the effluent discharge is clear, odorless and can be chlorinated if necessary, to meet Health Dept. standards. Local Health Departments often insist on aeration type plants instead of septic tanks, especially where the water table is high or where the soil has shown poor percolation. The treatment process, called extended aeration, is a speeded up version of what happens in nature when a river tumbles through rapids and over waterfalls, purifying itself by capturing oxygen.

Does this type plant really work? The plant in Fig. 2 has been proven in the field and tested by private and government organizations and has been accepted by the National Sanitation Foundation.

How does it actually work? This plant employs a biochemical action in which aerobic bacteria, using the oxygen solution, break down and oxidize household sewage. The three separate compartments in the plant each perform a specific function in the total purification process. The primary treatment compartment receives the household sewage and holds it long enough to allow solid matter to settle to the sludge layer at the bottom of the tank. Here an aerobic bacterial action continuously breaks down the sewage solids, both physically and biochemically, pretreating and conditioning them for passage into the second or aeration chamber.

In the aeration chamber the finely divided pretreated sewage from the primary treatment compartment is mixed with activated sludge and aerated. The *Jet* Aerator circulates and mixes the entire content while injecting ample air to meet the oxygen demand of the aerobic digestion process. The *Jet* control panel, furnished as part of the plant package, is set to automatically cycle the running time of the aerator each day. The final phase of the operation takes place in the settling/clarifying compartment. In this compartment, a tube settler eliminates currents and enhances the settling of any remaining suspended material which is returned, via the tanks sloping end wall, to the aeration chamber for further treatment. A nonmechanical surface skimmer, operated by tank roll, continuously skims any floating material from the surface of the settling compartment and returns it to the aeration compartment. The odorless clarified liquid flows into the final discharge line through the baffled outlet. Effluent disposal must conform to the requirements of the Health Authorities having jurisdiction. Normally, the highly-treated *Jet* effluent eliminates the need for leaching fields or finger systems, or subsurface filters. In most areas, *Jet* effluent is discharged to a flowing stream, a storm sewer or any well-defined line of drainage. An upflow filter and a chlorinator can be installed in this treatment plant if local conditions require it. Fig. 3 gives dimension of the *Jet* aeration sewage system.

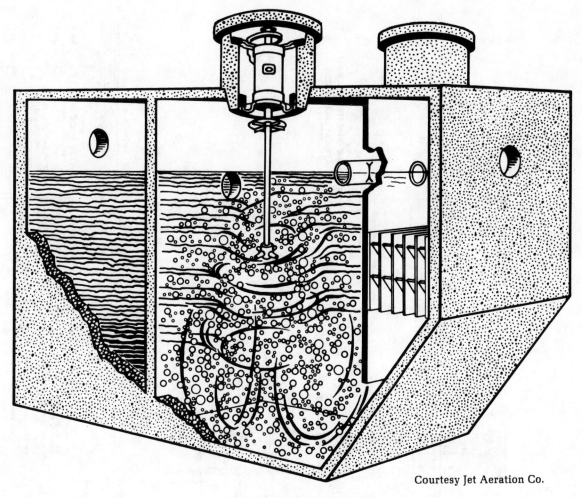

Courtesy Jet Aeration Co.

Fig. 2. *Jet* aeration sewage tank.

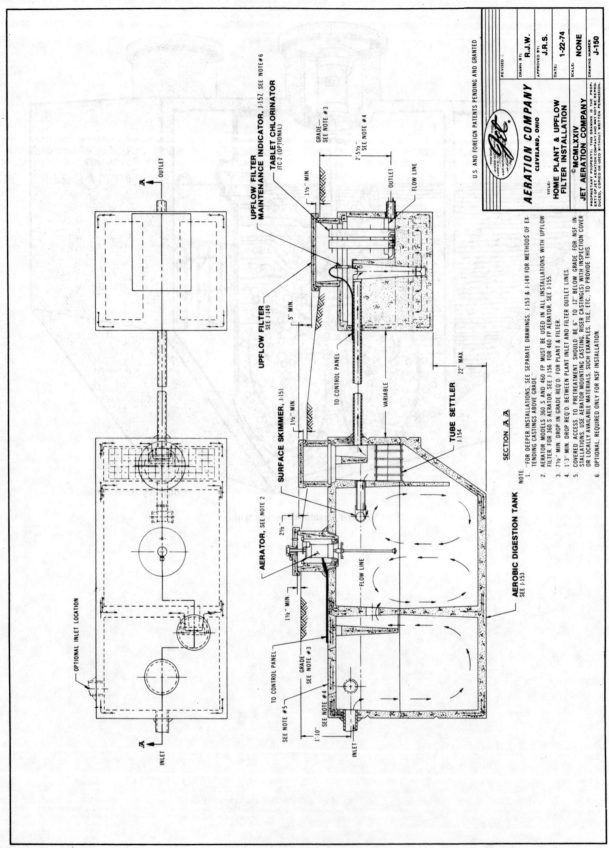

Fig. 3. Dimensions of the *Jet* Aeration tank.

Chapter 20

Private Water Systems

A good well or other source of water is a necessity for residents of suburban or rural areas not served by a water utility. The water furnished by a public or private utility is subject to inspections by health authorities. Water obtained from any other source such as wells, springs, lakes, etc., should be tested for purity before the water is used for human consumption. State and/or local Boards of Health or private laboratories will furnish sterile containers for water samples and test the water for purity. If the water is contaminated, filters, chlorinators, or iodine feeders can be installed in the private water system to make the water safe for human consumption.

Shallow well pumps can be used when the water does not have to be lifted more than 20 feet. Reciprocating (or piston) type pumps and shallow-well ejector (jet) type pumps are best suited for shallow well usage. Where water must be lifted more than 20 feet, a deep well pump should be used. "Convertible" type jet pumps can be used in either shallow wells or deep wells. The ejector can be mounted on the pump for use in shallow wells, or installed in the well casing for use in deep wells. Submersible-type pumps can be used on either shallow or deep wells. Submersible-type pumps do not have to be protected from freezing conditions; they are installed in a well below the water level.

Pressure/Storage Tanks

For proper operation, the water level in a pressure tank should be on the basis of ⅔ water, ⅓ air. With the top ⅓ of the tank filled with air, water is pumped into the tank, the air is compressed until the top (or high) setting of the pressure switch is reached, usually 40 lbs. The pressure switch then turns off the pump. Water can then be used from the tank. As the water is pushed out of the tank, the pressure drops until the low setting of the pressure switch is reached (usually 20 lbs.) and the pump

starts, beginning the cycle over again. Occasionally, due to the failure of an air volume control, or a leak in the pressure tank, the air cushion (the top ⅓ of the tank) is lost or diminished.

If there is only a very small area of air at the top of the tank, the pressure will drop from 40 lbs. to 20 lbs. and immediately upon opening a faucet or valve, the pump will start. If there is no air at the top of the tank, the pump will operate continuously because the pump will be unable to build up enough pressure to cause the pressure switch to turn the pump off. When these conditions occur, the tank is "water-logged". A water-logged tank should be corrected as soon as possible; the frequent on-off operation of the pressure switch will result in burned contacts on the switch and possible damage to the pump motor.

The air volume controls should be checked for proper operation; if a float operated air volume control is used, check to see that the float is not sticking and that the float ball has not lost its buoyancy. *Schrader* air valves, or snifter valves, should be checked to see that they open and close freely. The pressure tank can be checked for leakages at any openings when pressure is built up by covering the openings with soap suds; any air leakage will then show up as soap bubbles.

Correct a water-logged tank condition by turning off the electric power to the pump. There should be a drain valve at the bottom of the tank. Open this valve and drain the tank. In order to drain the tank properly, the air volume control, or a plug at or near the top of the tank, should be removed. Water draining from the bottom of the tank will cause a partial vacuum to be formed in the tank; removing the air volume control or the plug will relieve this vacuum.

The tank must be completely drained; the tank will then be filled with air at atmospheric pressure. (14.7 lbs., at sea level.) When the tank has been completely drained, replace the plug or the air volume control. These fittings must be tight in the tank with no air leakage when the system is in operation. Close the drain valve and turn on the electric power to the pump. If the pump lost its prime in this operation, it may have to be primed at this point. When the pump starts pumping water, the air in the tank will be compressed and forced into the top ⅓ of the tank. If there is no air leak in the tank, and if the air volume controls are working correctly, this ratio of ⅓ air and ⅔ water, will be maintained.

The operation instructions included with the pump will show the location of the priming plug. The pressure tank can also become air-bound. This condition can be caused when there is a leak in the suction pipe to the well — when the water drops below the end of the suction pipe — when there is a leak in the tubing between the pump and the air volume control on the tank — or by a leak in the air volume control device.

Air Volume Controls

There are several methods of maintaining the proper air volume in a storage tank. Snifter valves, such as a *Schrader* air valve, will draw in air when the pump runs. Diaphragm-type air volume controls, such as the *Brady* air volume control will deliver a charge of air each time the pump turns on. Float-type air volume controls regulate the volume of air in the tank by opening to admit air into the tank when the water level raises the float, and expelling air when the water level in the tank causes the float to drop. Float type controls have the disadvantage of sticking in one position, or the float can leak and lose its buoyancy.

Pressure Tanks

In addition to the standard type pressure tank, pressure tanks are now made with a sealing-type float inside. The tank is charged with 10 lbs. of air when it is installed and under normal operating conditions the proper air volume will be maintained indefinitely.

Piping Connections At The Well Casing

A sanitary well seal is a device (usually using bolts around the outside top side) to compress two plates together and expand a rubber seal against the inside of a well casing. The seal will have two holes in the top for use with an ejector (jet) installed in the casing or one hole in the top for use with a shallow well jet or submersible pump. A sanitary well seal should always be installed where the piping enters the well casing to prevent contamination from surface water of the well. The water systems of most of our cities today are supplied with water obtained from reservoirs. This "raw" water is processed in filtration plants and chemically treated. Chlorine, aluminum sulphate, and activated carbon are added to complete the purification process.

Many rural residents, living on lakeside property, now have their own individual water systems based on the same theory as the large city water systems. If the lake water has not been polluted by discharges from septic tanks or run-off from stock feeding and grazing areas, this type of water system is very satisfactory. The first step in construction of this type system is to dig a hole near the edge of the lake. The hole should be 8 to 10 feet in diameter and approximately 10 feet deep. A 12 inch layer of clean sand and small gravel is then placed in the bottom of the excavation. Concrete pipe, 24 to 30 inches in diameter is then set down in the hole, extended up to 2 feet above high water level of the lake. A mixture of clean sand and small gravel is then filled in around the concrete pipe. A 3 wire/220 volt submersible-type pump is the best type of pump to use with this type system. The pump, with 1 inch discharge pipe, is set down in the well and the discharge pipe and the electric wiring to the pump is taken out of the side of the well. (See Figs. 1 and 2.) The trench for the discharge piping and the electric wiring for the pump, must be deep enough to be below freezing level. A minimum depth of 54 inches (4½ feet) is recommended. One inch diameter plastic pipe, available in 100 feet rolls, and made expecially for this purpose, is excellent for use as the discharge piping from the pump to the tank location.

A final filtration system including a charcoal filter and an iodine feeder, or a chlorinator, can be installed at the tank location. The top of the 'well' must be sealed with a concrete top, set in mortar, to prevent surface contamination.

A well with a high iron content in the water will cause deposits of iron oxide to form in the piping, tanks, water heaters and water closet tanks. Once these deposits have formed it is virtually impossible to remove them. Water pressure tanks, water heaters and piping may have to be replaced. An iron filter should be installed in any piping system when the water has a high iron content. Filters can also be installed to remove the objectionable taste and odor of so-called 'sulfur' water.

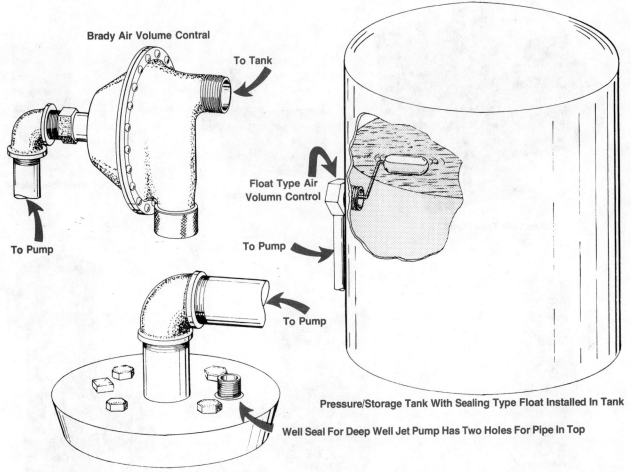

Brady Air Volume Contral

To Tank

To Pump

Float Type Air
Volumn Control

To Pump

To Pump

Pressure/Storage Tank With Sealing Type Float Installed In Tank

Well Seal For Deep Well Jet Pump Has Two Holes For Pipe In Top

Sanitary Well Seal For Shallow Well Pump

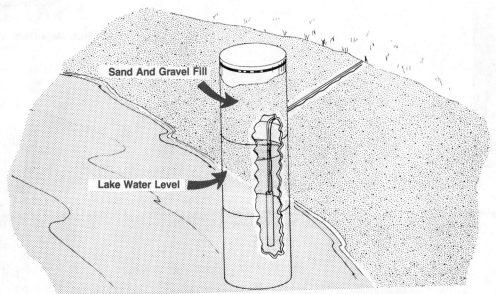

Sand And Gravel Fill

Lake Water Level

**The Pipe From Any Well Or Source Of Water Should Always Slope
Up From The Well To The Pump Or The Tank**

Fig. 1. Various inner working parts of a well water system.

Convertible Type Pump

Shallow Well Ejector

Convertible Ejector

Submersible Pump

Air-E-Tainer Float Type Tank

Tank Float Can Be Installed In Exisiting Tank To Prevent Waterlogging

Multi-stage Deep Well Pump

Iodizer

Courtesy Flint and Walling. Inc.

Fig. 2. Various pumps used in individual well water systems.

Chapter 21

Water Conditioning

There are many areas where the installation of a water conditioner is very desirable. While all water contains (among other elements,) calcium, in certain areas the water has a very high calcium content. This condition, called 'hard' water, is easily overcome by the use of a water softener. Iron is often present in the water obtained from wells, in some cases the iron content is extremely high. When the water is pumped into a tank and exposed to air, the iron content is changed to an iron oxide. The iron oxide will form thick deposits in pressure tanks, piping, water closet tanks, etc., and eventually clog the piping. Where this condition is encountered, an iron filter should be installed. The water treatment sequence would be (1) through the iron filter, (2) through the water softener. It is not necessary to soften all the water used in the home. The water used for drinking, cooking, water closets and sprinkling purposes need not be softened. Common practice is to soften only the hot water.

Note that a bypass connection is shown in Fig. 1. This is desirable even if the water softener has a built-in bypass connection. In the event the water softener needs repairing or has to be replaced, hot water can still be obtained if a bypass connection is correctly installed in the piping system.

Water is softened by circulating it through *Zeolite*. *Zeolite* is a manufactured resin, made in very small beads which resemble an orange colored sand. *Zeolite* catches the magnesium and calcium compounds which make the water hard. Periodic backwashing with salt water restores the *Zeolite* to its original condition. Well water with a high hydrogen sulphide (sulfur) content, causing a 'rotten-egg' smell, can be treated to remove the sulfur and the odor. Several methods are used to treat this type of condition depending on the concentration of hydrogen sulphide and iron in the water. Neutralizing filters can be used to correct acid water conditions. Companies that handle water softener equipment also handle other water treatment supplies.

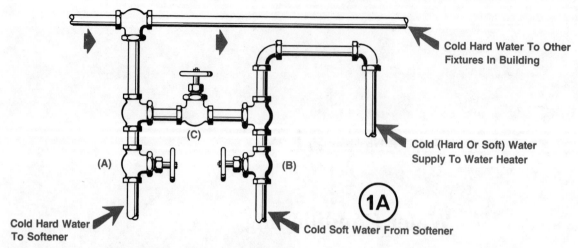

In Normal Operation: Valve A Is Open
Valve B Is Open
Valve C Is Closed

**Cold Hard Water Enters Softener Through Valve A
Leaves Through Valve B**

If Softener Needs Repair, Valve A Is Closed
Valve B Is Closed
Valve C Is Open

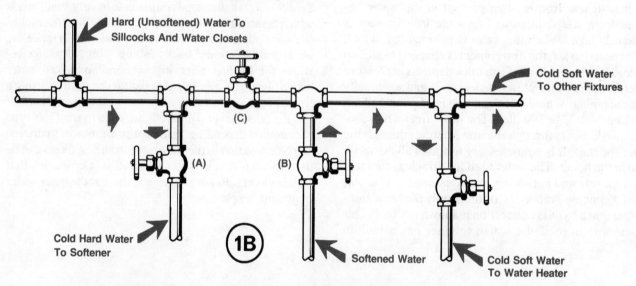

**If It Is Desirable To Use Hard (Unsoftened) Water To Supply The
Water Closets and Sillcocks And Use Softened Water For The
Other Fixtures, The Piping Connections Could Be As Shown In
Fig.1A.**
The Bypass Connection Operation Would Be: Valve A, Normally Open
Valve B, Normally Open
Valve C, Normally Closed
In The Event Of Repair Or Replacement Of The Water Softener,
Valve A Would Be Closed
Valve B Would Be Closed
Valve C Would Be Open
**And A Supply Of Hot Water (Unsoftened) Could Be Maintained
Until The Water Softener Is Repaired Or Replaced.**

Fig. 1. Various plumbing connections for a water softener.

Appendix

It helps to know the name of the fitting you need. Illustrated in Fig. 1 are all of the common steel pipe fittings. Elbows are available as straight pipe size and also as reducing elbows — ½ inch elbows or ½ x ⅜, ½ x ⅜, x ½ x ¼, etc. Tees are also abailable either as straight pipe size or as reducting tees — ½ x ½ x ½ (straight pipe size) or ½ x ⅜ x ⅜ or ½ x ⅜ x ½ or ½ x ½ x ⅜ etc. A tee is always 'read' described as end/end/side, thus, a tee which is ½ inch on one end, ½ inch on the other end, and ⅜ inch on the side is a ½ x ½ x ⅜ tee.

Couplings are often confused with unions. Couplings join pieces of pipe together as do unions, but unions must be used when joining two pieces of pipe together between pipes or fittings which cannot be turned or moved.

Pipe nipples are short lengths of threaded pipe, from 'close' or all-thread to 12 inch length. Street elbows, both 45° and 90° are not recommended, the male (outside threaded) end is restricted and street elbows are difficult to grasp with a pipe wrench. A 90° elbow and close nipple or a 45° elbow and close nipple will work much better than street 90° elbows or street 45° elbows.

Traps on fixtures are not there for the purpose of stopping or catching any object entering the drain of a fixture except as noted. Certain hospital-type fixtures, or fixtures used in connection with the making of casts have a plaster catching trap. The traps on the fixtures in the home are there for the purpose of preventing sewer gas from entering the home through the fixture drain piping.

Relief Valves

Relief valves on water heaters will often, after several years of service, start leaking or will open and discharge a quantity of water before shutting off again. This usually is not due to a defective relief valve; many communities are outgrowing the old water mains installed many years ago and the water utilities are raising the pressures on the water mains in order to force more water through the now undersized mains as water usage increases. Consult with your water utility if you have this problem, they can inform you on the correct pressure rating of a new relief valve.

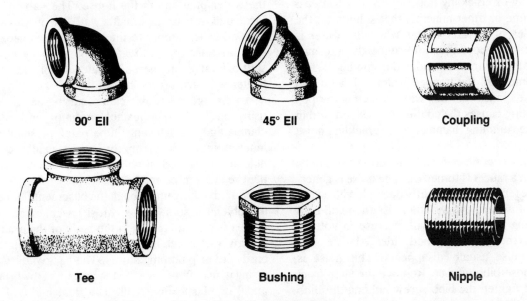

90° Ell   45° Ell   Coupling

Tee   Bushing   Nipple

Fig. 1. Various pipe fittings.

Condensation

Condensation forming on the outside of the water closet tank in the bathroom is a problem often experienced, especially in the winter season. When the warm moist air of the bathroom comes in contact with the cold surface of the water closet tank, the moisture in the air is condensed and forms as water droplets on the surface of the closet tank. If the water in the closet tank can be held at approximately the air temperature of the bathroom, no condensation will form. A thermostatic supply valve, which mixes hot and cold water to supply the water closet tank, can be obtained from your local plumber. Follow the installation instructions and condensation on the closet tank will be eliminated.

Thawing Frozen Piping

If the water pipes freeze and are in an accessible location, they can be safely thawed by pouring hot water on the frozen pipe or wrapping rags dipped in hot water around the frozen areas. A heat lamp can also be used for thawing frozen pipes if the lamp is a safe distance away from combustible materials. An electric hair dryer, aimed at the frozen areas will often thaw the pipe. An electric heating tape wrapped around the frozen areas will also thaw the pipe. If none of these methods do the job, your local plumber has special tools, electric thawing machinery and steamers that *will* solve the problem.

An ounce of prevention is worth a pound of cure! If there is a possibility that water pipes may freeze, open the faucet (hot and cold sides) and let a small stream of water flow. Water moving through the piping will not freeze.

Noise in the Piping System

A hammering or crackling noise in a water heater is caused from lime or other minerals that solidify on the bottom or sides of a water heater when the water is heated. This mineral coating is not water-tight and small droplets of water are trapped under this coating. When the burner comes on, the trapped water is heated and expands. The expansion of this water which is trapped under the coating causes the coating to flake off and in the process the rumbling, hammering or crackling noise is produced.

It is a common occurence for a hammering noise to be produced when a faucet is turned on. The noise is generally heard if the faucet is only slightly opened, with a very small stream of water flowing. The noise then will be a slow hammering noise, which will increase in volume and intensity as the faucet is opened wider. At wide open position, the noise usually disappears. This noise is caused by a loose bibb washer. Remove the stem from the faucet and tighten the bibb screw holding the bibb washer. A whistling type noise is often heard in the water closet tank. This noise is caused when an old-fashioned type ballcock, shown in Fig. 1 Chapter 4, starts to shut off the water flow as the proper water level in the tank is reached. The rate of water flow, through an old-fashioned ballcock varies. The water flows at a fast rate when the tank starts to refill, but as the float ball rises with the water level and in turn pinches off the water flow, a whistling noise is often produced. The chapter on repairing water closet tanks explains the operation and installation of a new type water fill valve. The installation of this new type valve will eliminate this source of noise.

Noise in the pipe is also caused by the sudden closing of a valve. This causes a shock wave to travel through the water pipe, creating a hammer-blow type noise. This noise is very common during usage of an automatic electric dishwasher or an automatic clothes washer. These appliances have electrically operated solenoid valves installed in the water pipe and when these valves close, they close very quickly, literally slamming closed and creating a shock wave in the water pipe. There is a device manufactured to cure this problem. This is a water hammer arrestor, or shock absorber. The ½ inch I.P.S. (iron pipe size) shock absorber is large enough for the average clothes washer or dishwasher. Best results will be obtained if a shock absorber is installed in both the hot and cold water pipes. Shock absorbers are not expensive and are available through plumbing shops.

General Information

Every home owner should know where the main shut-off valve is located on the water supply to the home. Knowing where this valve is located and how to operate it may prevent costly damage. Follow the cold water connection from the water heater back to the point where the water pipe enters the house. The valve should be located at or near the outside wall. The main valve controls all water supply to the house. If it is necessary to shut off only the hot-water supply to the house, the shut-off valve on the *cold* water supply to the water heater will turn off the hot water.

An emergency repair clamp can be easily made by using one or two adjustable radiator or heater hose clamps and a small inner tube patch or small piece of inner tubing. If the clamp (or clamps) is placed over the hole and tightened, it will hold pressure until a permanent repair can be made.

Water backing up through the basement floor drains is a problem in some areas. Most basement floor drains have a two-inch tapping in the body of the drain, under the cover. A 2 inch backwater valve can be purchased at hardware or plumbing stores. The backwater valve is a fitting with a rubber seat on the bottom of the fitting and a metal or plastic float ball. The float ball will drop to permit the passage of water through the inlet side of the floor drain, but the ball will raise and seat itself against the rubber seat of the fitting, preventing water from backing up through the floor drain.

A *vent* when applied to a plumbing system is a pipe which provides a flow of air to and from the plumbing drainage piping. Proper venting is necessary to permit good drainage flow and prevent siphonage in a drainage system.

To eliminate the possibility of accidental scalding, the thermostat on a water heater should be set to deliver a maximum hot-water temperature of 120°F. The hot-water temperature should be checked using a thermometer at the water heater or at a hot-water faucet nearest to the heater.

Bathtubs are now being made with slip-resistant surfaces. A great many injuries are caused by slipping and falling in a bathtub. Bathtubs now in use, which do not have 'slip-resistant' bottoms, can be made much safer by the application of non-slip tapes.

How To Figure Capacities Of Round Tanks

There are occasions when it is necessary or desirable to be able to compute the contents in gallons of round tanks, cisterns or wells. A tank 12 inches in diameter and 5 feet in length will hold 29.3760 gallons of water. A well casing 4 inches in diameter with 10 feet (120 ins.) of water standing in the casing has 6.528 gallons of water standing in the casing.

Shown are the two old standard methods of figuring the contents of round tanks. Also shown are two shorter and easier methods. The shorter methods are not only easier but there is less chance for error since fewer steps are used in obtaining the results.

C = capacity in gallons .7854 = area of circle
D = diameter   231 = cubic inches in gallons
L = length   7.48 = gallons in cubic foot

The old standard methods
1) when measurements are in inches:
   $$\frac{D^2 \times .7854 \times L}{231} = \text{capacity in gallons}$$

2) when measurements are in feet:
   $D^2 \times .7854 \times L \times 7.48 = \text{capacity in gallons}$
   PROBLEM: A tank is 12 inches in diameter and 5 feet in length. How many gallons will the tank hold?
   using method 1–

1) 12   2) 144   3) 113.0976
    x 12   x .7854   x 60
    144   576   6785.8560
                720
                1152
                1008
                113.0976

4)   29.376
    231) 6785.8560
          462
          2165
          2079
          868
          693
          1755
          1617
          1386
          1386   answer: 29.3760 gallons

Note: four steps are necessary in arriving at the answer.

using method 2 –
1) 1   2) .7854   3) .7854   4) 3.927
    x 1   x 1   x 5   x 7.48
    1   .7854   .39270   3.1416
                                     15708
                                     27489
                                     29.37396

answer: 29.373 gallons

Note: four steps are mecessary in arriving at the answer, also, the answer does not agree with answer shown on method 1. The difference in the answers is minor

The shorter methods in figuring tank capacities are:
using method 3 – (C = D² (inches) x L (feet) x .0408)
when measurements are in feet and inches
1) 12   2) 144   3) 720
    x 12   x 5   x .0408
    144   720   5760
                         28800
                         29.3760

answer: 29.3760 gallons

Note: three steps are necessary using this method.

using method 4 – (C = D² x L x .0034 = gallons)
when measurements are in inches
1) 12   2) 144   3) 8640
    x 12   x 60   x .0034
    144   8640   34560
                         25920
                         29.3760

Note: only three steps are necessary using this method, also methods 3 and 4 arrive at the same exact answer when carried to four decimal points.

PROBLEMS: A well casing is 4 inches in diameter, 120 feet deep and has 10 feet of water standing in the casing. How many gallons of water are in the casing?

using method 4—

1)
```
      4
    x 4
     16
```
2)
```
     16
   x 120
    320
     16
   1920
```
3)
```
    1920
  x .0034
    7680
    5760
  6.5280
```

using method 3 —

1)
```
      4
    x 4
     16
```
2)
```
     16
   x 10
    160
```
3)
```
     160
  x .0408
    1280
    6400
  6.5280
```

STEM INDEX BY MANUFACTURER

MANUFACTURER	FITS O.E.M. NUMBER:	PART NO.
American Brass	118 Concealed Sink Faucet	FS1-1
American Brass	104, 1104, 105, 1105 Sink	FS1-2
American Brass	100 Sink	FS4-1
American Kitchens	5771-C, 5772-H, LF-110, 120, 210	FS3-1
American Kitchens	4271-H, 4272-C, LF-100-C, 200-E	FS5-3
American Standard	36536-02 Built-in Bath & Shwr.	FS10-3
American Standard	21674-02(LH), 21186-02(RH) B-892, B-872	FS10-23
American Standard	25509-02 Tract Line Bath	FS10-25
American Standard	19276-02, K-370, K-371 (¾"x½") Exp. Bath Fitting	FS11-3
American Standard	21698-02 Diverter Stem Transfer Valve Fitting	FS11-66
American Standard	(Re-Nu Barrel) 20336-08 RH, 20563-08 LH	FS1-3
American Standard	64804-07 RH, 50637-07 LH, C-530-560 Colony Lav. Fitting	FS1-23
American Standard	6079-04-H, 6080-04-C, R-4100-1-2-3	FS2-1
American Standard	853-14-R, 856-14-L, R-4100-1-2-3	FS2-2
American Standard	54156-04-H, 54157-04-C, Crosley CF-100-200, Schaible 922-26, SCH-1622, Youngstn-Mull. 5505, 60954-55	FS2-3
American Standard	72950-07-RH, 72951-07-LH	FS2-11
American Standard	7617-07-LH, 7616-07-RH, B-876-8	FS3-2
American Standard	711-17-LH, 712-27-RH, B-876-8	FS3-3
American Standard	25535-02-RH, 25536-02-LH, TL-30-37 Tract Line Lav. Fitting	FS3-17
American Standard	21658-02-LH, 21659-02-RH, B906-4211	FS4-4
American Standard	64703-07-RH, 50668-07-LH, Colony Trim	FS4-27
American Standard	21451-02-LH, 20731-02-RH, B-900-2, B-901	FS5-4
American Standard	21883-02, F-305, P-4100-S Lavatory	FS6-1
American Standard	19376-02, F-105, F-115 Lav., B-787 Lavatory, B-912S Sink	FS7-1
American Standard	21668-02-LH, 21195-02-RH, B-874	FS7-2
American Standard	21706-02-LH, 21705-02-RH, P-3905 Lavatory	FS7-4
American Standard	55815-04-RH, 55816-04-LH, R-4046-48	FS8-2
American Standard	55838-04-RH, 55839-04-LH	FS7-13
Briggs	22268 Stem, 22269 Bonnet, T-8832-33 Trim Line Lavatory	FS1-21
Briggs	22291 Stem, 22292 Bonnet, For T-9153-59 Deck Fitting	FS4-73
Briggs	22097-RH-C, 22098-LH-H, T-8802-03 Slant-Back	FS5-7
Briggs	22033-C-RH, 22034-H-LH, T-8805-15 Shelf Back Lavatory	FS5-8
Briggs	6612-H, 6613-C, T-8200, T-9205	FS7-5
Briggs	5857: T-8105-15-25, T-8205-10 Bath & Shower	FS10-6
Briggs	67: 1170, 1200, 1224, 1230 Bath Fittings	FS11-5
Central Brass	SU-357-K Model Lavatory, Bath, Laundry and Sink	FS1-5
Central Brass	SU-1855-47 (R or L), 21, 23, 26, 28, 21-S, 26-S, 47½, 48½	FS3-15
Central Brass	SU-1994-L, SU-1994-R, 70-S, 71-S, 76-S	FS7-6
Central Brass	SU-1548R, 7868, 8868, 9868 Diverters	FS10-7
Chicago Faucets	217-X-LH, 217-X-RH, 889	FS6-3
Crane	F12537 Magic-Close Faucet	FS3-6
Crane	FB-8035-H, FB-8034-C, All Dial-Eze Faucets	FS4-7

MANUFACTURER	FITS O.E.M. NUMBER:	PART NO.
Crane	F12536-H, F12535-C, All Dial-Eze Faucets	FS4-8
Crane	New Sleeve Unit (25A-25 ⅜), C-32180-85-86, C-32165-66-67 Lav., C-32279-81-82 Lav., C-32835 Sink, C-32746-S Wall Sink	FS4-24
Crane	Sleeve Unit (25A-10¾) C-31806 Telsa, C-32267 Securo Jr.	FS5-16
Crane	F13167-RH, F13168-LH, Hospital Stem	FS6-5
Crane	New Sleeve Unit (25A-14¾) Imp. Telsa Wall Mount Sink Faucet	FS6-13
Crane	FB-7674, C-32475-450 Panel Back Lav., Criterion & Rival Lav., C-32820-450 Sinks. Complete Assembly. Pre-War Trim	FS9-23
Crane	FB-1341 (Pre-War) Complete Assembly	FS10-17
Crane	FB-1341 (Pre-War) Criterion & Rival Concealed Bath and By-Pass Valves	FS10-60
Crane	FB-7673, No. 1 Line Lav. (Pre-War) Complete Assembly	FS11-15
Crane	FB-1077 Complete Assembly (Pre-War) No. 1 Line Bath Valves, ½" and ¾"	FS12-1
Crosley	CF-100-200	FS2-3
Dick Bros.	Fits Dick Bros. 2011-H, 2011-C, D-4008-10 Wall Faucet	FS3-13
Dick Bros.	Fits Dick Bros. 3057 Deck Faucet	FS4-60
Elkay		FS4-28
Eljer	5288-1-RH, 5288-2-LH, No. 4 Unit	FS1-22
Eljer	4788 No. 3 Unit	FS4-25
Eljer	2807, 9575R, 9576R	FS4-72
Eljer	5182	FS4-61
Eljer	4650	FS4-62
Eljer	5186	FS5-61
Eljer	4290, E-9560, E-9562-R	FS5-67
Eljer	2733, E-9340-R-41R Concealed Lavatory Fittings	FS6-65
Eljer	2788, E-9564-5-6	FS7-64
Eljer	3045 Bath Fittings	FS8-66
Eljer	5259-1-RH, 5259-2-LH, Tub Valve Assembly, No. 5 Unit	FS9-29
Empire Brass	611 Ledge Type	FS1-1
Empire Brass	800 Ledge Type	FS1-2
Gerber	RPA-28-1, 29-1	FS1-6
Gerber	29-1: 250	FS1-61
Gerber	260, 225-40, 53, RPA-28-2, 29-2	FS2-4
Gerber	29-2, 260, 22540, 53	FS2-61
Gerber	393-1: 365, 368 Ledge Type Sink Faucets Since '61	FS4-12
Gerber	7-2 Tub Stem	FS10-8
Gerber	7-2, 12-1, 98-672	FS10-65
Gerber	RPA-13-1, 37-1, 98-722	FS11-64
Gopher-St. Paul	(See Union Brass)	
Harcraft Brass	3247, 10-100 through 10-171, 12-300 through 12-302	FS1-13
Harcraft Brass	99-3128-H, 99-3129-C, A-50, A-51-A, A-52-A, A-160, A-180, A-190, A-540, A-541 Swing Spouts, Center Sets, Exposed Ledge Types	FS1-14
Harcraft Brass	3140: Valve Stem for Diverters, Shower and Tub Fillers prior to '60	FS10-18
Harcraft Brass	3141: Diverter Stem for Diverters & Tub Fillers prior to '60	FS10-19

STEM INDEX BY MANUFACTURER

MANUFACTURER	FITS O.E.M. NUMBER:	PART NO.
Indiana Brass	631-C & D	FS1-7
Indiana Brass	552-C, w/582-D Gland Nut	FS8-61
Kohler	32462 RH, 39717 LH, (Renew Barrel)	FS1-8
Kohler	32473 (Renew Barrel) No Thread	FS1-62
Kohler	34070-C, 34071-H, K-8610-A, K-8005A-06-09	FS3-7
Kohler	31871-H, 31872-C, K-8686, K-8692, K-8655, K-8660	FS4-14
Kohler	20655-H, 20656-C, Valve Unit, Aquaric	FS4-63
Kohler	32491, K-8178, Single Lavatory Fitting	FS5-11
Kohler	22932-H, 22917-C Constellation, Galaxy, Triton, Aquaric	FS5-22
Kohler	31584-C, 31585-H, Hampton, Taughton, Marston, Grammercy, Strand; K-8100-15-32-33	FS7-10
Kohler	31591-H, 31592-C, K-8634-36, K-8638, K-8650	FS8-8
Kohler	37645-C (Cold), 37646-H (Hot) Concealed Lavatory	FS9-7
Kohler	20654 Valvet Unit Aquaric Bath & Shower	FS10-24
Kohler	20242, K-7030-32, K-7100, K-7210, K-7240-42-45-47 Bath & Shower Valves	FS11-10
Milwaukee	5012-H, 5012-C, K-4009 Lavatory Center Set	FS3-14
Noland (Sayco)	45D-12 for 280 Fitting	FS1-11
Price-Pfister	03205-01	FS3-63
Price-Pfister	635-645	FS1-9
Price-Pfister	03156-01-H, 03156-02-C	FS2-6
Price-Pfister	3147, 460-61 Single Lavatory	FS2-7
Price-Pfister	03175-01-00Z, 43-010 through 43-124 Fittings	FS2-14
Price-Pfister	03206-01	FS4-70
Price-Pfister	03220-01	FS4-71
Price-Pfister	03151-01-RH, 03151-02-LH	FS4-15
Price-Pfister	03155-01-RH, 03155-02-LH (Fits 703 Ledge Faucet)	FS6-7
Price-Pfister	3108, 10 & 12 DLH Series Crown Imperial Bath	FS8-9
Price-Pfister	3109	FS8-10
Price-Pfister	3110, 10, 12, 50, 60 Bath & Shower	FS10-10
Price-Pfister	03172-01-00-A	FS10-11
Price-Pfister	3111	FS10-68
Repcal Brass	14147 RH, 14148 LH, Old No. 16-5	FS1-10
Repcal Brass	FB-9173 RH, FB-9174 LH, Also 16-3A	FS1-19
Repcal Brass	F14143H, F14145C, 13-511, 205 Wall Sink, 123 Laundry	FS4-16
Repcal Brass	515-L-5	FS9-9
Repcal Brass	1112-5A, B-625P, B-626P, B-1137-P Bath Fittings	FS9-10
Repcal Brass	F14153	FS10-13
Repcal Brass	110-5	FS11-65
Republic Brass	(See Briggs)	
Richmond	18 Serration Broach (New Strle	FS6-63
Richmond		FS5-18
Savoy Brass	A-17-H, A-17-C, 18 Serration Broach (New Style)	FS6-63
Sayco (Noland)	45-D-12 Stem, 57-D-13 Bonnet For 280 Fitting	FS1-11
Sayco (Noland)	45-D-12 Stem for 280 Fitting	FS1-63
Sayco	255-R, 255-L, 1100 Sink Faucet	FS6-10
Sayco	214-R (Stem), 572-R (Bonnet)	FS8-64
Sayco	LOS-1: 206, 208, 308, 311, 406 Bath & Shower Valves (New Style)	FS9-13
Sayco	1-1116-R: 308, A-308-C, 208, 206, 20811 Bath & Shower (New Style)	FS9-64
Sayco	LOS-1D: Diverter Stem	FS9-65
Sears-Roebuck	Sears, Universal-Rundle Lav.	FS1-64
Sears-Roebuck	Sears-Homart	FS2-10

MANUFACTURER	FITS O.E.M. NUMBER:	PART NO.
Sears-Roebuck	Sears-Milwaukee 5077RH, 5078-LH	FS2-65
Sears-Roebuck	Sears, Elkay, Universal-Rundle: P-1077, 32008-L	FS4-28
Sears-Roebuck	Sears, Homart, Universal-Rundle	FS4-66
Sears-Roebuck	Sears, Homart, Universal-Rundle	FS5-64
Schaible	56628-H, 56629-C, (Old) 1932, 1936, 1941	FS6-11
Schaible	55838-04-H, 55839-04-C, New 932	FS7-13
Schaible	Schaible: 922-26, 1622	FS2-3
Schaible	957	FS5-5
Schaible	6618-H, 6619-C, 936	FS8-12
Speakman	3-336-C, 3-337-H, 3-291 (G-3-180) H, 3-292 (G-3-181) C, SK-561	FS2-9
Speakman	G3-158H, G3-159C, S-4760-61, S-4770-71, S-4700	FS3-9
Speakman	G3-140H, G3-139C, S-4700	FS5-14
Speakman	3-250H, 3-251C, S-4095, Diamond Lavatory	FS5-18
Speakman	G3-122 Bath Fitting	FS10-15
Speakman	3-258 Lavatory Fitting Stem	FS10-63
Sterling Faucet	99S-8079-H, 99S-8080-C, 20-310 Series Lavatory Fittings	FS1-17
Sterling Faucet	99S-8109-H, 99S-8110-C	FS1-18
Sterling Faucet	99S-8153-H, 99S-8154-C	FS2-12
Sterling Faucet	99S-8131-H, 99S-8132-C	FS2-66
Sterling Faucet	99S-8125-H, 99S-8126-C	FS3-19
Sterling Faucet	99S-8062-H, 99S-8063-C, 19-050, 19-060 Bath Fitting	FS3-20
Sterling Faucet	99S-8171-H, 99S-8172-C	FS3-21
Sterling Faucet	99S-0064-H, 99S-0294-C	FS4-21
Sterling Faucet	99S-0148-H, 99S-0648-C, S-200, S-1120-22	FS4-22
Sterling Faucet	99S-8021-H, 99S-8020-C, 20-10 (N-300) Lavatory Fitting	FS4-23
Sterling Faucet	99S-8128-H, 99S-8129-C	FS4-67
Sterling Faucet	99S-3193-H, 99S-3194-C	FS5-15
Sterling Faucet	99S-1080-H, 99S-1079-C	FS6-12
Sterling Faucet	99S-0235 (S-1100 Old)	FS7-15
Sterling Faucet	99S-8000 (Series 10-000) Tub Valve Assembly	FS9-19
Sterling Faucet	99S-8001 (Series 10-000) Diverter Assembly	FS9-20
Sterling Faucet	99S-0160-H, 99S-0316-C	FS9-30
Sterling Faucet	99S-8136	FS9-66
Sterling Faucet	99S-1023	FS9-67
Sterling Faucet	99S-1025	FS9-68
Sterling Faucet	99S-8005	FS10-16
Sterling Faucet	99S-0174: 10-200 (S-600 Series) Bath Fitting	FS10-64
Sterling Faucet	99S-8006	FS11-13
Streamway	118, 1118, 218, 1218 Concealed Ledge Type Sink Faucet	FS1-1
Streamway	108, 1108, 208, 1208 Sink Faucet; 104, 1104, 105, 1105 Lavatory; 103, 1103 Tub & Shower	FS1-2
Tracy		FS2-3
Union Brass	1837-A	FS1-12
Union Brass	3402, 1837-A, L-7370 (Also Gopher-St. Paul)	FS2-68
Union Brass	P-107-H, P-107-C, L-7335-41, L-7315-21 (Also Gopher-St. Paul)	FS3-11
Union Brass	1840-A-RH, 1840-A-LH, 30-2-3-4-5 Bath Fitting	FS7-17
Union Brass	P-106-H, P-106-C (Also Gopher-St. Paul)	FS8-67
Union Brass	Stem & Seat Holder	FS9-69
Union Brass	P-19: L-6300-06, L-6320-25, L-6404-05-06-08 Bath Fitting	FS9-70
Universal-Rundle	Sears, Elkay, Universal-Rundle P-1077, 32008-L	FS4-28
Universal-Rundle	Sears, Homart, Universal-Rundle	FS4-66
Universal-Rundle	Sears, Homart, Universal-Rundle	FS5-64
Universal-Rundle	R-39-H, R-39-C, O-310-25	FS7-18
Youngstown-Mullins	5505, 60954-55	FS2-3
Youngstown-Mullins	5546-H, 5547-C	FS5-5

FAUCET STEMS

The faucet stems shown here are all actual size drawings, giving you the easiest, quickest solution to replacing old or worn faucet stems. All stems are grouped into 12 basic sizes, from No. 1 (smallest) to No. 12 (largest).

To identify your stem, simply place it on the drawings until an exact duplicate is found. This is your stem number. The text accompanying each drawing includes original manufacturer, OEM part no. and if hot or cold sides are different. If number includes "C" (FS1-9C), this is cold side. If number includes "H" (FS1-9H), this is hot side. If number includes "HC" (FS1-13HC), stem fits either side.

Stem packages are color-coded: Red for HOT side (H); Green for COLD side (C); Brown for Hot or Cold (HC). Order by COMPLETE part number.

When replacing stems, it is often advisable to replace old seats also. The replacement seat numbers required for each stem are shown (in most cases) along with each stem drawing.

BRASS BIBB SCREW: S-553
MONEL BIBB SCREW: S-1258

SEAT NO. S-2001

WASHER: W-145A (O-FLAT)

FS1-1

FS1-1C   FS1-1H
Fits AMER. BRASS, EMPIRE BRASS, STREAMWAY 118 Concealed Sink Faucet

BRASS BIBB SCREW: S-555
MONEL BIBB SCREW: S-1262

WASHER: W-145A (O-FLAT)

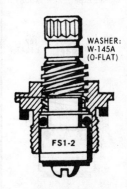

FS1-2

FS1-2C   FS1-2H
Fits AMER. BRASS, EMPIRE BRASS, STREAMWAY: 104, 1104, 105, 1105 Sink

---

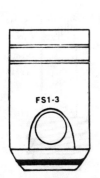

FS1-3

FS1-3C   FS1-3H
Fits AMER. STANDARD (Re-Nu Barrel) OEM 20336-08 RH, 20563-08 LH

BRASS BIBB SCREW: S-555
MONEL BIBB SCREW: S-1262

SEAT NO. S-1091-C

WASHER: W-146 (¼ S-FLAT)

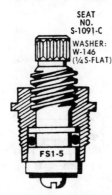

FS1-5

FS1-5C   FS1-5H
Fits CENTRAL SU-357-K Model Lav., Bath, Laundry and Sink

BRASS BIBB SCREW: S-552
MONEL BIBB SCREW: S-1263

SEAT NO. S-1091C

WASHER: W-147 (¼ FLAT)

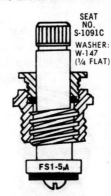

FS1-5A

FS1-5AH   FS1-5AC
For all bath, lavatory, laundry and sink faucets. Interchanges with FS1-5.

BRASS BIBB SCREW: S-553
MONEL BIBB SCREW: S-1258

SEAT NO. S-2007

WASHER: W-148 (¼ L-FLAT)

FS1-6

FS1-6C   FS1-6H
Fits GERBER OEM RPA-28-1, 29-1

BRASS BIBB SCREW: S-552
MONEL BIBB SCREW: S-1263

SEAT NO. S-2008

WASHER: W-147 (¼ FLAT)

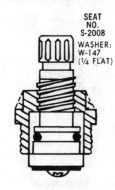

FS1-7C   FS1-7H
Fits INDIANA BRASS OEM 631-C & D

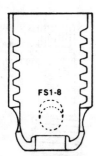

FS1-8

FS1-8C   FS1-8H
Fits KOHLER OEM 32462 RH, OEM 39717 LH

FS1-8A

FS1-8AC   FS1-8AH
Same as OEM 32462 RH (FS1-8C) Same as OEM 39717 LH (FS1-8H) but with "O" Ring

BRASS BIBB SCREW: S-553
MONEL BIBB SCREW: S-1258

SEAT NO. S-2012

WASHER: W-145A (O-FLAT)

FS1-9

FS1-9C   FS1-9H
Fits PRICE-PFISTER 635-645

BRASS BIBB SCREW: S-553
MONEL BIBB SCREW: S-1258

WASHER: W-147 (¼ FLAT)

SEAT NO. S-2024

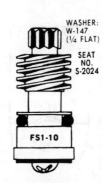

FS1-10

FS1-10C   FS1-10H
Fits REPCAL OEM 14147 RH, OEM 14148 LH Old No. 16-5

BRASS BIBB SCREW: S-555
MONEL BIBB SCREW: S-1262

SEAT NO. S-2012

WASHER: W-145 (OO FLAT)

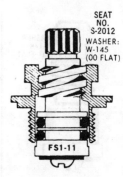

FS1-11

FS1-11C   FS1-11H
Fits SAYCO-NOLAND OEM 45-D-12 Stem, 57-D-13 Bonnet for 280 Fitting

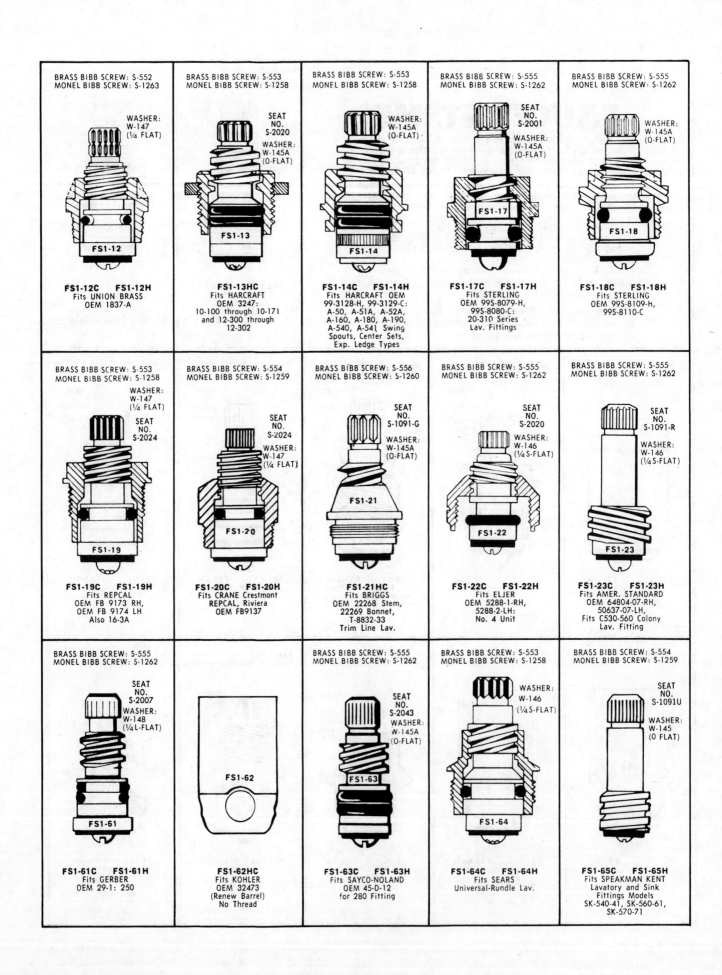

BRASS BIBB SCREW: S-552
MONEL BIBB SCREW: S-1263

WASHER:
W-147
(¼ FLAT)

FS1-12

FS1-12C   FS1-12H
Fits UNION BRASS
OEM 1837-A

BRASS BIBB SCREW: S-553
MONEL BIBB SCREW: S-1258

SEAT
NO.
S-2020

WASHER:
W-145A
(O-FLAT)

FS1-13

FS1-13HC
Fits HARCRAFT
OEM 3247:
10-100 through 10-171
and 12-300 through
12-302

BRASS BIBB SCREW: S-553
MONEL BIBB SCREW: S-1258

WASHER:
W-145A
(O-FLAT)

FS1-14

FS1-14C   FS1-14H
Fits HARCRAFT OEM
99-3128-H, 99-3129-C:
A-50, A-51A, A-52A,
A-160, A-180, A-190,
A-540, A-541 Swing
Spouts, Center Sets,
Exp. Ledge Types

BRASS BIBB SCREW: S-555
MONEL BIBB SCREW: S-1262

SEAT
NO.
S-2001

WASHER:
W-145A
(O-FLAT)

FS1-17

FS1-17C   FS1-17H
Fits STERLING
OEM 99S-8079-H,
99S-8080-C:
20-310 Series
Lav. Fittings

BRASS BIBB SCREW: S-555
MONEL BIBB SCREW: S-1262

WASHER:
W-145A
(O-FLAT)

FS1-18

FS1-18C   FS1-18H
Fits STERLING
OEM 99S-8109-H,
99S-8110-C

BRASS BIBB SCREW: S-553
MONEL BIBB SCREW: S-1258

WASHER:
W-147
(¼ FLAT)

SEAT
NO.
S-2024

FS1-19

FS1-19C   FS1-19H
Fits REPCAL
OEM FB 9173 RH,
OEM FB 9174 LH
Also 16-3A

BRASS BIBB SCREW: S-554
MONEL BIBB SCREW: S-1259

SEAT
NO.
S-2024

WASHER:
W-147
(¼ FLAT)

FS1-20

FS1-20C   FS1-20H
Fits CRANE Crestmont
REPCAL, Riviera
OEM FB9137

BRASS BIBB SCREW: S-556
MONEL BIBB SCREW: S-1260

SEAT
NO.
S-1091-G

WASHER:
W-145A
(O-FLAT)

FS1-21

FS1-21HC
Fits BRIGGS
OEM 22268 Stem,
22269 Bonnet,
T-8832-33
Trim Line Lav.

BRASS BIBB SCREW: S-555
MONEL BIBB SCREW: S-1262

SEAT
NO.
S-2020

WASHER:
W-146
(¼ S-FLAT)

FS1-22

FS1-22C   FS1-22H
Fits ELJER
OEM 5288-1-RH,
5288-2-LH:
No. 4 Unit

BRASS BIBB SCREW: S-555
MONEL BIBB SCREW: S-1262

SEAT
NO.
S-1091-R

WASHER:
W-146
(¼ S-FLAT)

FS1-23

FS1-23C   FS1-23H
Fits AMER. STANDARD
OEM 64804-07-RH,
50637-07-LH,
Fits C530-560 Colony
Lav. Fitting

BRASS BIBB SCREW: S-555
MONEL BIBB SCREW: S-1262

SEAT
NO.
S-2007

WASHER:
W-148
(¼ L-FLAT)

FS1-61

FS1-61C   FS1-61H
Fits GERBER
OEM 29-1: 250

FS1-62

FS1-62HC
Fits KOHLER
OEM 32473
(Renew Barrel)
No Thread

BRASS BIBB SCREW: S-555
MONEL BIBB SCREW: S-1262

SEAT
NO.
S-2043

WASHER:
W-145A
(O-FLAT)

FS1-63

FS1-63C   FS1-63H
Fits SAYCO-NOLAND
OEM 45-D-12
for 280 Fitting

BRASS BIBB SCREW: S-553
MONEL BIBB SCREW: S-1258

WASHER:
W-146
(¼ S-FLAT)

FS1-64

FS1-64C   FS1-64H
Fits SEARS
Universal-Rundle Lav.

BRASS BIBB SCREW: S-554
MONEL BIBB SCREW: S-1259

SEAT
NO.
S-1091U

WASHER:
W-145
(O FLAT)

FS1-65

FS1-65C   FS1-65H
Fits SPEAKMAN KENT
Lavatory and Sink
Fittings Models
SK-540-41, SK-560-61,
SK-570-71

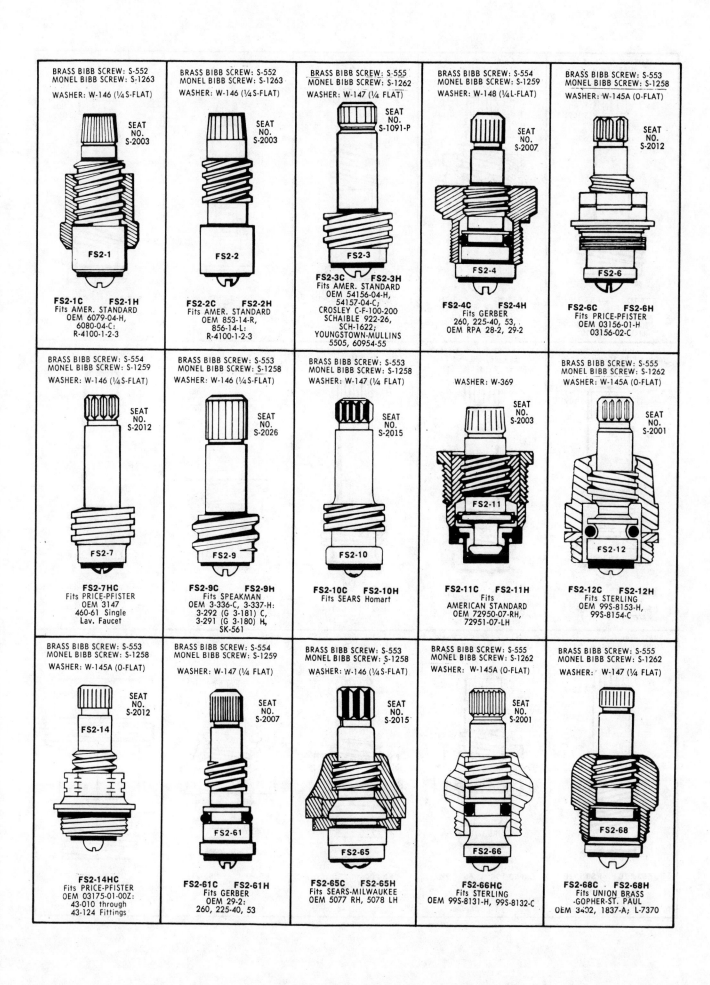

BRASS BIBB SCREW: S-552
MONEL BIBB SCREW: S-1263

WASHER: W-146 (¼S-FLAT)

SEAT NO. S-2003

FS2-1

FS2-1C   FS2-1H
Fits AMER. STANDARD
OEM 6079-04-H,
6080-04-C;
R-4100-1-2-3

BRASS BIBB SCREW: S-552
MONEL BIBB SCREW: S-1263

WASHER: W-146 (¼S-FLAT)

SEAT NO. S-2003

FS2-2

FS2-2C   FS2-2H
Fits AMER. STANDARD
OEM 853-14-R,
856-14-L;
R-4100-1-2-3

BRASS BIBB SCREW: S-555
MONEL BIBB SCREW: S-1262

WASHER: W-147 (¼ FLAT)

SEAT NO. S-1091-P

FS2-3

FS2-3C   FS2-3H
Fits AMER. STANDARD
OEM 54156-04-H,
54157-04-C;
CROSLEY C-F-100-200
SCHAIBLE 922-26,
SCH-1622;
YOUNGSTOWN-MULLINS
5505, 60954-55

BRASS BIBB SCREW: S-554
MONEL BIBB SCREW: S-1259

WASHER: W-148 (¼L-FLAT)

SEAT NO. S-2007

FS2-4

FS2-4C   FS2-4H
Fits GERBER
260, 225-40, 53,
OEM RPA 28-2, 29-2

BRASS BIBB SCREW: S-553
MONEL BIBB SCREW: S-1258

WASHER: W-145A (O-FLAT)

SEAT NO. S-2012

FS2-6

FS2-6C   FS2-6H
Fits PRICE-PFISTER
OEM 03156-01-H
03156-02-C

BRASS BIBB SCREW: S-554
MONEL BIBB SCREW: S-1259

WASHER: W-146 (¼S-FLAT)

SEAT NO. S-2012

FS2-7

FS2-7HC
Fits PRICE-PFISTER
OEM 3147
460-61 Single
Lav. Faucet

BRASS BIBB SCREW: S-553
MONEL BIBB SCREW: S-1258

WASHER: W-146 (¼S-FLAT)

SEAT NO. S-2026

FS2-9

FS2-9C   FS2-9H
Fits SPEAKMAN
OEM 3-336-C, 3-337-H:
3-292 (G 3-181) C,
3-291 (G 3-180) H,
SK-561

BRASS BIBB SCREW: S-553
MONEL BIBB SCREW: S-1258

WASHER: W-147 (¼ FLAT)

SEAT NO. S-2015

FS2-10

FS2-10C   FS2-10H
Fits SEARS Homart

WASHER: W-369

SEAT NO. S-2003

FS2-11

FS2-11C   FS2-11H
Fits
AMERICAN STANDARD
OEM 72950-07-RH,
72951-07-LH

BRASS BIBB SCREW: S-555
MONEL BIBB SCREW: S-1262

WASHER: W-145A (O-FLAT)

SEAT NO. S-2001

FS2-12

FS2-12C   FS2-12H
Fits STERLING
OEM 99S-8153-H,
99S-8154-C

BRASS BIBB SCREW: S-553
MONEL BIBB SCREW: S-1258

WASHER: W-145A (O-FLAT)

SEAT NO. S-2012

FS2-14

FS2-14HC
Fits PRICE-PFISTER
OEM 03175-01-00Z:
43-010 through
43-124 Fittings

BRASS BIBB SCREW: S-554
MONEL BIBB SCREW: S-1259

WASHER: W-147 (¼ FLAT)

SEAT NO. S-2007

FS2-61

FS2-61C   FS2-61H
Fits GERBER
OEM 29-2:
260, 225-40, 53

BRASS BIBB SCREW: S-553
MONEL BIBB SCREW: S-1258

WASHER: W-146 (¼S-FLAT)

SEAT NO. S-2015

FS2-65

FS2-65C   FS2-65H
Fits SEARS-MILWAUKEE
OEM 5077 RH, 5078 LH

BRASS BIBB SCREW: S-555
MONEL BIBB SCREW: S-1262

WASHER: W-145A (O-FLAT)

SEAT NO. S-2001

FS2-66

FS2-66HC
Fits STERLING
OEM 99S-8131-H, 99S-8132-C

BRASS BIBB SCREW: S-555
MONEL BIBB SCREW: S-1262

WASHER: W-147 (¼ FLAT)

FS2-68

FS2-68C   FS2-68H
Fits UNION BRASS
-GOPHER-ST. PAUL
OEM 3402, 1837-A; L-7370

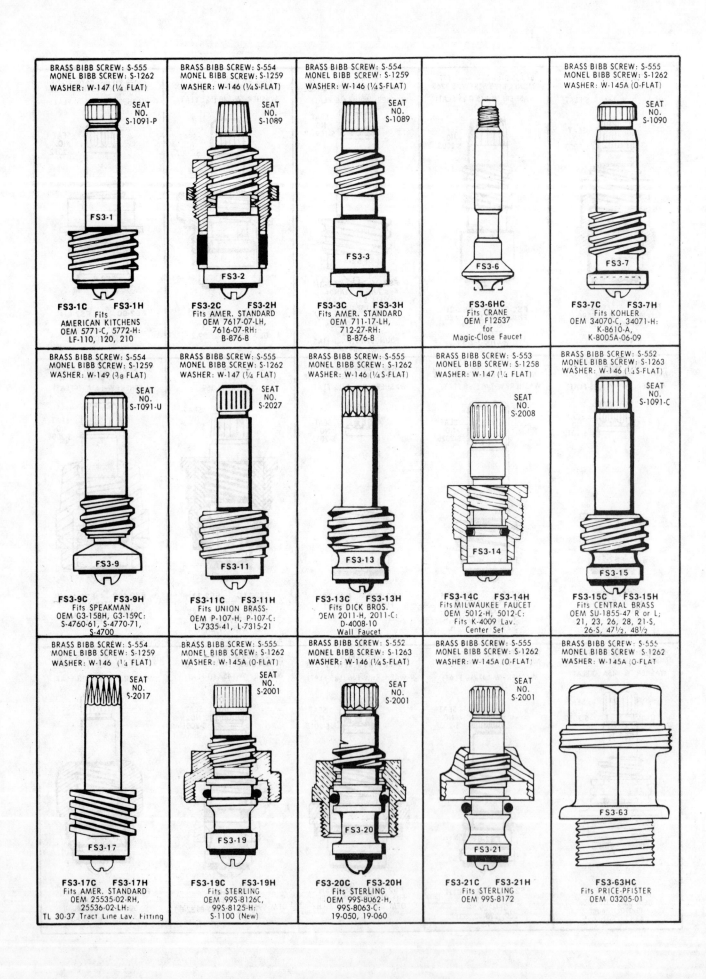

BRASS BIBB SCREW: S-555
MONEL BIBB SCREW: S-1262
WASHER: W-147 (¼ FLAT)

SEAT
NO.
S-1091-P

FS3-1

FS3-1C   FS3-1H
Fits
AMERICAN KITCHENS
OEM 5771-C, 5772-H:
LF-110, 120, 210

BRASS BIBB SCREW: S-554
MONEL BIBB SCREW: S-1259
WASHER: W-146 (¼S-FLAT)

SEAT
NO.
S-1089

FS3-2

FS3-2C   FS3-2H
Fits AMER. STANDARD
OEM 7617-07-LH,
7616-07-RH:
B-876-8

BRASS BIBB SCREW: S-554
MONEL BIBB SCREW: S-1259
WASHER: W-146 (¼S-FLAT)

SEAT
NO.
S-1089

FS3-3

FS3-3C   FS3-3H
Fits AMER. STANDARD
OEM 711-17-LH,
712-27-RH:
B-876-8

FS3-6

FS3-6HC
Fits CRANE
OEM F12537
for
Magic-Close Faucet

BRASS BIBB SCREW: S-555
MONEL BIBB SCREW: S-1262
WASHER: W-145A (O-FLAT)

SEAT
NO.
S-1090

FS3-7

FS3-7C   FS3-7H
Fits KOHLER
OEM 34070-C, 34071-H:
K-8610-A,
K-8005A-06-09

BRASS BIBB SCREW: S-554
MONEL BIBB SCREW: S-1259
WASHER: W-149 (⅜ FLAT)

SEAT
NO.
S-T091-U

FS3-9

FS3-9C   FS3-9H
Fits SPEAKMAN
OEM G3-158H, G3-159C:
S-4760-61, S-4770-71,
S-4700

BRASS BIBB SCREW: S-555
MONEL BIBB SCREW: S-1262
WASHER: W-147 (¼ FLAT)

SEAT
NO.
S-2027

FS3-11

FS3-11C   FS3-11H
Fits UNION BRASS-
OEM P-107-H, P-107-C:
L-7335-41, L-7315-21

BRASS BIBB SCREW: S-555
MONEL BIBB SCREW: S-1262
WASHER: W-146 (¼S-FLAT)

FS3-13

FS3-13C   FS3-13H
Fits DICK BROS.
OEM 2011-H, 2011-C:
D-4008-10
Wall Faucet

BRASS BIBB SCREW: S-553
MONEL BIBB SCREW: S-1258
WASHER: W-147 (¼ FLAT)

SEAT
NO.
S-2008

FS3-14

FS3-14C   FS3-14H
Fits MILWAUKEE FAUCET
OEM 5012-H, 5012-C:
Fits K-4009 Lav.
Center Set

BRASS BIBB SCREW: S-552
MONEL BIBB SCREW: S-1263
WASHER: W-146 (¼S-FLAT)

SEAT
NO.
S-1091-C

FS3-15

FS3-15C   FS3-15H
Fits CENTRAL BRASS
OEM SU-1855-47 R or L:
21, 23, 26, 28, 21-S,
26-S, 47½, 48½

BRASS BIBB SCREW: S-554
MONEL BIBB SCREW: S-1259
WASHER: W-146 (¼ FLAT)

SEAT
NO.
S-2017

FS3-17

FS3-17C   FS3-17H
Fits AMER. STANDARD
OEM 25535-02-RH,
25536-02-LH:
TL 30-37 Tract Line Lav. Fitting

BRASS BIBB SCREW: S-555
MONEL BIBB SCREW: S-1262
WASHER: W-145A (O-FLAT)

SEAT
NO.
S-2001

FS3-19

FS3-19C   FS3-19H
Fits STERLING
OEM 99S-8126C,
99S-8125-H:
S-1100 (New)

BRASS BIBB SCREW: S-552
MONEL BIBB SCREW: S-1263
WASHER: W-146 (¼S-FLAT)

SEAT
NO.
S-2001

FS3-20

FS3-20C   FS3-20H
Fits STERLING
OEM 99S-8062-H,
99S-8063-C:
19-050, 19-060

BRASS BIBB SCREW: S-555
MONEL BIBB SCREW: S-1262
WASHER: W-145A (O-FLAT)

SEAT
NO.
S-2001

FS3-21

FS3-21C   FS3-21H
Fits STERLING
OEM 99S-8172

BRASS BIBB SCREW: S-555
MONEL BIBB SCREW: S-1262
WASHER: W-145A (O-FLAT)

FS3-63

FS3-63HC
Fits PRICE-PFISTER
OEM 03205-01

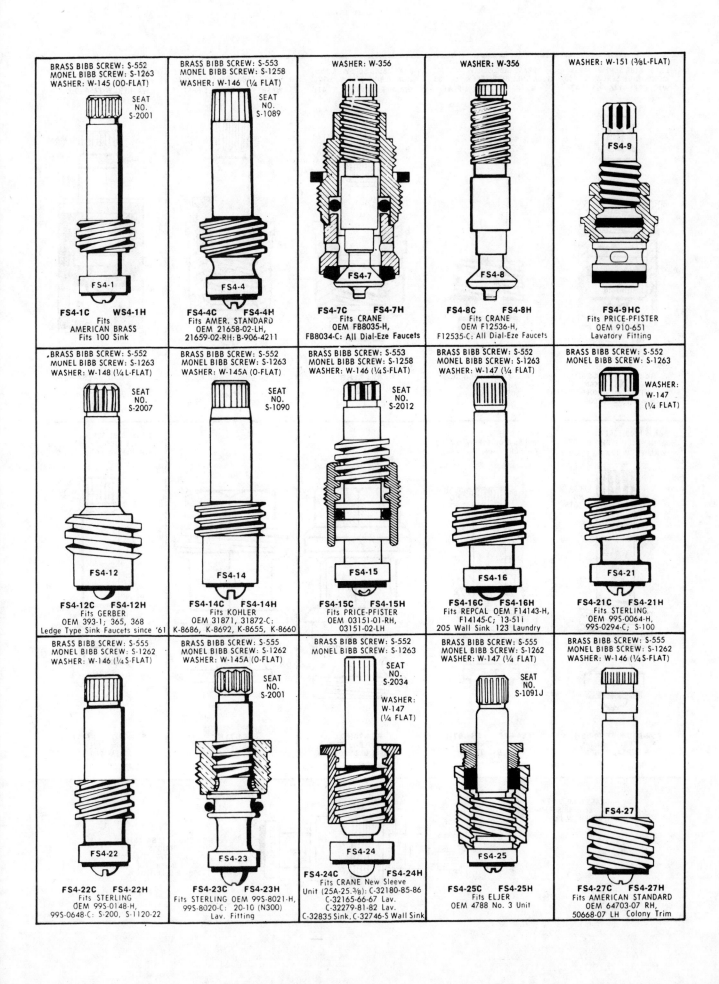

BRASS BIBB SCREW: S-552
MONEL BIBB SCREW: S-1263
WASHER: W-145 (OO-FLAT)

SEAT NO. S-2001

FS4-1

FS4-1C   WS4-1H
Fits
AMERICAN BRASS
Fits 100 Sink

BRASS BIBB SCREW: S-553
MONEL BIBB SCREW: S-1258
WASHER: W-146 (¼ FLAT)

SEAT NO. S-1089

FS4-4

FS4-4C   FS4-4H
Fits AMER. STANDARD
OEM 21658-02-LH,
21659-02-RH: B-906-4211

WASHER: W-356

FS4-7

FS4-7C   FS4-7H
Fits CRANE
OEM FB8035-H,
FB8034-C: All Dial-Eze Faucets

WASHER: W-356

FS4-8

FS4-8C   FS4-8H
Fits CRANE
F12536-H,
F12535-C: All Dial-Eze Faucets

WASHER: W-151 (⅜L-FLAT)

FS4-9

FS4-9HC
Fits PRICE-PFISTER
OEM 910-651
Lavatory Fitting

BRASS BIBB SCREW: S-552
MUNEL BIBB SCREW: S-1263
WASHER: W-148 (¼L-FLAT)

SEAT NO. S-2007

FS4-12

FS4-12C   FS4-12H
Fits GERBER
OEM 393-1; 365, 368
Ledge Type Sink Faucets since '61

BRASS BIBB SCREW: S-552
MONEL BIBB SCREW: S-1263
WASHER: W-145A (O-FLAT)

SEAT NO. S-1090

FS4-14

FS4-14C   FS4-14H
Fits KOHLER
OEM 31871, 31872-C:
K-8686, K-8692, K-8655, K-8660

BRASS BIBB SCREW: S-553
MONEL BIBB SCREW: S-1258
WASHER: W-146 (¼S-FLAT)

SEAT NO. S-2012

FS4-15

FS4-15C   FS4-15H
Fits PRICE-PFISTER
OEM 03151-01-RH,
03151-02-LH

BRASS BIBB SCREW: S-552
MONEL BIBB SCREW: S-1263
WASHER: W-147 (¼ FLAT)

FS4-16

FS4-16C   FS4-16H
Fits REPCAL OEM F14143-H,
F14145-C; 13-51i
205 Wall Sink 123 Laundry

BRASS BIBB SCREW: S-552
MONEL BIBB SCREW: S-1263

WASHER:
W-147
(¼ FLAT)

FS4-21

FS4-21C   FS4-21H
Fits STERLING
OEM 99S-0064-H,
99S-0294-C; S-100

BRASS BIBB SCREW: S-555
MONEL BIBB SCREW: S-1262
WASHER: W-146 (¼S-FLAT)

FS4-22

FS4-22C   FS4-22H
Fits STERLING
OEM 99S-0148-H,
99S-0648-C: S-200, S-1120-22

BRASS BIBB SCREW: S-555
MONEL BIBB SCREW: S-1262
WASHER: W-145A (O-FLAT)

SEAT NO. S-2001

FS4-23

FS4-23C   FS4-23H
Fits STERLING OEM 99S-8021-H,
99S-8020-C: 20-10 (N300)
Lav. Fitting

BRASS BIBB SCREW: S-552
MONEL BIBB SCREW: S-1263

SEAT NO. S-2034

WASHER:
W-147
(¼ FLAT)

FS4-24

FS4-24C   FS4-24H
Fits CRANE New Sleeve
Unit (25A-25.³⁄₈): C-32180-85-86
C-32165-66-67 Lav.
C-32279-81-82 Lav.
C-32835 Sink, C-32746-S Wall Sink

BRASS BIBB SCREW: S-555
MONEL BIBB SCREW: S-1262
WASHER: W-147 (¼ FLAT)

SEAT NO. S-1091J

FS4-25

FS4-25C   FS4-25H
Fits ELJER
OEM 4788 No. 3 Unit

BRASS BIBB SCREW: S-555
MONEL BIBB SCREW: S-1262
WASHER: W-146 (¼S-FLAT)

FS4-27

FS4-27C   FS4-27H
Fits AMERICAN STANDARD
OEM 64703-07 RH,
50668-07 LH Colony Trim

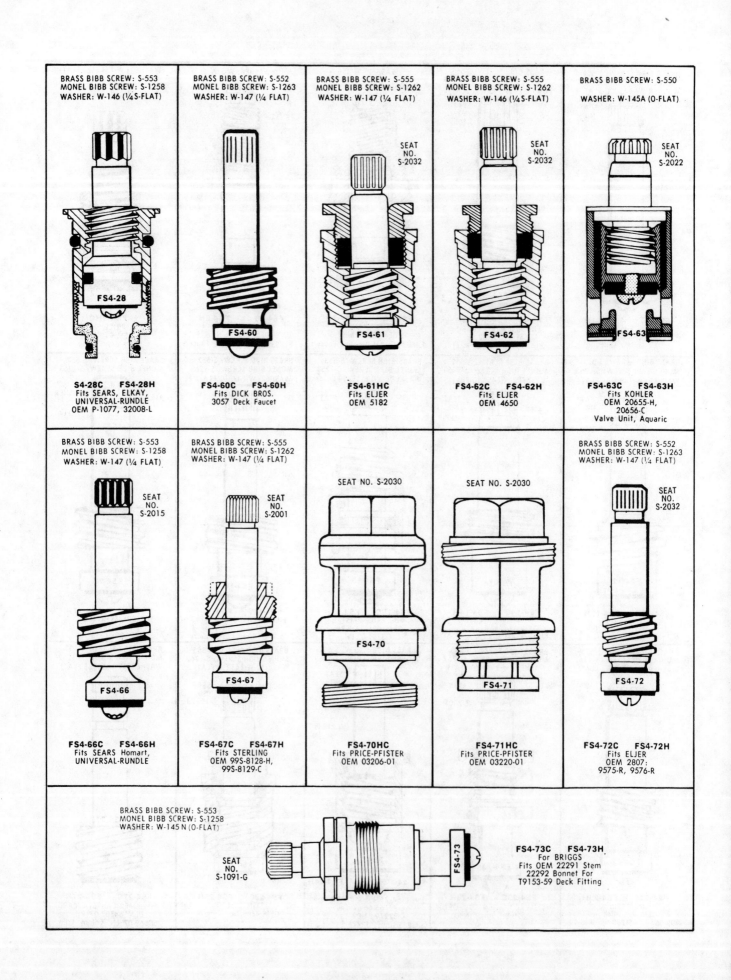

BRASS BIBB SCREW: S-553
MONEL BIBB SCREW: S-1258
WASHER: W-146 (¼S-FLAT)

BRASS BIBB SCREW: S-552
MONEL BIBB SCREW: S-1263
WASHER: W-147 (¼ FLAT)

BRASS BIBB SCREW: S-555
MONEL BIBB SCREW: S-1262
WASHER: W-147 (¼ FLAT)

BRASS BIBB SCREW: S-555
MONEL BIBB SCREW: S-1262
WASHER: W-146 (¼S-FLAT)

BRASS BIBB SCREW: S-550
WASHER: W-145A (O-FLAT)

SEAT NO. S-2032

SEAT NO. S-2032

SEAT NO. S-2022

FS4-28

FS4-60

FS4-61

FS4-62

FS4-63

S4-28C   FS4-28H
Fits SEARS, ELKAY,
UNIVERSAL-RUNDLE
OEM P-1077, 32008-L

FS4-60C   FS4-60H
Fits DICK BROS.
3057 Deck Faucet

FS4-61HC
Fits ELJER
OEM 5182

FS4-62C   FS4-62H
Fits ELJER
OEM 4650

FS4-63C   FS4-63H
Fits KOHLER
OEM 20655-H,
20656-C
Valve Unit, Aquaric

BRASS BIBB SCREW: S-553
MONEL BIBB SCREW: S-1258
WASHER: W-147 (¼ FLAT)

BRASS BIBB SCREW: S-555
MONEL BIBB SCREW: S-1262
WASHER: W-147 (¼ FLAT)

SEAT NO. S-2015

SEAT NO. S-2001

SEAT NO. S-2030

SEAT NO. S-2030

BRASS BIBB SCREW: S-552
MONEL BIBB SCREW: S-1263
WASHER: W-147 (¼ FLAT)

SEAT NO. S-2032

FS4-66

FS4-67

FS4-70

FS4-71

FS4-72

FS4-66C   FS4-66H
Fits SEARS Homart,
UNIVERSAL-RUNDLE

FS4-67C   FS4-67H
Fits STERLING
OEM 99S-8128-H,
99S-8129-C

FS4-70HC
Fits PRICE-PFISTER
OEM 03206-01

FS4-71HC
Fits PRICE-PFISTER
OEM 03220-01

FS4-72C   FS4-72H
Fits ELJER
OEM 2807:
9575-R, 9576-R

BRASS BIBB SCREW: S-553
MONEL BIBB SCREW: S-1258
WASHER: W-145N (O-FLAT)

SEAT NO. S-1091-G

FS4-73

FS4-73C   FS4-73H
For BRIGGS
Fits OEM 22291 Stem
22292 Bonnet For
T9153-59 Deck Fitting

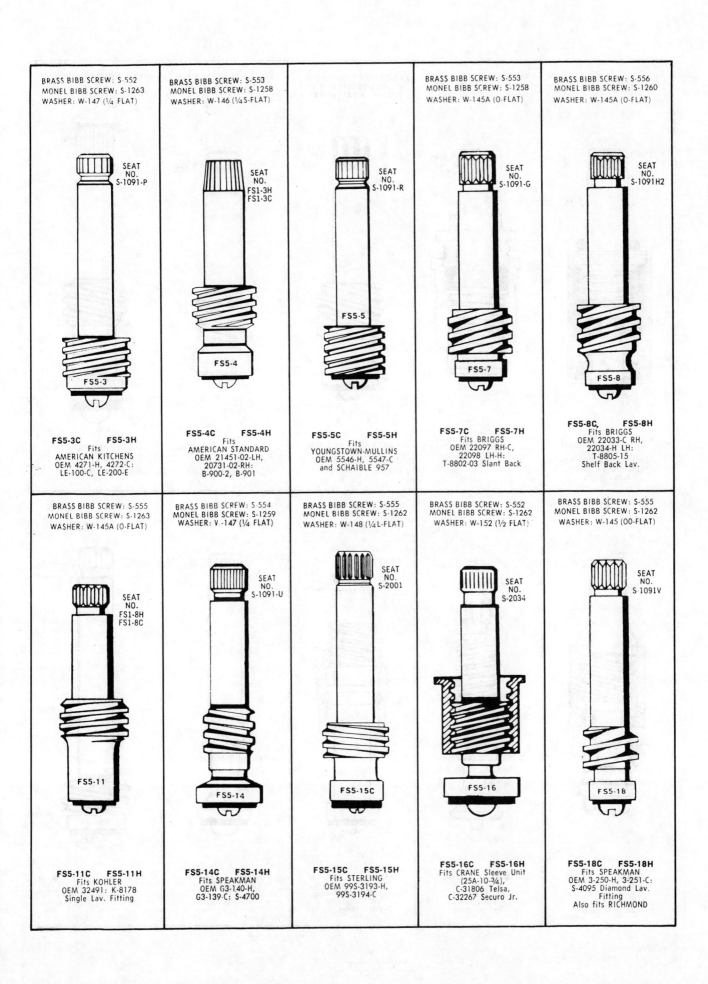

BRASS BIBB SCREW: S-552
MONEL BIBB SCREW: S-1263
WASHER: W-147 (¼ FLAT)

SEAT NO. S-1091-P

FS5-3

FS5-3C   FS5-3H
Fits
AMERICAN KITCHENS
OEM 4271-H, 4272-C:
LE-100-C, LE-200-E

BRASS BIBB SCREW: S-553
MONEL BIBB SCREW: S-1258
WASHER: W-146 (¼S-FLAT)

SEAT NO. FS1-3H FS1-3C

FS5-4

FS5-4C   FS5-4H
Fits
AMERICAN STANDARD
OEM 21451-02-LH,
20731-02-RH:
B-900-2, B-901

BRASS BIBB SCREW: S-553
MONEL BIBB SCREW: S-1258
WASHER: W-145A (O-FLAT)

SEAT NO. S-1091-R

FS5-5

FS5-5C   FS5-5H
Fits
YOUNGSTOWN-MULLINS
OEM 5546-H, 5547-C
and SCHAIBLE 957

BRASS BIBB SCREW: S-553
MONEL BIBB SCREW: S-1258
WASHER: W-145A (O-FLAT)

SEAT NO. S-1091-G

FS5-7

FS5-7C   FS5-7H
Fits BRIGGS
OEM 22097 RH-C,
22098 LH-H:
T-8802-03 Slant Back

BRASS BIBB SCREW: S-556
MONEL BIBB SCREW: S-1260
WASHER: W-145A (O-FLAT)

SEAT NO. S-1091H2

FS5-8

FS5-8C,   FS5-8H
Fits BRIGGS
OEM 22033-C RH,
22034-H LH:
T-8805-15
Shelf Back Lav.

BRASS BIBB SCREW: S-555
MONEL BIBB SCREW: S-1263
WASHER: W-145A (O-FLAT)

SEAT NO. FS1-8H FS1-8C

FS5-11

FS5-11C   FS5-11H
Fits KOHLER
OEM 32491: K-8178
Single Lav. Fitting

BRASS BIBB SCREW: S-554
MONEL BIBB SCREW: S-1259
WASHER: V-147 (¼ FLAT)

SEAT NO. S-1091-U

FS5-14

FS5-14C   FS5-14H
Fits SPEAKMAN
OEM G3-140-H,
G3-139-C: S-4700

BRASS BIBB SCREW: S-555
MONEL BIBB SCREW: S-1262
WASHER: W-148 (¼L-FLAT)

SEAT NO. S-2001

FS5-15C

FS5-15C   FS5-15H
Fits STERLING
OEM 99S-3193-H,
99S-3194-C

BRASS BIBB SCREW: S-552
MONEL BIBB SCREW: S-1262
WASHER: W-152 (½ FLAT)

SEAT NO. S-2034

FS5-16

FS5-16C   FS5-16H
Fits CRANE Sleeve Unit
(25A-10-¾),
C-31806 Telsa,
C-32267 Securo Jr.

BRASS BIBB SCREW: S-555
MONEL BIBB SCREW: S-1262
WASHER: W-145 (00-FLAT)

SEAT NO. S-1091V

FS5-18

FS5-18C   FS5-18H
Fits SPEAKMAN
OEM 3-250-H, 3-251-C:
S-4095 Diamond Lav.
Fitting
Also fits RICHMOND

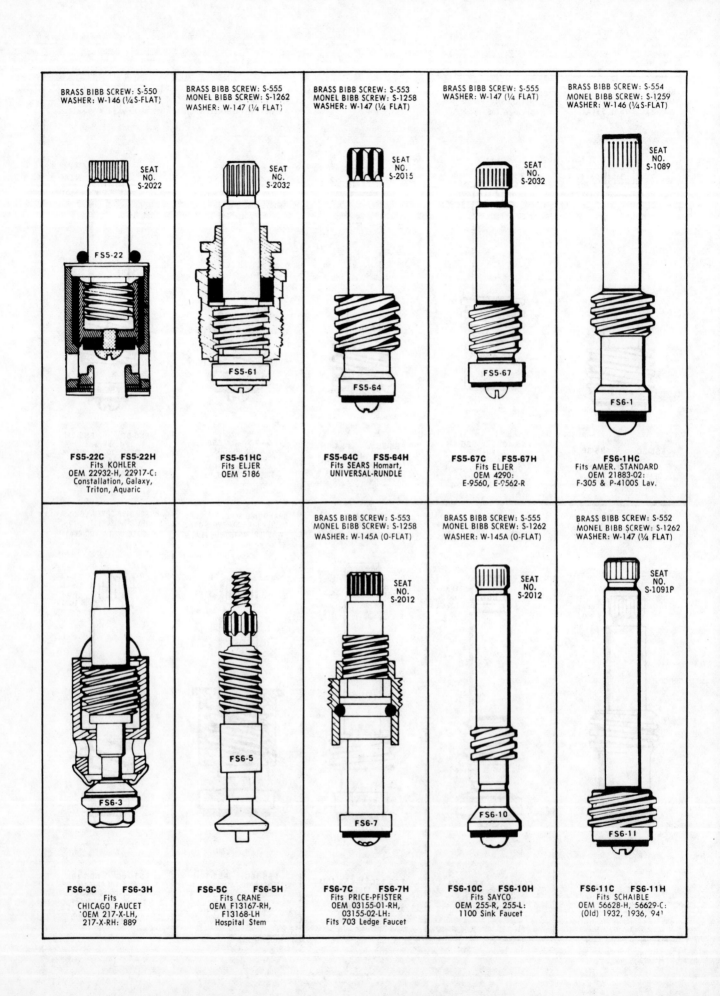

BRASS BIBB SCREW: S-550
WASHER: W-146 (¼S-FLAT)

SEAT NO. S-2022

FS5-22

FS5-22C   FS5-22H
Fits KOHLER
OEM 22932-H, 22917-C:
Constallation, Galaxy,
Triton, Aquaric

BRASS BIBB SCREW: S-555
MONEL BIBB SCREW: S-1262
WASHER: W-147 (¼ FLAT)

SEAT NO. S-2032

FS5-61

FS5-61HC
Fits ELJER
OEM 5186

BRASS BIBB SCREW: S-553
MONEL BIBB SCREW: S-1258
WASHER: W-147 (¼ FLAT)

SEAT NO. S-2015

FS5-64

FS5-64C   FS5-64H
Fits SEARS Homart,
UNIVERSAL-RUNDLE

BRASS BIBB SCREW: S-555
WASHER: W-147 (¼ FLAT)

SEAT NO. S-2032

FS5-67

FS5-67C   FS5-67H
Fits ELJER
OEM 4290:
E-9560, E-9562-R

BRASS BIBB SCREW: S-554
MONEL BIBB SCREW: S-1259
WASHER: W-146 (¼S-FLAT)

SEAT NO. S-1089

FS6-1

FS6-1HC
Fits AMER. STANDARD
OEM 21883-02:
F-305 & P-4100S Lav.

FS6-3

FS6-3C   FS6-3H
Fits
CHICAGO FAUCET
OEM 217-X-LH,
217-X-RH: 889

FS6-5

FS6-5C   FS6-5H
Fits CRANE
OEM F13167-RH,
F13168-LH
Hospital Stem

BRASS BIBB SCREW: S-553
MONEL BIBB SCREW: S-1258
WASHER: W-145A (O-FLAT)

SEAT NO. S-2012

FS6-7

FS6-7C   FS6-7H
Fits PRICE-PFISTER
OEM 03155-01-RH,
03155-02-LH:
Fits 703 Ledge Faucet

BRASS BIBB SCREW: S-555
MONEL BIBB SCREW: S-1262
WASHER: W-145A (O-FLAT)

SEAT NO. S-2012

FS6-10

FS6-10C   FS6-10H
Fits SAYCO
OEM 255-R, 255-L:
1100 Sink Faucet

BRASS BIBB SCREW: S-552
MONEL BIBB SCREW: S-1262
WASHER: W-147 (¼ FLAT)

SEAT NO. S-1091P

FS6-11

FS6-11C   FS6-11H
Fits SCHAIBLE
OEM 56628-H, 56629-C:
(Old) 1932, 1936, 941

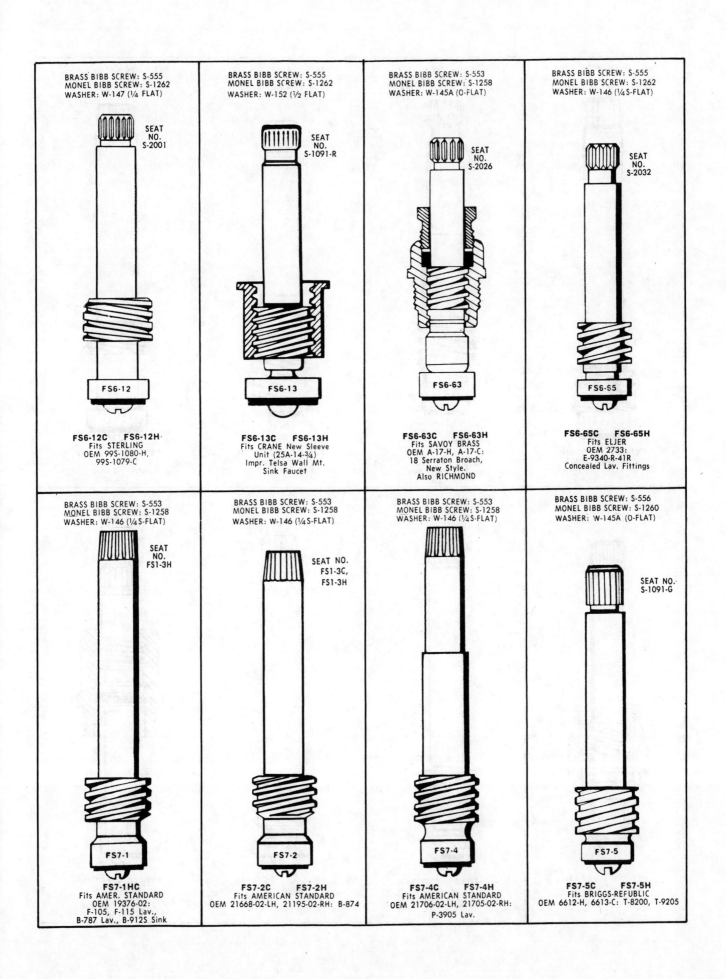

BRASS BIBB SCREW: S-555
MONEL BIBB SCREW: S-1262
WASHER: W-147 (¼ FLAT)

SEAT NO. S-2001

FS6-12

FS6-12C   FS6-12H
Fits STERLING
OEM 99S-1080-H,
99S-1079-C

BRASS BIBB SCREW: S-555
MONEL BIBB SCREW: S-1262
WASHER: W-152 (½ FLAT)

SEAT NO. S-1091-R

FS6-13

FS6-13C   FS6-13H
Fits CRANE New Sleeve
Unit (25A-14-¾)
Impr. Telsa Wall Mt.
Sink Faucet

BRASS BIBB SCREW: S-553
MONEL BIBB SCREW: S-1258
WASHER: W-145A (O-FLAT)

SEAT NO. S-2026

FS6-63

FS6-63C   FS6-63H
Fits SAVOY BRASS
OEM A-17-H, A-17-C:
18 Serraton Broach,
New Style.
Also RICHMOND

BRASS BIBB SCREW: S-555
MONEL BIBB SCREW: S-1262
WASHER: W-146 (¼S-FLAT)

SEAT NO. S-2032

FS6-65

FS6-65C   FS6-65H
Fits ELJER
OEM 2733:
E-9340-R-41R
Concealed Lav. Fittings

BRASS BIBB SCREW: S-553
MONEL BIBB SCREW: S-1258
WASHER: W-146 (¼S-FLAT)

SEAT NO. FS1-3H

FS7-1

FS7-1HC
Fits AMER. STANDARD
OEM 19376-02:
F-105, F-115 Lav.,
B-787 Lav., B-912S Sink

BRASS BIBB SCREW: S-553
MONEL BIBB SCREW: S-1258
WASHER: W-146 (¼S-FLAT)

SEAT NO.
FS1-3C,
FS1-3H

FS7-2

FS7-2C   FS7-2H
Fits AMERICAN STANDARD
OEM 21668-02-LH, 21195-02-RH: B-874

BRASS BIBB SCREW: S-553
MONEL BIBB SCREW: S-1258
WASHER: W-146 (¼S-FLAT)

FS7-4

FS7-4C   FS7-4H
Fits AMERICAN STANDARD
OEM 21706-02-LH, 21705-02-RH:
P-3905 Lav.

BRASS BIBB SCREW: S-556
MONEL BIBB SCREW: S-1260
WASHER: W-145A (O-FLAT)

SEAT NO. S-1091-G

FS7-5

FS7-5C   FS7-5H
Fits BRIGGS-REFUBLIC
OEM 6612-H, 6613-C: T-8200, T-9205

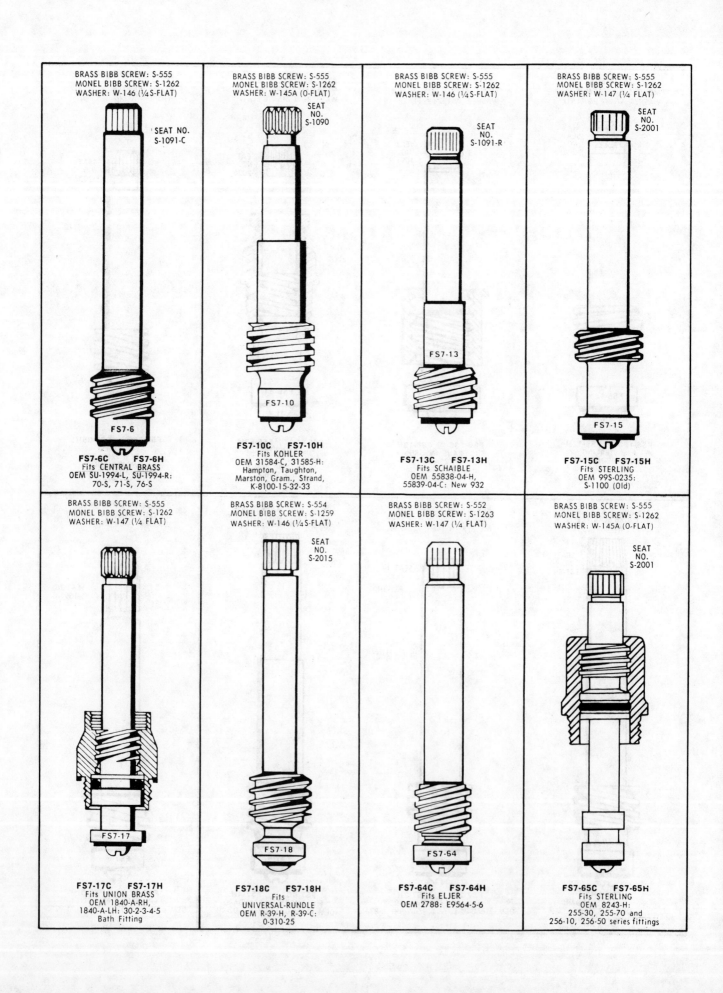

BRASS BIBB SCREW: S-555
MONEL BIBB SCREW: S-1262
WASHER: W-146 (¼S-FLAT)

SEAT NO.
S-1091-C

FS7-6

FS7-6C   FS7-6H
Fits CENTRAL BRASS
OEM SU-1994-L, SU-1994-R:
70-S, 71-S, 76-S

BRASS BIBB SCREW: S-555
MONEL BIBB SCREW: S-1262
WASHER: W-145A (O-FLAT)

SEAT
NO.
S-1090

FS7-10

FS7-10C   FS7-10H
Fits KOHLER
OEM 31584-C, 31585-H:
Hampton, Taughton,
Marston, Gram., Strand,
K-8100-15-32-33

BRASS BIBB SCREW: S-555
MONEL BIBB SCREW: S-1262
WASHER: W-146 (¼S-FLAT)

SEAT
NO.
S-1091-R

FS7-13

FS7-13C   FS7-13H
Fits SCHAIBLE
OEM 55838-04-H,
55839-04-C: New 932

BRASS BIBB SCREW: S-555
MONEL BIBB SCREW: S-1262
WASHER: W-147 (¼ FLAT)

SEAT
NO.
S-2001

FS7-15

FS7-15C   FS7-15H
Fits STERLING
OEM 99S-0235:
S-1100 (Old)

BRASS BIBB SCREW: S-555
MONEL BIBB SCREW: S-1262
WASHER: W-147 (¼ FLAT)

FS7-17

FS7-17C   FS7-17H
Fits UNION BRASS
OEM 1840-A-RH,
1840-A-LH: 30-2-3-4-5
Bath Fitting

BRASS BIBB SCREW: S-554
MONEL BIBB SCREW: S-1259
WASHER: W-146 (¼S-FLAT)

SEAT
NO.
S-2015

FS7-18

FS7-18C   FS7-18H
Fits
UNIVERSAL-RUNDLE
OEM R-39-H, R-39-C:
0-310-25

BRASS BIBB SCREW: S-552
MONEL BIBB SCREW: S-1263
WASHER: W-147 (¼ FLAT)

FS7-64

FS7-64C   FS7-64H
Fits ELJER
OEM 2788: E9564-5-6

BRASS BIBB SCREW: S-555
MONEL BIBB SCREW: S-1262
WASHER: W-145A (O-FLAT)

SEAT
NO.
S-2001

FS7-65C   FS7-65H
Fits STERLING
OEM 8243-H:
255-30, 255-70 and
256-10, 256-50 series fittings

BRASS BIBB SCREW: S-552
MONEL BIBB SCREW: S-1263
WASHER: W-146 (¼ S-FLAT)

SEAT
NO.
S-1091-R

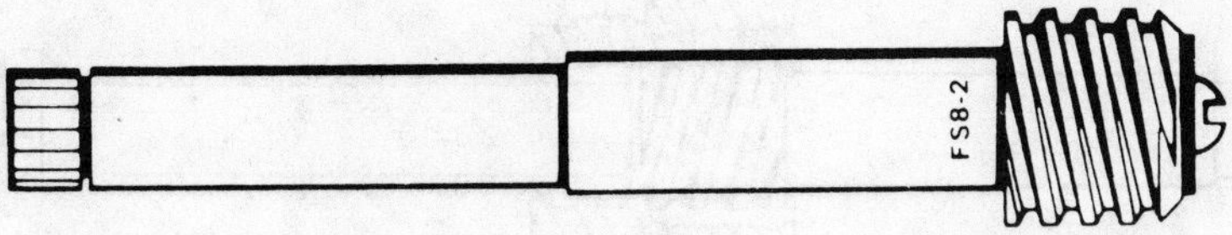

FS8-2

FS8-2C   FS8-2H
Fits AMER. STANDARD
OEM 55815-04-RH,
55816-04-LH:
R-4046-48

BRASS BIBB SCREW: S-552
MONEL BIBB SCREW: S-1263
WASHER: W-145A (0-FLAT)

SEAT
NO.
S-1091

FS8-8

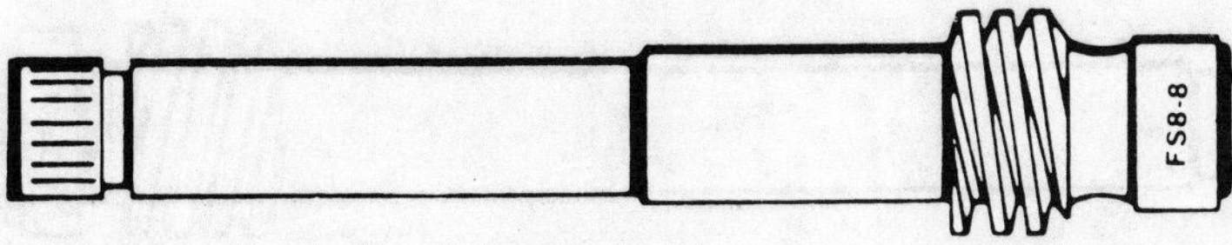

FS8-8C   FS8-8H
Fits KOHLER
OEM 31591-H,
31591-C: K-8634-36
K-8638, -K8650

BRASS BIBB SCREW: S-555
MONEL BIBB SCREW: S-1262
WASHER: W-149 (⅜ FLAT)

SEAT
NO.
S-2030

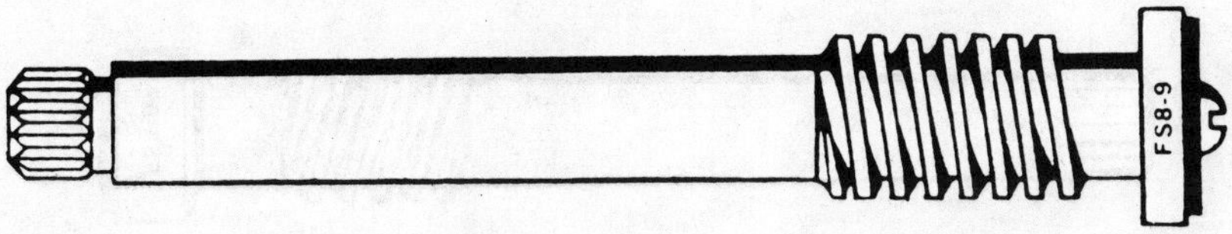

FS8-9

FS8-9HC
Fits PRICE-PFISTER
OEM 3108: 10 & 12
DLH Series Crown-Imp.
Bath Fittings

BRASS BIBB SCREW: S-553
MONEL BIBB SCREW: S-1258
WASHER: W-151 (⅜L-FLAT)

SEAT
NO.
S-2030

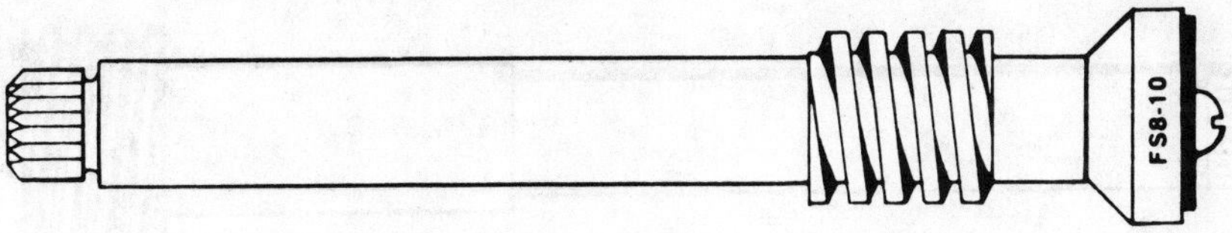

FS8-10

FS8-10HC
Fits PRICE-PFISTER
OEM 3109

BRASS BIBB SCREW: S-552
MONEL BIBB SCREW: S-1263
WASHER: W-147 (¼ FLAT)

SEAT
NO.
S-1091-P

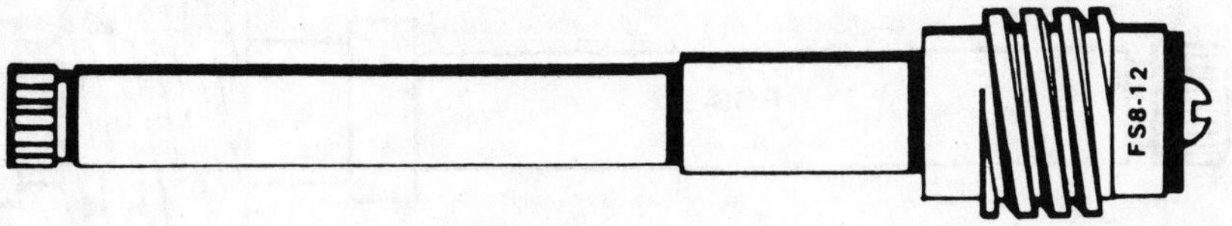

FS8-12

FS8-12C   FS8-12H
Fits SCHAIBLE
OEM 6618-H, 6619-C:
936

BRASS BIBB SCREW: S-555
MONEL BIBB SCREW: S-1262
WASHER: W-147 (¼ FLAT)

SEAT
NO.
S-2051

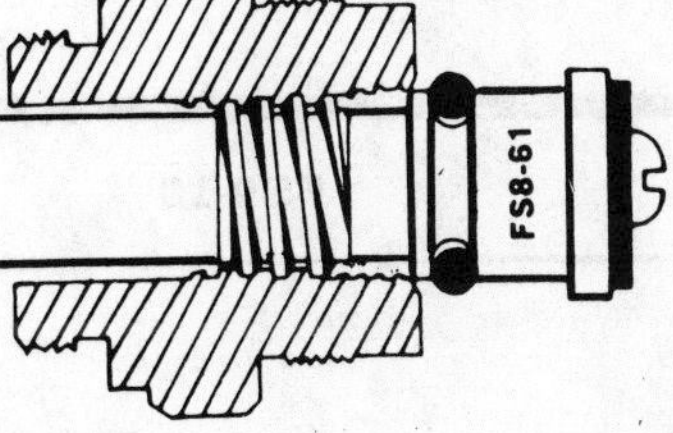

FS8-61

FS8-61C   FS8-61H
Fits INDIANA BRASS OEM 552-H
w/582-D Gland Nut

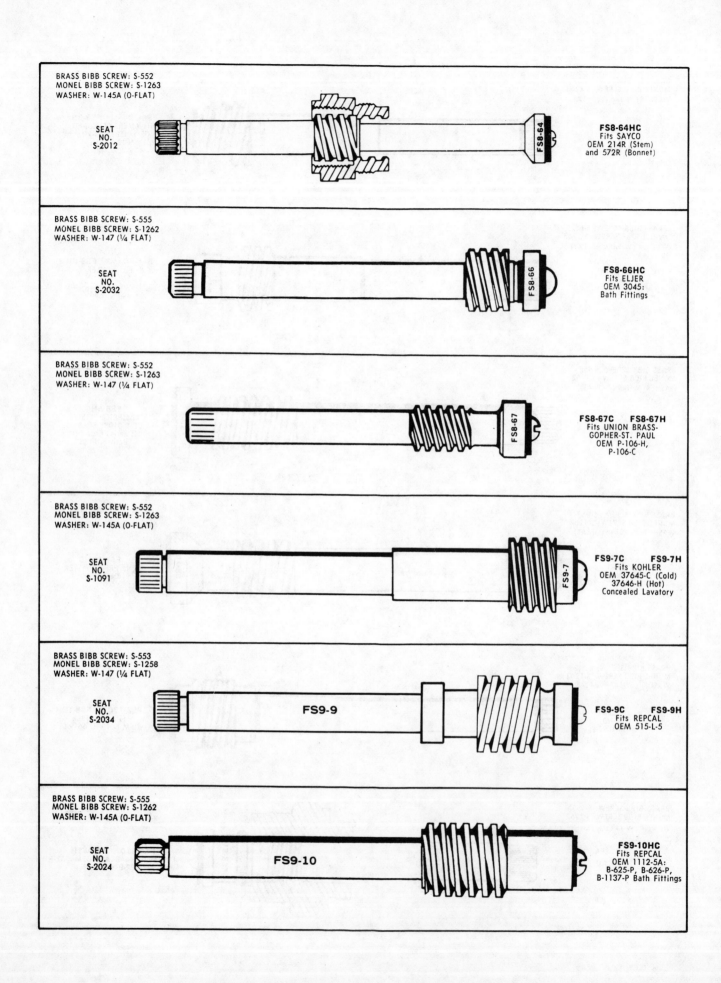

BRASS BIBB SCREW: S-552
MONEL BIBB SCREW: S-1263
WASHER: W-145A (0-FLAT)

SEAT
NO.
S-2012

FS8-64

FS8-64HC
Fits SAYCO
OEM 214R (Stem)
and 572R (Bonnet)

BRASS BIBB SCREW: S-555
MONEL BIBB SCREW: S-1262
WASHER: W-147 (¼ FLAT)

SEAT
NO.
S-2032

FS8-66

FS8-66HC
Fits ELJER
OEM 3045:
Bath Fittings

BRASS BIBB SCREW: S-552
MONEL BIBB SCREW: S-1263
WASHER: W-147 (¼ FLAT)

FS8-67

FS8-67C   FS8-67H
Fits UNION BRASS-
GOPHER-ST. PAUL
OEM P-106-H,
P-106-C

BRASS BIBB SCREW: S-552
MONEL BIBB SCREW: S-1263
WASHER: W-145A (0-FLAT)

SEAT
NO.
S-1091

FS9-7

FS9-7C   FS9-7H
Fits KOHLER
OEM 37645-C (Cold)
37646-H (Hot)
Concealed Lavatory

BRASS BIBB SCREW: S-553
MONEL BIBB SCREW: S-1258
WASHER: W-147 (¼ FLAT)

SEAT
NO.
S-2034

FS9-9

FS9-9C   FS9-9H
Fits REPCAL
OEM 515-L-5

BRASS BIBB SCREW: S-555
MONEL BIBB SCREW: S-1262
WASHER: W-145A (0-FLAT)

SEAT
NO.
S-2024

FS9-10

FS9-10HC
Fits REPCAL
OEM 1112-5A:
B-625-P, B-626-P,
B-1137-P Bath Fittings

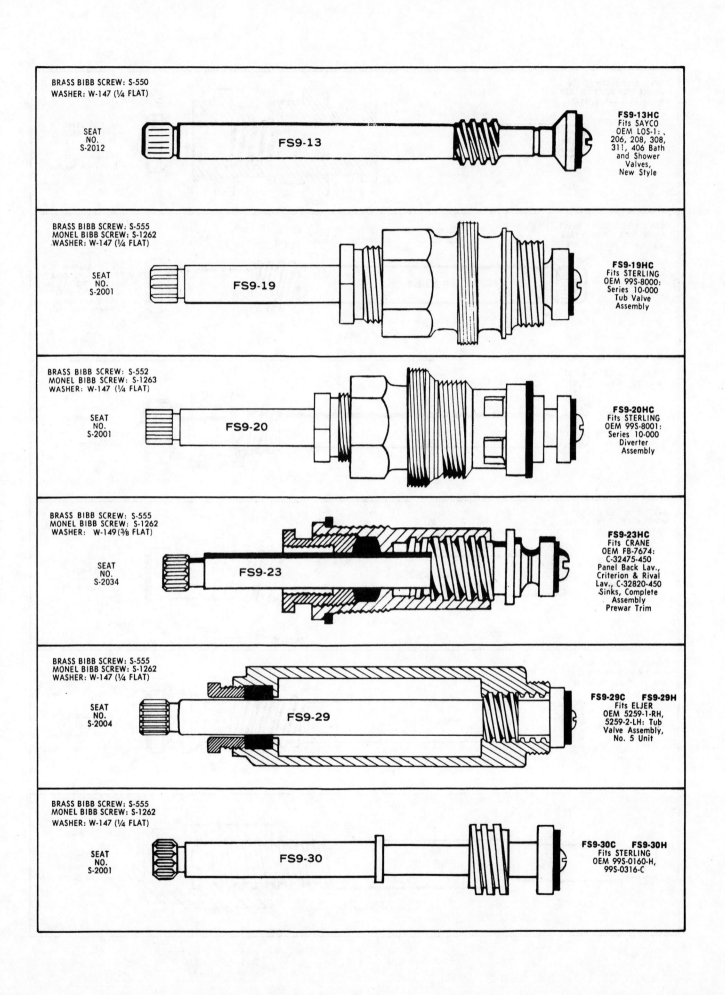

BRASS BIBB SCREW: S-550
WASHER: W-147 (¼ FLAT)

SEAT
NO.
S-2012

FS9-13

FS9-13HC
Fits SAYCO
OEM LOS-1,
206, 208, 308,
311, 406 Bath
and Shower
Valves,
New Style

BRASS BIBB SCREW: S-555
MONEL BIBB SCREW: S-1262
WASHER: W-147 (¼ FLAT)

SEAT
NO.
S-2001

FS9-19

FS9-19HC
Fits STERLING
OEM 99S-8000:
Series 10-000
Tub Valve
Assembly

BRASS BIBB SCREW: S-552
MONEL BIBB SCREW: S-1263
WASHER: W-147 (¼ FLAT)

SEAT
NO.
S-2001

FS9-20

FS9-20HC
Fits STERLING
OEM 99S-8001:
Series 10-000
Diverter
Assembly

BRASS BIBB SCREW: S-555
MONEL BIBB SCREW: S-1262
WASHER: W-149 (⅜ FLAT)

SEAT
NO.
S-2034

FS9-23

FS9-23HC
Fits CRANE
OEM FB-7674:
C-32475-450
Panel Back Lav.,
Criterion & Rival
Lav., C-32820-450
Sinks, Complete
Assembly
Prewar Trim

BRASS BIBB SCREW: S-555
MONEL BIBB SCREW: S-1262
WASHER: W-147 (¼ FLAT)

SEAT
NO.
S-2004

FS9-29

FS9-29C   FS9-29H
Fits ELJER
OEM 5259-1-RH,
5259-2-LH: Tub
Valve Assembly,
No. 5 Unit

BRASS BIBB SCREW: S-555
MONEL BIBB SCREW: S-1262
WASHER: W-147 (¼ FLAT)

SEAT
NO.
S-2001

FS9-30

FS9-30C   FS9-30H
Fits STERLING
OEM 99S-0160-H,
99S-0316-C

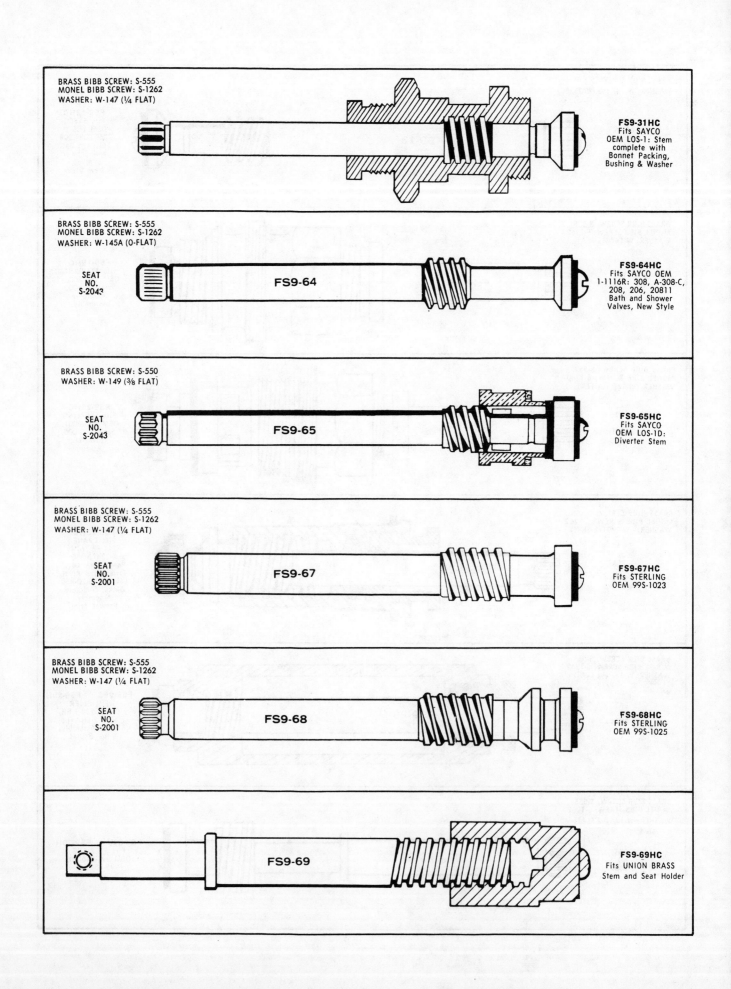

BRASS BIBB SCREW: S-555
MONEL BIBB SCREW: S-1262
WASHER: W-147 (¼ FLAT)

FS9-31HC
Fits SAYCO
OEM LOS-1: Stem
complete with
Bonnet Packing,
Bushing & Washer

BRASS BIBB SCREW: S-555
MONEL BIBB SCREW: S-1262
WASHER: W-145A (O-FLAT)

SEAT
NO.
S-2043

FS9-64

FS9-64HC
Fits SAYCO OEM
1-1116R: 308, A-308-C,
208, 206, 20811
Bath and Shower
Valves, New Style

BRASS BIBB SCREW: S-550
WASHER: W-149 (⅜ FLAT)

SEAT
NO.
S-2043

FS9-65

FS9-65HC
Fits SAYCO
OEM LOS-1D:
Diverter Stem

BRASS BIBB SCREW: S-555
MONEL BIBB SCREW: S-1262
WASHER: W-147 (¼ FLAT)

SEAT
NO.
S-2001

FS9-67

FS9-67HC
Fits STERLING
OEM 99S-1023

BRASS BIBB SCREW: S-555
MONEL BIBB SCREW: S-1262
WASHER: W-147 (¼ FLAT)

SEAT
NO.
S-2001

FS9-68

FS9-68HC
Fits STERLING
OEM 99S-1025

FS9-69

FS9-69HC
Fits UNION BRASS
Stem and Seat Holder

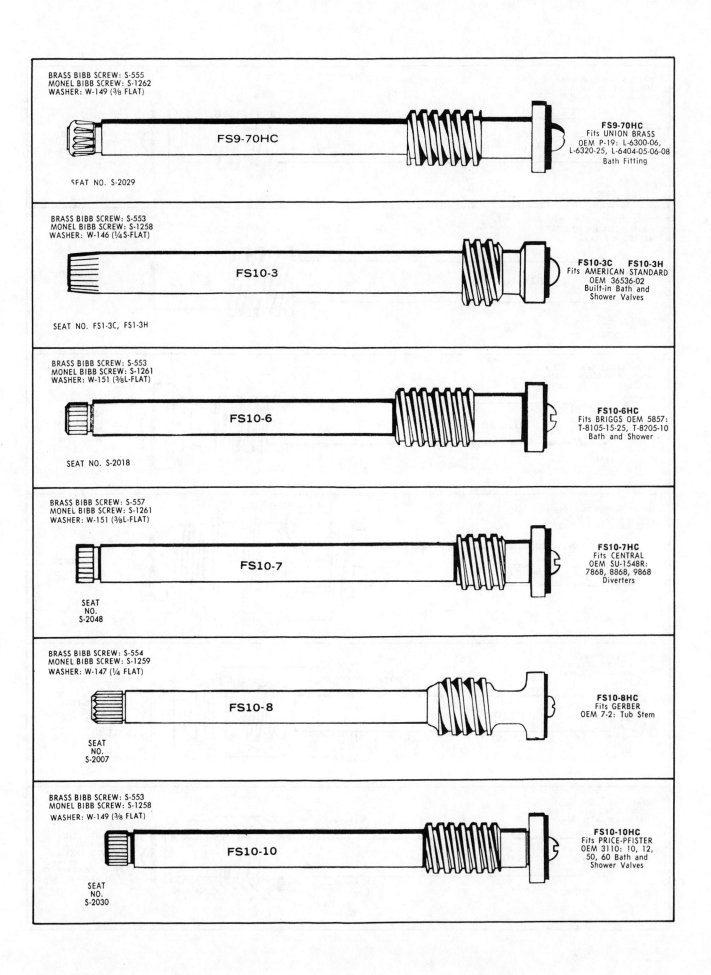

BRASS BIBB SCREW: S-555
MONEL BIBB SCREW: S-1262
WASHER: W-149 (3/8 FLAT)

FS9-70HC

SEAT NO. S-2029

FS9-70HC
Fits UNION BRASS
OEM P-19: L-6300-06,
L-6320-25, L-6404-05-06-08
Bath Fitting

BRASS BIBB SCREW: S-553
MONEL BIBB SCREW: S-1258
WASHER: W-146 (1/4 S-FLAT)

FS10-3

SEAT NO. FS1-3C, FS1-3H

FS10-3C   FS10-3H
Fits AMERICAN STANDARD
OEM 36536-02
Built-in Bath and
Shower Valves

BRASS BIBB SCREW: S-553
MONEL BIBB SCREW: S-1261
WASHER: W-151 (3/8L-FLAT)

FS10-6

SEAT NO. S-2018

FS10-6HC
Fits BRIGGS OEM 5857:
T-8105-15-25, T-8205-10
Bath and Shower

BRASS BIBB SCREW: S-557
MONEL BIBB SCREW: S-1261
WASHER: W-151 (3/8L-FLAT)

FS10-7

SEAT
NO.
S-2048

FS10-7HC
Fits CENTRAL
OEM SU-1548R:
7868, 8868, 9868
Diverters

BRASS BIBB SCREW: S-554
MONEL BIBB SCREW: S-1259
WASHER: W-147 (1/4 FLAT)

FS10-8

SEAT
NO.
S-2007

FS10-8HC
Fits GERBER
OEM 7-2: Tub Stem

BRASS BIBB SCREW: S-553
MONEL BIBB SCREW: S-1258
WASHER: W-149 (3/8 FLAT)

FS10-10

SEAT
NO.
S-2030

FS10-10HC
Fits PRICE-PFISTER
OEM 3110: 10, 12,
50, 60 Bath and
Shower Valves

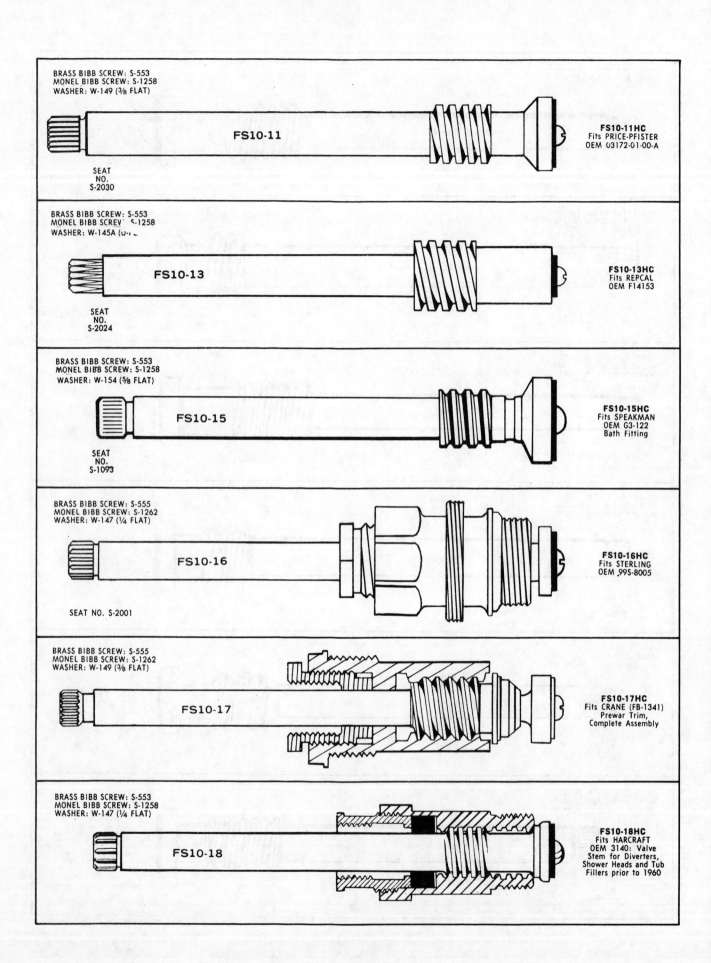

BRASS BIBB SCREW: S-553
MONEL BIBB SCREW: S-1258
WASHER: W-149 (3/8 FLAT)

FS10-11

SEAT
NO.
S-2030

FS10-11HC
Fits PRICE-PFISTER
OEM 03172-01-00-A

BRASS BIBB SCREW: S-553
MONEL BIBB SCREW: S-1258
WASHER: W-145A (U-1

FS10-13

SEAT
NO.
S-2024

FS10-13HC
Fits REPCAL
OEM F14153

BRASS BIBB SCREW: S-553
MONEL BIBB SCREW: S-1258
WASHER: W-154 (5/8 FLAT)

FS10-15

SEAT
NO.
S-1093

FS10-15HC
Fits SPEAKMAN
OEM G3-122
Bath Fitting

BRASS BIBB SCREW: S-555
MONEL BIBB SCREW: S-1262
WASHER: W-147 (1/4 FLAT)

FS10-16

SEAT NO. S-2001

FS10-16HC
Fits STERLING
OEM 99S-8005

BRASS BIBB SCREW: S-555
MONEL BIBB SCREW: S-1262
WASHER: W-149 (3/8 FLAT)

FS10-17

FS10-17HC
Fits CRANE (FB-1341)
Prewar Trim,
Complete Assembly

BRASS BIBB SCREW: S-553
MONEL BIBB SCREW: S-1258
WASHER: W-147 (1/4 FLAT)

FS10-18

FS10-18HC
Fits HARCRAFT
OEM 3140: Valve
Stem for Diverters,
Shower Heads and Tub
Fillers prior to 1960

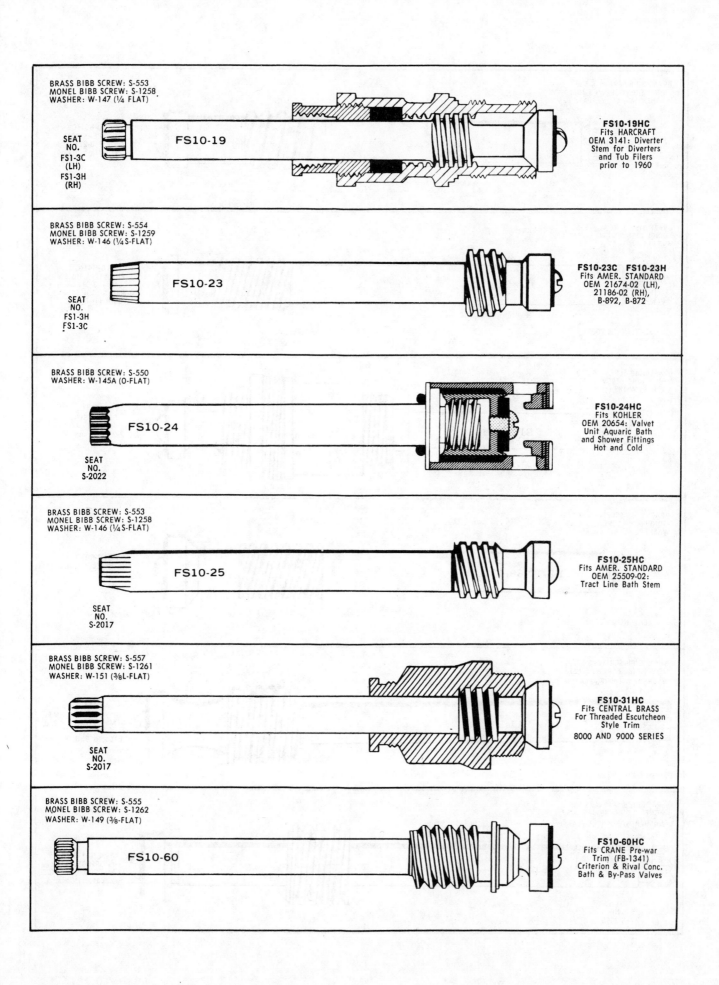

BRASS BIBB SCREW: S-553
MONEL BIBB SCREW: S-1258
WASHER: W-147 (¼ FLAT)

SEAT NO.
FS1-3C (LH)
FS1-3H (RH)

FS10-19

FS10-19HC
Fits HARCRAFT
OEM 3141: Diverter
Stem for Diverters
and Tub Filers
prior to 1960

BRASS BIBB SCREW: S-554
MONEL BIBB SCREW: S-1259
WASHER: W-146 (¼ S-FLAT)

SEAT NO.
FS1-3H
FS1-3C

FS10-23

FS10-23C FS10-23H
Fits AMER. STANDARD
OEM 21674-02 (LH),
21186-02 (RH),
B-892, B-872

BRASS BIBB SCREW: S-550
WASHER: W-145A (0-FLAT)

SEAT NO.
S-2022

FS10-24

FS10-24HC
Fits KOHLER
OEM 20654: Valvet
Unit Aquaric Bath
and Shower Fittings
Hot and Cold

BRASS BIBB SCREW: S-553
MONEL BIBB SCREW: S-1258
WASHER: W-146 (¼ S-FLAT)

SEAT NO.
S-2017

FS10-25

FS10-25HC
Fits AMER. STANDARD
OEM 25509-02:
Tract Line Bath Stem

BRASS BIBB SCREW: S-557
MONEL BIBB SCREW: S-1261
WASHER: W-151 (⅜L-FLAT)

SEAT NO.
S-2017

FS10-31

FS10-31HC
Fits CENTRAL BRASS
For Threaded Escutcheon
Style Trim
8000 AND 9000 SERIES

BRASS BIBB SCREW: S-555
MONEL BIBB SCREW: S-1262
WASHER: W-149 (⅜-FLAT)

FS10-60

FS10-60HC
Fits CRANE Pre-war
Trim (FB-1341)
Criterion & Rival Conc.
Bath & By-Pass Valves

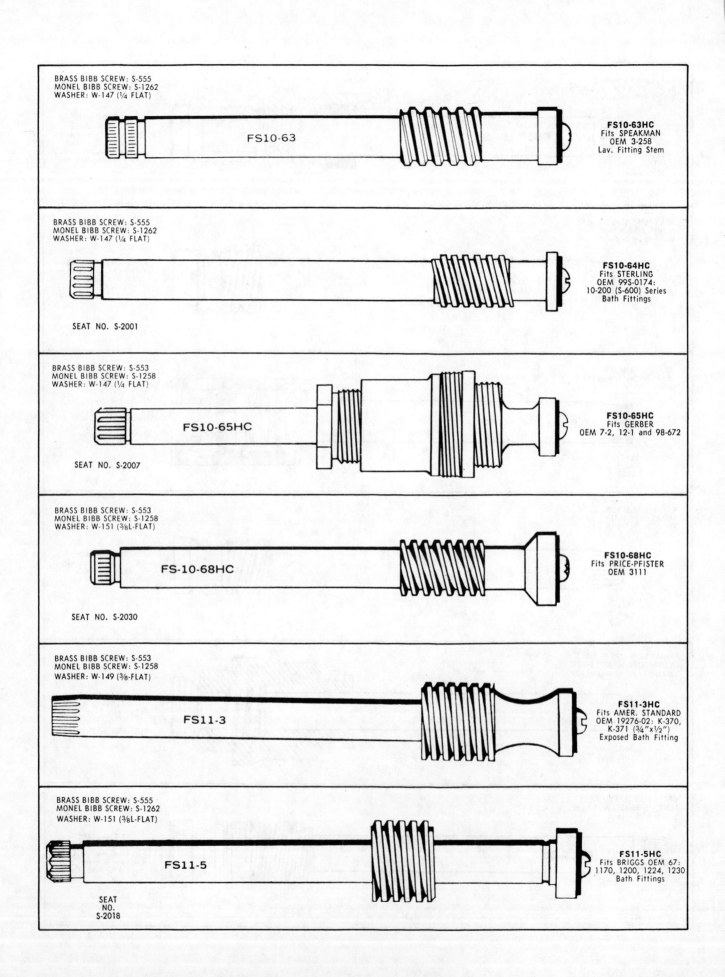

BRASS BIBB SCREW: S-555
MONEL BIBB SCREW: S-1262
WASHER: W-147 (1/4 FLAT)

FS10-63

FS10-63HC
Fits SPEAKMAN
OEM 3-258
Lav. Fitting Stem

BRASS BIBB SCREW: S-555
MONEL BIBB SCREW: S-1262
WASHER: W-147 (1/4 FLAT)

SEAT NO. S-2001

FS10-64HC
Fits STERLING
OEM 99S-0174:
10-200 (S-600) Series
Bath Fittings

BRASS BIBB SCREW: S-553
MONEL BIBB SCREW: S-1258
WASHER: W-147 (1/4 FLAT)

FS10-65HC

SEAT NO. S-2007

FS10-65HC
Fits GERBER
OEM 7-2, 12-1 and 98-672

BRASS BIBB SCREW: S-553
MONEL BIBB SCREW: S-1258
WASHER: W-151 (3/8L-FLAT)

FS-10-68HC

SEAT NO. S-2030

FS10-68HC
Fits PRICE-PFISTER
OEM 3111

BRASS BIBB SCREW: S-553
MONEL BIBB SCREW: S-1258
WASHER: W-149 (3/8-FLAT)

FS11-3

FS11-3HC
Fits AMER. STANDARD
OEM 19276-02: K-370,
K-371 (3/4"x1/2")
Exposed Bath Fitting

BRASS BIBB SCREW: S-555
MONEL BIBB SCREW: S-1262
WASHER: W-151 (3/8L-FLAT)

FS11-5

SEAT
NO.
S-2018

FS11-5HC
Fits BRIGGS OEM 67:
1170, 1200, 1224, 1230
Bath Fittings

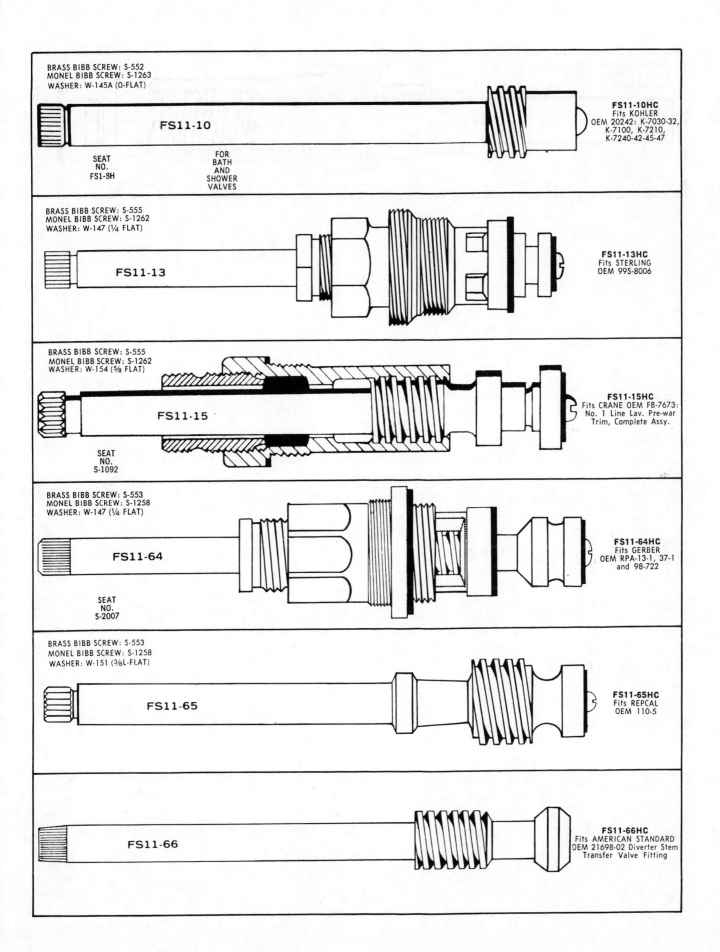

BRASS BIBB SCREW: S-552
MONEL BIBB SCREW: S-1263
WASHER: W-145A (O-FLAT)

FS11-10

SEAT NO. FS1-8H

FOR BATH AND SHOWER VALVES

FS11-10HC
Fits KOHLER
OEM 20242: K-7030-32,
K-7100, K-7210,
K-7240-42-45-47

BRASS BIBB SCREW: S-555
MONEL BIBB SCREW: S-1262
WASHER: W-147 (¼ FLAT)

FS11-13

FS11-13HC
Fits STERLING
OEM 99S-8006

BRASS BIBB SCREW: S-555
MONEL BIBB SCREW: S-1262
WASHER: W-154 (⅝ FLAT)

FS11-15

SEAT NO. S-1092

FS11-15HC
Fits CRANE OEM FB-7673:
No. 1 Line Lav. Pre-war
Trim, Complete Assy.

BRASS BIBB SCREW: S-553
MONEL BIBB SCREW: S-1258
WASHER: W-147 (¼ FLAT)

FS11-64

SEAT NO. S-2007

FS11-64HC
Fits GERBER
OEM RPA-13-1, 37-1
and 98-722

BRASS BIBB SCREW: S-553
MONEL BIBB SCREW: S-1258
WASHER: W-151 (⅜L-FLAT)

FS11-65

FS11-65HC
Fits REPCAL
OEM 110-5

FS11-66

FS11-66HC
Fits AMERICAN STANDARD
OEM 21698-02 Diverter Stem
Transfer Valve Fitting

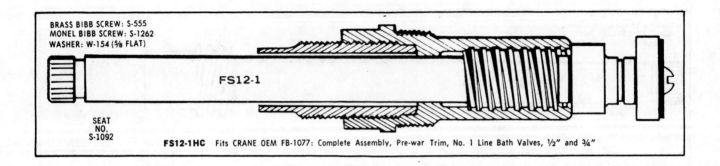

BRASS BIBB SCREW: S-555
MONEL BIBB SCREW: S-1262
WASHER: W-154 (⅝ FLAT)

FS12-1

SEAT
NO.
S-1092

FS12-1HC Fits CRANE OEM FB-1077: Complete Assembly, Pre-war Trim, No. 1 Line Bath Valves, ½" and ¾"

SWING SPOUT PACKINGS (ALL REG. WALLS) (ACTUAL SIZE ILLUSTRATIONS)

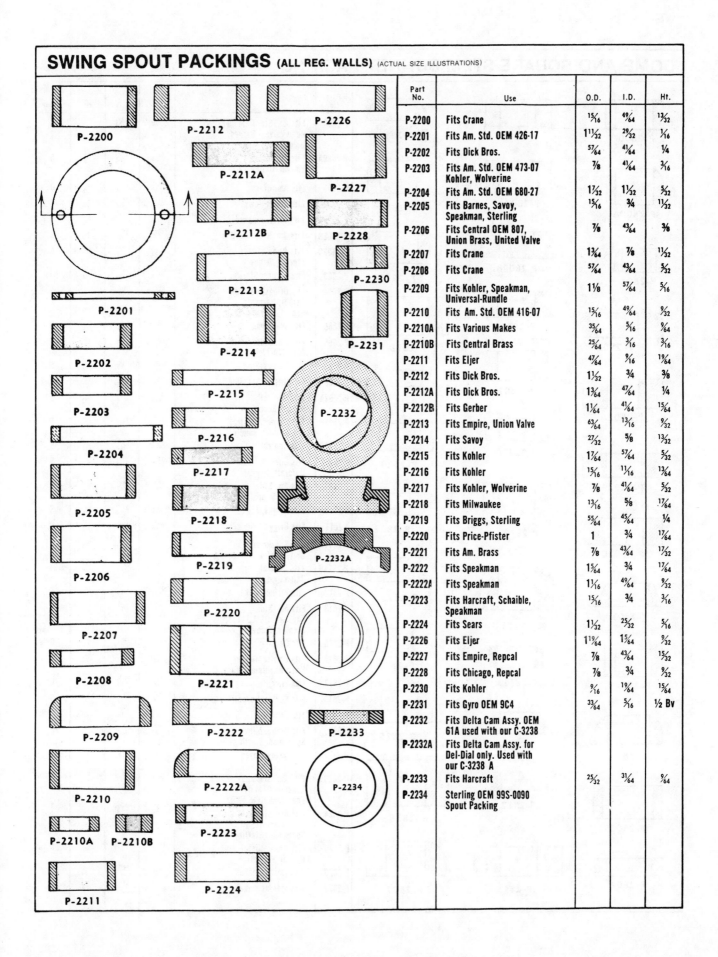

Part No.	Use	O.D.	I.D.	Ht.
P-2200	Fits Crane	15/16	49/64	13/32
P-2201	Fits Am. Std. OEM 426-17	1 11/32	29/32	1/16
P-2202	Fits Dick Bros.	57/64	41/64	1/4
P-2203	Fits Am. Std. OEM 473-07 Kohler, Wolverine	7/8	41/64	3/16
P-2204	Fits Am. Std. OEM 680-27	1 7/32	1 1/32	5/32
P-2205	Fits Barnes, Savoy, Speakman, Sterling	15/16	3/4	11/32
P-2206	Fits Central OEM 807, Union Brass, United Valve	7/8	43/64	3/8
P-2207	Fits Crane	1 3/64	7/8	11/32
P-2208	Fits Crane	57/64	43/64	5/32
P-2209	Fits Kohler, Speakman, Universal-Rundle	1 1/8	57/64	5/16
P-2210	Fits Am. Std. OEM 416-07	15/16	49/64	9/32
P-2210A	Fits Various Makes	35/64	5/16	9/64
P-2210B	Fits Central Brass	25/64	3/16	3/16
P-2211	Fits Eljer	47/64	9/16	19/64
P-2212	Fits Dick Bros.	1 1/32	3/4	3/8
P-2212A	Fits Dick Bros.	1 3/64	47/64	1/4
P-2212B	Fits Gerber	1 1/64	41/64	15/64
P-2213	Fits Empire, Union Valve	63/64	13/16	9/32
P-2214	Fits Savoy	27/32	5/8	13/32
P-2215	Fits Kohler	1 1/64	57/64	5/32
P-2216	Fits Kohler	15/16	11/16	13/64
P-2217	Fits Kohler, Wolverine	7/8	41/64	5/32
P-2218	Fits Milwaukee	13/16	5/8	11/64
P-2219	Fits Briggs, Sterling	55/64	45/64	1/4
P-2220	Fits Price-Pfister	1	3/4	17/64
P-2221	Fits Am. Brass	7/8	43/64	17/32
P-2222	Fits Speakman	1 5/64	3/4	17/64
P-2222A	Fits Speakman	1 1/16	49/64	9/32
P-2223	Fits Harcraft, Schaible, Speakman	15/16	3/4	3/16
P-2224	Fits Sears	1 1/32	25/32	5/16
P-2226	Fits Eljer	1 19/64	1 5/64	9/32
P-2227	Fits Empire, Repcal	7/8	43/64	15/32
P-2228	Fits Chicago, Repcal	7/8	3/4	9/32
P-2230	Fits Kohler	9/16	19/64	15/64
P-2231	Fits Gyro OEM 9C4	33/64	5/16	1/2 Bv
P-2232	Fits Delta Cam Assy. OEM 61A used with our C-3238			
P-2232A	Fits Delta Cam Assy. for Del-Dial only. Used with our C-3238 A			
P-2233	Fits Harcraft	25/32	31/64	9/64
P-2234	Sterling OEM 99S-0090 Spout Packing			

DOME AND SQUARE STEM BONNET PACKING

"S"—SQUARE "D"—DOME (ACTUAL SIZE ILLUSTRATIONS)

P-2800
P-2800A
P-2801
P-2802
P-2802A
P-2802B
P-2802C
P-2803
P-2804
P-2805
P-2806
P-2807

P-2808
P-2808A
P-2808B
P-2808C
P-2809
P 2810
P-2810A
P-2811
P-2811A
P-2811B
P-2812
P-2813

P-2815
P-2816
P-2817
P-2818
P-2819
P-2820
P-2822
P-2823
P-2824
P-2825
P-2826
P-2827

Part No.	Use	O.D.	I.D.	Ht.	Shape
P-2800	Fits Am. Std. OEM 397 Renu Valves, Barnes, Union Brass	$\frac{41}{64}$	$\frac{27}{64}$	$\frac{5}{16}$	S
P-2800A	Fits Am. Brass, Empire, United Valve	$\frac{41}{64}$	$\frac{27}{64}$	$\frac{3}{8}$	D
P-2801	Fits Am. Brass	$\frac{17}{32}$	$\frac{13}{32}$	$\frac{1}{4}$	D
P-2802	Fits Youngstown	$\frac{63}{64}$	$\frac{29}{64}$	$\frac{1}{4}$	S
P-2802A	Fits various makes	$\frac{7}{8}$	$\frac{11}{32}$	$\frac{5}{16}$	S
P-2802B	Fits Am. Kit., Schaible	$\frac{7}{8}$	$\frac{25}{64}$	$\frac{1}{4}$	S
P-2802C	Fits Price-Pfister, Sterling, Tracy	$\frac{55}{64}$	$\frac{7}{16}$	$\frac{9}{32}$	S
P-2803	Fits Barnes, Sterling	$\frac{49}{64}$	$\frac{27}{64}$	$\frac{1}{4}$	Bev
P-2804	Fits Briggs, Central, Gerber, Harcraft, Repcal, Sterling, Wolv.	$\frac{41}{64}$	$\frac{25}{64}$	$\frac{11}{32}$	S
P-2805	Fits Speakman	$\frac{3}{4}$	$\frac{29}{64}$	$\frac{5}{16}$	D
P-2806	Fits Am. Std. OEM 679-27	$\frac{3}{4}$	$\frac{19}{32}$	$\frac{11}{32}$	S
P-2707	Fits Barnes	$\frac{3}{4}$	$\frac{1}{2}$	$\frac{3}{16}$	S
P-2808	Fits Briggs, Schaible	$\frac{3}{4}$	$\frac{13}{32}$	$\frac{9}{32}$	S
P-2808A	Fits Harcraft, Kohler, Repcal	$\frac{47}{64}$	$\frac{7}{16}$	$\frac{11}{32}$	S
P-2808B	Fits Barnes, Central, Crane, Univ.-Rundle	$\frac{13}{16}$	$\frac{13}{32}$	$\frac{9}{32}$	S
P-2808C	Fits Central	$\frac{25}{32}$	$\frac{27}{64}$	$\frac{9}{32}$	S
P-2809	Fits Eljer, Harcraft, Repcal	$\frac{11}{16}$	$\frac{27}{64}$	$\frac{5}{16}$	S
P-2810	Fits Am. Std., Gerber, Repcal	$\frac{53}{64}$	$\frac{13}{32}$	$\frac{29}{64}$	D
P-2810A	Fits Briggs, Eljer	$\frac{3}{4}$	$\frac{27}{64}$	$\frac{7}{16}$	S
P-2811	Fits Briggs	$\frac{45}{64}$	$\frac{13}{32}$	$\frac{1}{4}$	D
P-2811A	Fits Speakman, United Valve	$\frac{39}{64}$	$\frac{7}{16}$	$\frac{3}{16}$	S
P-2811B	Fits Central, Union Brass	$\frac{23}{32}$	$\frac{27}{64}$	$\frac{15}{64}$	Bev
P-2812	Fits various makes	$\frac{41}{64}$	$\frac{27}{64}$	$\frac{7}{64}$	S
P-2813	Fits Am. Std., Yngstwn. OEM 54111, Crosley, Tracy	1	$\frac{7}{16}$	$\frac{5}{16}$	S
P-2815	Fits Chicago	$\frac{43}{64}$	$\frac{13}{32}$	$\frac{15}{64}$	D
P-2816	Fits Empire, Gerber, Price-Pf., Union Brass	$\frac{43}{64}$	$\frac{13}{32}$	$\frac{11}{32}$	S
P-2817	Fits Eljer	$\frac{11}{16}$	$\frac{27}{64}$	$\frac{11}{32}$	D
P-2818	Fits Kohler, Repcal	$\frac{57}{64}$	$\frac{29}{64}$	$\frac{13}{32}$	S
P-2819	Fits Kohler	$\frac{47}{64}$	$\frac{7}{16}$	$\frac{11}{64}$	D
P-2820	Fits Crane, Milwkee., Speakman, Sterling	$\frac{47}{64}$	$\frac{7}{16}$	$\frac{5}{16}$	D
P-2822	Fits Savoy, Sayco, Union Brass	$\frac{9}{16}$	$\frac{3}{8}$	$\frac{15}{64}$	S
P-2823	Fits Am. Std. and Schaible OEM 57283, Epeakman OEM 49-92	$\frac{3}{4}$	$\frac{25}{64}$	$\frac{1}{4}$	S
P-2824	Fits Am. Std. OEM 397, Harcraft, Speakman, Sterling	$\frac{5}{8}$	$\frac{3}{8}$	$\frac{1}{4}$	S
P-2825	Fits Price-Pfister, Repcal, Wolverine	$\frac{23}{32}$	$\frac{13}{32}$	$\frac{11}{32}$	D
P-2826	Fits various makes	$\frac{29}{32}$	$\frac{13}{32}$	$\frac{31}{64}$	S
P-2827	Fits various makes	$\frac{7}{8}$	$\frac{7}{16}$	$\frac{3}{8}$	D

DOME AND SQUARE STEM BONNET PACKING "S"—SQUARE "D"—DOME (ACTUAL SIZE ILLUSTRATIONS)

Part No.	Use	O.D.	I.D.	Ht.	Shape
P-2828	Fits various makes	$\frac{21}{32}$	$\frac{27}{64}$	$\frac{21}{64}$	Bev
P-2829	Fits Am. Std. OEM 1311-07, Dick, Sears	$\frac{47}{64}$	$\frac{25}{64}$	$\frac{3}{8}$	D
P-2830	Fits Sterling OEM 99S-0509				
P-2831	Fits Sterling OEM 99S-0386				
P-2832	Fits Sterling OEM 99S-0028				
P-2833	Fits various makes	$\frac{5}{8}$	$\frac{5}{16}$	$\frac{9}{32}$	D
P-2834	Fits Am. Std.	$\frac{11}{16}$	$\frac{7}{16}$	$\frac{11}{32}$	D
P-2835	Fits Kohler	$\frac{11}{16}$	$\frac{7}{16}$	$\frac{5}{32}$	S
P-2836	Fits Briggs, Savoy, Sayco, Union Brass	$\frac{39}{64}$	$\frac{3}{8}$	$\frac{5}{16}$	S
P-2838	"V" Packing for Am. Std. Push-Pull Lav. Single Lev. Fct. OEM 29312-07				
P-2839	Fits Am. Std. OEM 391-07	$\frac{29}{32}$	$\frac{9}{16}$	$\frac{15}{32}$	S
P-2840	Fits Briggs	$\frac{19}{32}$	$\frac{25}{64}$	$\frac{25}{64}$	S
P-2841	Fits Kohler	$\frac{3}{4}$	$\frac{7}{16}$	$\frac{7}{32}$	D
P-2844	Fits Sterling OEM 99S-0427	$\frac{49}{64}$	$\frac{25}{64}$	$\frac{5}{16}$	D
P-2845	Fits Union Brass OEM 1903	$\frac{15}{16}$	$\frac{23}{64}$	$\frac{9}{32}$	D
P-3001V	Graphite-Coated, Self-Forming Packing, 24" length (Blister card)	$\frac{3}{32}$			
P-3002V	Graphited Asbestos, Stem and Valve Packing (Blister card)				
P-3003V	Teflon-Coated Asbestos Stem and Valve Packing (Blister card)				

P-2828 P-2835 P-2844
P-2829 P-2836 P-2845
P-2830 P-2837
P-2831 P-2838
P-2832 P-2839
P-2833 P-2840 P-3001V
P-2834 P-2841 P-3002V P-3003V

SEALMASTER FORM A SEAL PACKING
SEALMASTER STEM and VALVE PACKING

"O" RINGS PACKING RINGS FOR SWING SPOUT, SLIP-JOINTS, PUMP AND HYDRAULIC APPLICATIONS (ACTUAL SIZE ILLUSTRATIONS)

Part No.	Use	O.D.	I.D.	Wall
R-201		$\frac{1}{4}$	$\frac{1}{8}$	$\frac{1}{16}$
R-202		$\frac{9}{32}$	$\frac{5}{32}$	$\frac{1}{16}$
R-203		$\frac{5}{16}$	$\frac{3}{16}$	$\frac{1}{16}$
R-204		$\frac{11}{32}$	$\frac{7}{32}$	$\frac{1}{16}$
R-205	Fits Central, Kohler, Schaible, Tracy	$\frac{3}{8}$	$\frac{1}{4}$	$\frac{1}{16}$
R-206	Fits Central, Crane, Eljer, Kohler, Milwaukee, Sears, Sterling	$\frac{7}{16}$	$\frac{5}{16}$	$\frac{1}{16}$
R-207	Fits Crane, Eljer, Milwaukee, Speakman, Universal-Rundle, Elkay OEM 30453	$\frac{1}{2}$	$\frac{3}{8}$	$\frac{1}{16}$
R-208	Fits Am. Kit., Crane, Gerber, Harcraft, Indiana Brass, Milwaukee, Price-Pfister, Repcal, Sterling, Sayco, Union Brass	$\frac{9}{16}$	$\frac{3}{8}$	$\frac{3}{32}$
R-209	Fits Gyro, Kohler OEM 34263, Moen, Sears, Savoy	$\frac{5}{8}$	$\frac{7}{16}$	$\frac{3}{32}$
R-210	Fits Central, Gerber, Indiana Brass, Moen, Univ.-Rundle	$\frac{11}{16}$	$\frac{1}{2}$	$\frac{3}{32}$

R-201 R-202 R-203
R-204 R-205 R-206
R-207 R-208 R-209 R-210

"O" RINGS PACKING RINGS FOR SWING SPOUT, SLIP-JOINTS, PUMP AND HYDRAULIC APPLICATIONS (ACTUAL SIZE ILLUSTRATIONS)

Part No.	Use	O.D.	I.D.	Wall
R-211	Fits Am. Std., Barnes, Central, Crane, Dick, Eljer, Indiana Brass, Milwaukee, Repcal, Moen, Price-Pfister, Schaible, Sears, Sterling	$\frac{3}{4}$	$\frac{9}{16}$	$\frac{3}{32}$
R-212	Fits Am. Std., Moen, Speakman, Union Brass	$\frac{13}{16}$	$\frac{5}{8}$	$\frac{3}{32}$
R-212A	Fits various makes	$\frac{13}{16}$	$\frac{11}{16}$	$\frac{1}{16}$
R-213	Fits Am. Kit., Barnes, Briggs, Gyro, Indiana Brass, Kohler, Moen, Price-Pfister, Repcal, Schaible, Sears, Tracy, Union Brass, Youngstown	$\frac{7}{8}$	$\frac{11}{16}$	$\frac{3}{32}$
R-214	Fits Crane	$\frac{15}{16}$	$\frac{3}{4}$	$\frac{3}{32}$
R-215	Fits Crane	1	$\frac{3}{4}$	$\frac{1}{8}$
R-216	Fits Eljer, Gyro, Moen	$1\frac{1}{16}$	$\frac{13}{16}$	$\frac{1}{8}$
R-217	Fits Am. Std. OEM 507-37, Moen 137, Repcal	$1\frac{1}{8}$	$\frac{7}{8}$	$\frac{1}{8}$
R-218	Fits Speakman	$1\frac{3}{16}$	$\frac{15}{16}$	$\frac{1}{8}$
R-219	Fits various makes	$1\frac{1}{4}$	1	$\frac{1}{8}$
R-220	Fits various makes	$1\frac{5}{16}$	$1\frac{1}{16}$	$\frac{1}{8}$
R-221	Fits Moen Model 52	$1\frac{3}{8}$	$1\frac{1}{8}$	$\frac{1}{8}$
R-222	Fits various makes	$1\frac{7}{16}$	$1\frac{3}{16}$	$\frac{1}{8}$
R-223	Fits Sears, Moen Model 42E	$1\frac{1}{2}$	$1\frac{1}{4}$	$\frac{1}{8}$
R-224	For special applications	$1\frac{9}{16}$	$1\frac{5}{16}$	$\frac{1}{4}$
R-225	For special applications	$1\frac{5}{8}$	$1\frac{3}{8}$	$\frac{1}{8}$
R-226	For special applications	$1\frac{11}{16}$	$1\frac{7}{16}$	$\frac{1}{8}$
R-227	For special applications	$1\frac{3}{4}$	$1\frac{1}{2}$	$\frac{1}{8}$
R-228	Fits Crane Magic-Close	$\frac{9}{16}$	$\frac{5}{16}$	$\frac{1}{8}$
R-990	Fits Am. Std. OEM 246-37 Aqua-Seal, Delta	$\frac{5}{8}$	$\frac{1}{2}$	$\frac{1}{16}$
R-991	Fits Delta, Tracy	$1\frac{1}{2}$	$1\frac{5}{16}$	$\frac{3}{32}$
R-992	Fits Am. Std. OEM 55234, Am. Kit., Crane, Crosley, Eljer, Kohler, Repcal, Schaible, Sears, Tracy, Yngstwn., Kohler OEM 29464	$1\frac{1}{16}$	$\frac{7}{8}$	$\frac{3}{32}$
R-993	Fits Central, Gerber, Moen, Indiana Brass, Univ.-Rundle	$\frac{11}{16}$	$\frac{1}{2}$	$\frac{3}{32}$
R-994	Fits Kohler OEM 34264	$1\frac{1}{16}$	$\frac{15}{16}$	$\frac{1}{16}$
R-995	Fits Am. Std., Kohler OEM 34300	$\frac{7}{8}$	$\frac{3}{4}$	$\frac{1}{16}$
R-996	Fits Alamark	$1\frac{3}{16}$	1	$\frac{3}{32}$
R-997	Fits Eljer OEM 5290	$\frac{9}{16}$	$\frac{5}{16}$	$\frac{1}{8}$
R-998	Fits Moen Model 42E	$1\frac{7}{16}$	$1\frac{5}{16}$	$\frac{1}{16}$
R-999	Fits Am. Std. OEM 886-17	$\frac{5}{8}$	$\frac{1}{2}$	$\frac{1}{16}$
R-1004	Fits Crane OEM F12376	$\frac{45}{64}$	$\frac{27}{64}$	$\frac{9}{64}$
R-1005	Fits Delta	$1\frac{1}{4}$	$1\frac{1}{8}$	$\frac{1}{16}$
R-1006	Fits Am. Std. OEM 12035, Schaible, Sears, Youngstown	$1\frac{3}{16}$	$1\frac{1}{16}$	$\frac{1}{16}$
R-1008	Fits Harcraft	$\frac{11}{16}$	$\frac{9}{16}$	$\frac{1}{16}$
R-1011	Fits Barnes	$\frac{23}{32}$	$\frac{15}{32}$	$\frac{9}{64}$
R-1015	Fits Eljer OEM 5285 Lusterline, Burlington 68-9	$1\frac{1}{4}$	$1\frac{1}{16}$	$\frac{3}{32}$

BIBB SCREWS

BRASS (SMALL HEADS)

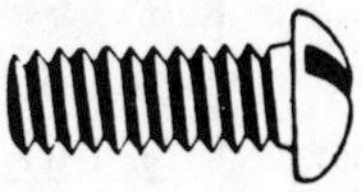

Part No.	Description
S-550	(¼ x 8-32)
S-552	(½ x 8-32)
S-553	(⅜ x 10-24)
S-554	(½ x 10-24)
S-555	(⅜ x 8-32 #7 Head)
S-556	(⅜ x 10-28)
S-557	(½ x 10-28)
S-558	(⅜ x 10-32)
S-559	(½ x 10-32)
S-560	(⅜ x 6-32)
S-561	(½ x 6-32)

MONEL (SELF-LOCKING)

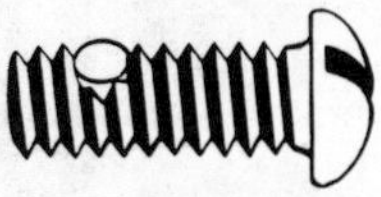

Will not corrode or crystalize. The head of MONEL screw is small so it won't touch seat of faucet. The nylon insert plug locks the screw in at proper depth. There is no chance for screw to come loose.

Part No.	Description
S-1256	(⅜x6-32) Gr.
S-1257	(½x6-32)
S-1258	(⅜x10-24)
S-1259	(½x10-24)
S-1260	(⅜x10-28)
S-1261	(½x10-28)
S-1262	(⅜x8-32)
S-1263	(½x8-32)
S-1264	(⅜x10-32)
S-1265	(½x10-32)

RENEW SEATS (ACTUAL SIZE ILLUSTRATIONS)

FS1-3C / FS1-3H, FS1-62HC, S-1091H, S-1091H2, S-1089, S-1091J, S-1090, S-1091P, S-1091, S-1091R, S-1091A, S-1091S, S-1091C, S-1091S1, FS1-8C / FS1-8H, FS1-8AC / FS1-8AH, S-1091G, S-1091U

Part No.	Manufacturer of Faucet	O.E.M. Part No.	Thd.	Specifications O.D. (Thd.)	Ht.
FS1-3C	Amer. Standard	20563-08	Barrel Seat Cold Side (L.H. Thread)w O-Rings		
FS1-3H	Amer. Standard	20336-08	Barrel Seat Hot Side (R.H. Thread)w O-Rings		
FS1-8H	Kohler	39717	Barrel Seat Hot Side (L.H. Thread)		
FS1-8C	Kohler	32462	Barrel Seat Cold Side (R.H. Thread)		
FS1-8AC	Kohler	32462	Barrel Seat Cold Side (R.H. Thread)		
FS1-8AH	Kohler	39717	Barrel Seat Hot Side (L.H. Thread)		
FS1-62HC	Kohler	32473	Barrel Seat (Not Threaded)		
S-1089	Amer. Standard	174-14	24 -	35/64 -	3/8
S-1090	Kohler	40602	27 -	1/2 -	25/64
S-1091	Kohler	33345	27 -	5/8 -	25/64
S-1091A	Dick Brothers	3055/4010	27 -	9/16 -	7/16
S-1091C	Central Brass	263-B	24 -	1/2 -	15/32
S-1091G	Briggs-Republic	22092	24 -	1/2 -	15/32
S-1091H	Briggs-Republic	8777	20 -	9/16 -	3/8
S-1091H2	Briggs-Republic	22040	20 -	1/2 -	3/4
S-1091J	Eljer	2734	27 -	1/2 -	1/2
S-1091P	Amer. Standard Amer. Kitchens Crosley Eljer Schaible Tracy Youngstown	56534-07 6534 6534 .... 6534 8662-8663 6534	20 -	1/2 -	3/8
S-1091R	Amer. Standard Schaible Youngstown	57281-07 5506 5506	20 -	1/2 -	7/16
S-1091S	Tracy	8435	20 -	5/8 -	7/16
S-1091S1	Amer. Standard Schaible Sears Youngstown	12002-07	20 -	35/64 -	5/16
S-1091U	Speakman	05-0305 Fmrly. S-5460	28 -	9/16 -	3/8

RENEW SEATS (ACTUAL SIZE ILLUSTRATIONS)

S-1092
S-2009
S-1093
S-2011
S-1094
S-2011A
S-2001
S-2021
S-2012
S-2001A
S-2022
S-2003
S-2015
S-2023
S-2004
S-2016
S-2024
S-2005
S-2017
S-2025
S-2006
S-2018
S-2026
S-2007
S-2019
S-2027
S-2008
S-2020
S-2028
S-2029
S-2030

Part No.	Manufacturer of Faucet	O.E.M. Part No.	Specifications Thd. - O.D. (Thd.) - Ht.		
			Thd.	O.D. (Thd.)	Ht.
S-1092	Crane	F-5914	22 - 43/64	-	27/64
S-1093	Burlington Br.	6-1	24 - 17/32	-	23/64
S-1094	Speakman	5-451	24 - 5/8	-	7/16
S-2001	Amer. Kitchen Sterling United Valve	802464 99S-0369	24 - 1/2	-	3/8
S-2001A	Empire Brass	20	24 - 1/2	-	11/32
S-2003	American Brass Amer. Standard	 862-14	24 - 7/16	-	13/32
S-2004	Barnes Eljer	7501 5257	20 - 1/2	-	11/32
S-2005	Barnes	7302	24 - 9/16	-	3/8
S-2006	Empire Brass	510	24 - 1/2	-	1¼
S-2007	Gerber	16	20 - 5/8	-	7/16
S-2008	Indiana Brass Milwaukee	638-0 964	27 - 1/2	-	3/8
S-2009	Milwaukee	3169	27 - 1/2	-	23/32
S-2011	Crane Savoy	 33	20 - 35/64	-	7/16
S-2011A	Savoy	524	20 - 1/2	-	13/32
S-2012	Price-Pfister Sayco	05716-01 6	20 - 1/2	-	3/8
S-2015	Univ.-Rundle	12R	24 - 17/32	-	3/8
S-2016	Wolverine	586S	24 - 1/2	-	3/8
S-2017	Amer. Standard	1848-07	18 - 17/32	-	3/8
S-2018	Briggs-Republic	91	20 - 5/8	-	27/32
S-2019	Gerber	98	20 - 5/8	-	1⅛
S-2020	Harcraft	6311	20 - 9/16	-	3/8
S-2021	Indiana	550-0	27 - 9/16	-	1⅝
S-2022	Kohler	23004	No Thd. - 25/32	-	21/64
S-2023	Harcraft	5377	20 - 3/4	-	3/8
S-2024	Repcal	1112-18A	No Thd. - 5/8	-	1/4
S-2025	Repcal	F14301 1112-18B	18 - 9/16	-	3/8
S-2026 (S-2010)	Speakman Richmond	05-0775 Fmrly. S-5465 K-12	27 - 7/16	-	11/32
S-2027 (S-2013)	Crane Union Brass	 2601	18 - 1/2	-	5/16
S-2028	Union Brass	2606	18 - 1/2	-	1 5/32
S-2029	Union Brass	2602	18 - 5/8	-	15/32
S-2030	Price-Pfister	05713-01	18 - 21/32	-	3/8

RENEW SEATS (ACTUAL SIZE ILLUSTRATIONS)

Part No.	Manufacturer of Faucet	O.E.M. Part No.	Thd.	O.D. (Thd.)	Ht.
S-2031	Eljer	2924	27	1/2	1 9/32
S-2032	Eljer	2750	27	35/64	15/16
S-2033	Eljer	4703	27	1/2	1 19/32
S-2034	Crane	F5913	22	9/16	13/32
S-2035	Crane	F5914	24	11/16	15/32
S-2036	Crane Speakman		24	11/16	11/32
S-2037	Crane Michigan Brass	F3364 13117	18	9/16	3/4
S-2038	Crane Eljer Price-Pfister Sayco	583 5257 5716-01 6A3	24	1/2	5/16
S-2039	Briggs-Republic	6762	24	9/16	7/16
S-2040	Briggs-Republic	22379	20	5/8	7/16
S-2041	Wolverine	587-S	24	19/32	25/64
S-2042	Kohler	22526	27	5/8	3/8
S-2043	Sayco	6-468-C-39	20	1/2	3/4
S-2044	Repcal	1112-18	28	9/16	11/32
S-2045	Union Brass	2632	20	39/64	5/16
S-2046	Central	59	27	9/16	13/32
S-2047	Central	263	24	1/2	47/64
S-2048	Central	165	24	5/8	23/32
S-2049	Central	165/6500	24	39/64	27/64
S-2051	Indiana Brass	590-0	27	9/16	5/8
S-2052	For use with Chicago, Economy, Sexauer or Skinner Tappg. Tools		27	3/8	3/8
S-2053	For use with Chicago, Economy, Sexauer or Skinner Tappg. Tools		27	3/8	19/64
S-2054	For use with Chicago, Economy, Sexauer or Skinner Tappg. Tools		27	7/16	3/8
S-2055	For use with Chicago, Economy, Sexauer or Skinner Tappg. Tools		27	7/16	19/64
S-2056	For use with Chicago, Economy, Sexauer or Skinner Tappg. Tools		27	1/2	19/64
S-2057	For use with Chicago, Economy, Sexauer or Skinner Tappg. Tools		27	1/2	1/4
S-2058	For use with Chicago, Economy, Sexauer or Skinner Tappg. Tools		27	1/2	3/8
S-2059	For use with Chicago, Economy, Sexauer or Skinner Tappg. Tools		27	1/2	19/64
S-2060	For use with Chicago, Economy, Sexauer or Skinner Tappg. Tools		27	9/16	3/8
S-2061	For use with Chicago, Economy, Sexauer or Skinner Tappg. Tools		27	9/16	19/64
S-2062	For use with Chicago, Economy, Sexauer or Skinner Tappg. Tools		27	5/8	3/8

S-2031, S-2032, S-2033, S-2034, S-2035, S-2036, S-2037, S-2038

S-2039, S-2040, S-2041, S-2042, S-2043, S-2044, S-2045, S-2046, S-2047, S-2048, S-2049

S-2051, S-2052, S-2053, S-2054, S-2055, S-2056, S-2057, S-2058, S-2059, S-2060, S-2061, S-2062

Pressure Fittings Solder-type for copper plumbing

Fig. No.	Nominal Size	Class	Pack Quantity	Pack Weight	Master Pack	Master Weight
600 COUPLING W/STOP — Copper x Copper						
	1/4	C	100	.9	—	—
	3/8	B	25	.4	—	—
	3/8 x 1/4	C	50	.7	—	—
	1/2	A	25	.7	2000	56.0
	1/2 x 1/4	C	25	.5		
	1/2 x 3/8	B	25	.7	1000	30.0
	3/4	A	25	1.7	1000	66.0
	3/4 x 5/8	C	50	3.3	—	—
	3/4 x 1/2	A	25	1.5	500	30.0
	3/4 x 3/8	C	50	2.7	—	—
	1	C	10	1.1	500	56.0
	1 x 3/4	C	25	3.2	250	33.0
	1 x 1/2	C	25	2.4	—	—
	1 1/4	C	25	4.1	250	41.0
600-2 FITTING REDUCER — Slip Fit x Copper						
	3/8 x 1/4	C	50	.7	—	—
	1/2 x 3/8	C	100	2.9	—	—
	3/4 x 1/2	B	25	1.4	500	28.0
	1 x 3/4	C	25	3.2	250	32.0
	1 x 1/2	C	25	2.1	—	—
601 REPAIR COUPLING LESS STOP — Copper x Copper						
	1/4	C	100	.8	—	—
	3/8	C	100	1.0	—	—
	1/2	B	100	2.8	—	—
	3/4	B	50	3.3	500	33.0
	1	C	25	2.7	—	—
701-D DRAIN COUPLING — Copper x Copper						
	3/8	C	10	.6	—	—
	1/2	C	10	.6	—	—
	3/4	C	10	1.4	—	—
701-2-D DRAIN COUPLING — Slip Fit x Copper						
	1/2	C	10	.9	—	—
	3/4	C	10	1.5	—	—
717-D DRAIN CAP — Copper						
	1/2	C	10	.6	—	—
	3/4	C	10	1.1	—	—
603 ADAPTER — Copper x Inside Thread						
	1/4	C	50	1.5	—	—
	3/8	C	25	1.1	—	—
	3/8 x 1/2	B	25	2.0	—	—
	1/2	A	25	2.1	1000	83.0
	1/2 x 3/4	B	25	3.3	500	66.0
	1/2 x 3/8	B	25	1.1	1000	44.0
	1/2 x 1/4	C	25	1.2	1000	48.0
	5/8 x 3/4	C	25	2.7	—	—
603 ADAPTER — Copper x Inside Thread (Continued)						
	3/4	B	25	3.3	500	67.0
	3/4 x 1	C	25	4.7	—	—
	3/4 x 1/2	C	25	2.5	—	—
	1	A	10	2.2	250	54.0
	1 x 3/4	—	25	4.7	—	—
603-2 ADAPTER — Slip Fit x Inside Thread						
	1/2	C	25	2.2	—	—
604 ADAPTER — Copper x Outside Thread						
	1/4	C	100	3.0	—	—
	3/8	B	25	1.0	—	—
	3/8 x 3/4	C	50	6.5	—	—
	3/8 x 1/2	B	25	2.2	1000	88.0
	1/2	A	25	1.9	1000	74.0
	1/2 x 1	C	25	5.7	—	—
	1/2 x 3/4	B	25	4.1	500	82.0
	1/2 x 3/8	B	25	1.2	1000	44.0
	3/4	B	25	3.7	500	75.0
	3/4 x 1	C	25	7.0	—	—
	3/4 x 1/2	C	50	5.2	—	—
	1	B	10	2.4	250	60.0
	1 x 3/4	C	25	4.2	—	—
	1 1/4	C	10	3.6	200	71.0
604-2 ADAPTER — Slip Fit x Outside Thread						
	1/2	C	25	2.2	—	—
705 TEE — BASEBOARD — Copper x Inside Thread x Copper						
	1/2 x 1/8 x 1/2	C	50	5.0	—	—
	3/4 x 1/8 x 3/4	C	25	4.6	250	46.5
606 ELL — 45° — Copper x Copper						
	1/4	C	50	.9	—	—
	3/8	C	25	.5	—	—
	1/2	A	25	1.0	1000	40.0
	3/4	A	25	2.3	500	46.0
	1	B	10	1.6	250	40.0
	1 1/4	C	10	2.5	200	50.0
	1 1/2	C	10	3.5	100	35.0
606-2 ELL — 45° — Slip Fit x Copper						
	3/8	C	100	2.5	—	—
	1/2	A	25	1.0	1000	44.0
	3/4	A	10	.9	250	24.0
	1	C	10	1.5	100	15.0

Pressure Fittings Solder-type for copper plumbing

Fig. No.	Nominal Size	Class	Pack Quantity	Pack Weight	Master Pack	Master Weight
607 ELL — 90° — Copper x Copper						
	¼	C	100	1.5	—	—
	⅜	Ⓐ	25	.7	2000	63.0
	½	Ⓐ	25	1.2	2000	84.0
	½ x ⅜	C	25	1.0	—	—
	¾	Ⓐ	25	2.6	500	51.0
	¾ x ½	Ⓐ	25	2.0	500	42.0
	1	B	10	2.9	200	42.0
	1 x ¾	B	25	3.4	250	35.0
	1¼	C	10	3.2	200	64.0
	1½	C	5	2.2	100	44.0
607-LT ELL — 90° LONG RADIUS — Copper x Copper						
	½	B	50	4.0	500	40.0
	½ x ⅜	C	50	2.8	—	—
	¾	C	25	4.1	250	41.0
	1	C	10	2.8	100	28.0
607-2 ELL — 90° — Slip Fit x Copper						
	⅜	C	25	.8	—	—
	½	Ⓐ	25	1.1	1000	43.0
	¾	Ⓐ	25	2.6	500	54.0
	1	C	25	5.0	250	50.0
607-3 ELL — 90° — Copper x Female						
	½	C	25	3.0	—	—
707-3 ELL — 90° — Copper x Inside Thread						
	½	Ⓑ	25	4.0	500	80.0
	½ x ¾	B	25	5.0	500	100.0
	½ x ⅜	B	25	3.3	1000	131.0
	¾	A	10	2.6	250	64.0
	¾ x ½	C	25	4.7	500	94.0
707-3-5 ELL — DROP EAR — 90° — Copper x Inside Thread						
	½	Ⓐ	25	5.2	250	52.0
	½ x ¾	C	25	9.1	250	91.0
	½ x ⅜	C	25	4.0	500	80.0
	¾	C	25	7.7	250	78.0
707-4 ELL — 90° — Copper x Outside Thread						
	½	B	25	2.8	500	56.0
	½ x ¾	C	25	2.0	250	40.0
	¾	C	10	2.1	—	—
707-4-5 ELL — DROP EAR — 90° — Copper x Outside Thread						
	½	C	25	5.2	—	—

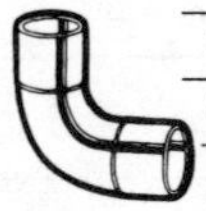

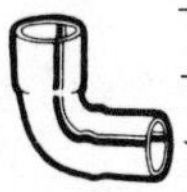

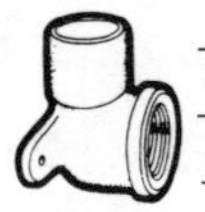

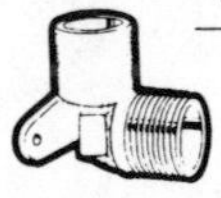

Fig. No.	Nominal Size	Class	Pack Quantity	Pack Weight	Master Pack	Master Weight
707-5 ELL — DROP EAR — 90° — Copper x Copper						
	½	B	50	5.8	500	58.0
	¾	C	10	2.4	—	—
708 90° SINK FITTINGS						
	½	C	25	5.0	—	—
611 TEE — Copper x Copper x Copper						
	¼	C	50	1.5	50	1.5
	⅜	Ⓑ	25	1.0	1000	42.0
	⅜ x ⅜ x ½	C	50	3.1	50	3.1
	⅜ x ⅜ x ¼	C	50	2.0	50	2.0
	½	Ⓐ	25	1.7	1000	68.0
	½ x ½ x ¾	B	25	4.1	250	41.0
	½ x ½ x ⅜	C	25	1.6	500	32.0
	½ x ⅜ x ½	C	50	3.1	—	—
	¾	Ⓐ	25	4.2	500	92.0
	¾ x ¾ x 1	B	10	2.8	100	28.0
	¾ x ¾ x ½	Ⓑ	10	1.4	10	1.4
	¾ x ½ x ¾	Ⓑ	10	1.6	10	1.6
	¾ x ½ x ½	Ⓑ	10	1.3	10	1.3
	1	B	10	3.2	200	64.0
	1 x 1 x ¾	C	10	2.7	200	56.0
	1 x ¾ x 1	C	10	3.2	100	32.0
	1 x ¾ x ¾	C	10	2.7	200	56.0
	1 x ¾ x ½	C	10	2.1	100	21.0
	1 x ½ x ¾	C	10	2.6	10	2.6
611-2 TEE — FITTING CONNECTION — Copper x Slip Fit x Copper						
	½	C	50	3.1	50	3.1
	¾	C	25	4.2	25	4.2
712 TEE — Copper x Copper x Inside Thread						
	⅜	C	50	5.0	—	—
	½	Ⓐ	25	4.3	500	86.0
	½ x ½ x ¾	C	25	3.7	—	—
	½ x ½ x ⅜	C	25	3.5	250	35.0
	¾	C	10	3.1	250	78.0
	¾ x ¾ x ½	C	25	6.5	250	65.0
712-5 TEE — DROP EAR — Copper x Copper x Inside Thread						
	½	C	25	5.6	250	56.0
	¾	C	10	3.7	100	37.0
	¾ x ¾ x ½	C	10	2.8	—	—
714 TEE — Copper x Inside Thread x Copper						
	½	C	25	4.0	500	80.0

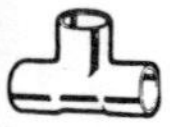

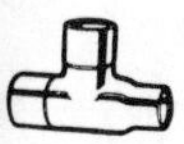

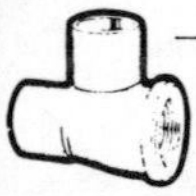

Pressure Fittings Solder-type for copper plumbing

Fig. No.	Nominal Size	Class	Pack Quantity	Pack Weight	Master Pack	Master Weight
616 FITTING PLUG						
	⅜	C	50	.5	—	—
	½	B	25	.5	—	—
	¾	C	50	2.0	—	—
617 CAP-TUBE — Fits Over End of Tube						
	⅜	C	50	.6	—	—
	½ ·	Ⓐ	25	.5	4000	74.0
	¾	Ⓐ	25	1.0	1000	40.0
	1	C	50	3.4	500	34.0
618 BUSHING — Slip Fit x Copper						
	⅜ x ¼	C	50	.6	—	—
	½ x ⅜	Ⓑ	25	.5	—	—
	¾ x ½	Ⓑ	25	1.7	—	—
	1 x ¾	C	25	3.0	—	—
619 AIR CHAMBER — Fits ½" Solder Cup						
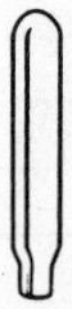	½ x 12	C	1	.4	72	29.0
	½ x 14*	C	1	.7	50	35.0

*For Wisconsin Code

Fig. No.	Nominal Size	Class	Pack Quantity	Pack Weight	Master Pack	Master Weight
624 TUBE STRAP						
	⅜	Ⓑ	100	1.2	—	—
	½	Ⓐ	100	1.5	2000	30.0
	¾	Ⓑ	100	2.6	2000	52.0
724-5-A HY-SET HANGER COUPLING — Copper						
	½	C	100	6.0	—	—
	¾	C	100	10.0	—	—
733 UNION — Copper x Copper						
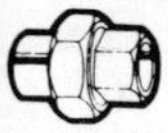	⅜	C	50	6.7	—	—
	½	Ⓐ	10	2.3	250	57.0
	¾	Ⓑ	10	3.4	250	85.0
	1	C	10	5.6	100	56.0
736 CROSS-OVER — Copper x Copper						
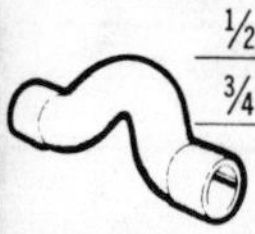	½	C	5	1.3	100	27.0
	¾	C	5	2.8	—	—
638 RETURN BEND — Copper x Copper						
	⅜	C	50	2.0	—	—
	½	C	25	3.2	—	—
	¾	C	10	2.6	—	—

Fig. No.	Size	Wall Thickness	Box Quantity	Box Weight	SWP* at 250° F
60 COPPER TUBING — Readi Cut™ Lengths					
	½ x 4'	0.018	25	13.3	245 PSI
	½ x 6'	0.018	25	19.9	245 PSI
	½ x 10'	0.018	25	33.2	245 PSI
	¾ x 4'	0.021	15	13.0	200 PSI
	¾ x 6'	0.021	15	19.6	200 PSI
	¾ x 10'	0.021	15	32.7	200 PSI

*Safe Working Pressure — ASME Rating.

Rigid Lengths — for general internal plumbing use — will withstand bursting pressure of 3700 PSI.

Flared Fittings for copper piping

Fig. No.	Nominal Size	Pack Quantity	Pack Weight	Master Pack	Master Weight
500 TUBE NUT					
	3/8	1	.1	—	—
	1/2	1	.1	—	—
	3/4	1	.2	—	—
	1	1	.3	—	—
501 COUPLING — Copper x Copper					
	3/8	25	8.7	—	—
	1/2	10	4.3	—	—
	3/4	10	6.8	—	—
	1	5	6.5	—	—
502 COUPLING — TWO PART W/RING — Copper x Copper					
	3/8	25	5.7	—	—
	1/2	10	3.9	—	—
	3/4	10	5.2	—	—
503 ADAPTER — Copper x Female					
	3/8	25	5.2	—	—
	1/2	25	9.0	—	—
	1/2 x 3/4	10	3.9	—	—
	3/4	10	5.0	—	—
	3/4 x 1/2	10	4.5	—	—
	1	5	5.1	—	—
504 ADAPTER — Copper x Male					
	3/8	25	6.5	—	—
	1/2	25	7.2	—	—
	1/2 x 3/8	10	2.8	—	—
	1/2 x 3/4	10	4.0	—	—
	3/4	10	5.7	—	—
	1	5	4.4	—	—
507 ELL — 90° — Copper x Copper					
	3/8	25	10.0	—	—
	1/2	10	5.5	—	→
	3/4	10	7.2	—	—
	1	5	8.1	—	—

Fig. No.	Nominal Size	Pack Quantity	Pack Weight	Master Pack	Master Weight
507-3 ELL — 90° — Copper x Female					
	3/8	25	9.0	—	—
	1/2	10	4.9	—	—
	3/4	10	7.0	—	—
507-4 ELL — 90° — Copper x Male					
	3/8	25	6.0	—	—
	1/2	10	3.4	—	—
	3/4	5	2.7	—	—
511 TEE — Copper x Copper x Copper					
	3/8	10	5.9	—	—
	1/2	10	6.9	—	—
	3/4	5	5.2	—	—
	1	1	2.1	—	—
512 TEE — Copper x Copper x Female					
	1/2	5	3.4	—	—
	3/4	5	5.0	—	—
516 CAP					
	3/8	50	8.0	—	—
	1/2	50	6.5	—	—
	3/4	1	.2	—	—
523 FITTING REDUCER — Nut Seat x Copper					
	3/8 x 1/4	1	.2	—	—
	1/2 x 3/8	1	.3	—	—
	3/4 x 1/2	10	4.7	—	—
	3/4 x 3/8	25	12.0	—	—

Flared Valves for copper piping

Fig. No.	Nominal Size	Pack Quantity	Pack Weight	Master Pack	Master Weight
577-17 FLARED ANGLE STOP VALVE — Copper x Female					
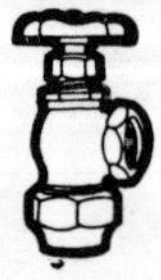	¾	5	4.6	—	—
	¾ x ½	5	4.9	—	—

Fig. No.	Nominal Size	Pack Quantity	Pack Weight	Master Pack	Master Weight
578-17 FLARED ANGLE STOP & WASTE VALVE — Copper x Female					
	¾	5	4.8	—	—
	¾ x ½	5	4.9	—	—

DWV Fittings Solder-type for copper drainage systems

Fig. No.	Nominal Size	Pack Quantity	Pack Weight	Master Pack	Master Weight
901 COUPLING — Copper x Copper					
	1¼	25	1.7	250	17.0
	1½	25	2.2	250	22.0
	1½ x 1¼	25	3.5	250	35.0
	2	10	1.4	100	14.0
	2 x 1½	10	2.0	100	20.0
	2 x 1¼	10	1.9	100	19.0
	3	5	1.5	50	15.0
	3 x 1½	5	1.5	50	15.0
	3 x 1¼	5	1.6	—	—

Fig. No.	Nominal Size	Pack Quantity	Pack Weight	Master Pack	Master Weight
901-RP COUPLING — REPAIR — Copper x Copper					
	1¼	25	2.0	250	20.0
	1½	25	2.7	250	27.8
	2	15	1.7	100	17.9
	3	5	1.7	50	17.7
901-2 BUSHING — EXTENDED — Fitting x Copper					
	1½ x 1¼	25	3.1	250	42.0
	3 x 1½	5	1.6	50	20.0

DWV Fittings — Solder-type for copper drainage systems

Fig. No.	Nominal Size	Pack Quantity	Pack Weight	Master Pack	Master Weight
801-2-T ADAPTER — TRAP — Fitting x O.D. Tube					
	1½ x 1½ OD	10	2.3	—	—
	1½ x 1¼ OD	10	2.6	—	—
901-7 ADAPTER — SLIP JOINT — TRAP — Copper x Slip Joint					
	1¼	1	.2	1	.2
	1½	10	2.7	100	27.0
	1½ x 1¼	10	2.6	100	26.0
802 ADAPTER — TRAP — Copper x Outside Thread					
	1¼ x 1½	10	4.1	100	41.0
	1½	25	6.0	250	60.0
903 ADAPTER — Copper x Inside Thread					
	1¼	25	4.6	250	46.0
	1½	10	2.5	100	25.0
	2	5	1.8	100	36.0
	3*	1	1.0	25	25.0
*Manufactured in Cast Bronze only. Specify Fig. No. 803					
903-2 ADAPTER — Fitting x Inside Thread					
	1¼	10	2.0	100	20.0
	1½	10	2.5	100	25.0
	2	5	1.9	100	38.0
	3*	1	1.1	25	28.0
*Manufactured in Cast Bronze only. Specify Fig. No. 803-2					
904 ADAPTER — Copper x Outside Thread					
	1¼	25	6.4	250	64.0
	1½	10	3.3	200	66.0
	2	5	2.4	100	48.0
	3*	1	1.5	25	38.0
*Manufactured in Cast Bronze only. Specify Fig. No. 804					
905 ADAPTER — SOIL PIPE — Copper x Spigot					
	1½ x 2	5	3.0	50	30.0
	2 x 2	5	2.4	50	24.0
	2 x 3*	2	2.4	20	24.0
	3 x 3*	2	1.9	30	29.0
	3 x 4*	1	1.9	15	29.0
*Manufactured in Cast Bronze only. Specify Fig. No. 805					
906 ELBOW — 45° — Copper x Copper					
	1¼	10	1.3	200	26.0
	1½	10	2.1	200	42.0
	2	5	2.0	50	20.0
	3	2	1.5	30	23.0
906-2 ELBOW — 45° — Fitting x Copper					
	1¼	10	1.4	100	14.0
	1½	10	2.1	100	21.0
	2	5	2.0	50	20.0
	3	2	1.5	30	23.0

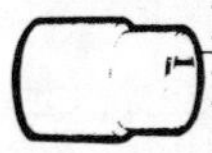

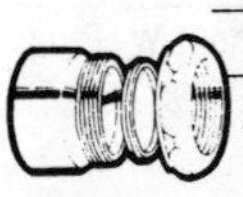

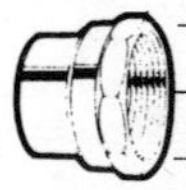

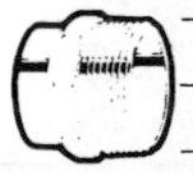

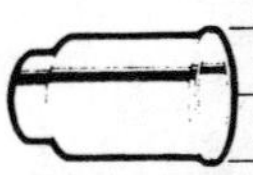

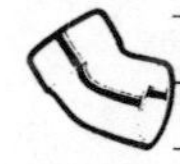

Fig. No.	Nominal Size	Pack Quantity	Pack Weight	Master Pack	Master Weight
907 ELBOW — 90° — Copper x Copper					
	1¼	10	1.9	200	38.0
	1½	10	3.2	100	32.0
	2	5	2.9	50	29.0
	3	2	2.5	20	25.0
907-2 ELBOW — 90° — Fitting x Copper					
	1¼	10	1.9	100	19.0
	1½	10	3.2	100	32.0
	2	5	2.9	50	29.0
	3	2	2.6	20	26.0
807-7 ELBOW — 90° — Copper x Slip Joint					
	1½	1	.7	50	35.0
	1½ x 1¼	1	.6	—	—
908 ELBOW — 22½° — Copper x Copper					
	1¼	25	2.7	—	—
	1½	10	1.6	100	16.0
	2	5	1.4	—	—
	3	2	1.2	—	—
810-Y — 45° — Copper x Copper x Copper					
	1¼	1	.6	50	30.0
	1½	1	.8	50	40.0
	2	1	1.2	50	60.0
	2 x 2 x 1½	1	1.2	50	60.0
	2 x 1½ x 1½	1	.8	50	40.0
	3	1	2.7	20	54.0
	3 x 3 x 1½	1	1.6	25	40.0
	3 x 3 x 2	1	1.6	25	40.0
	3 x 2 x 2	1	1.6	—	—
911 TEE — Copper x Copper x Copper					
	1¼	10	3.2	100	32.0
	1½	10	4.2	100	42.0
	2	5	2.1	60	25.0
	2 x 2 x 1½	5	2.8	60	34.0
	2 x 1½ x 2	5	3.6	60	43.0
	2 x 1½ x 1½	5	2.9	60	35.0
	3*	2	3.6	20	36.0
	3 x 3 x 2*	2	2.5	30	38.0
	3 x 3 x 1½*	2	2.1	30	32.0
*Manufactured in Cast Bronze only. Specify Fig. No. 811.					
816-S CLEANOUT — Fitting x Cleanout W/Plug					
	1¼	25	5.0	—	—
	1½ x 1	25	7.7	250	77.0
	2 x 1½	10	4.8	100	48.0
	3 x 2½	5	5.3	50	53.0

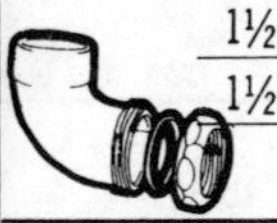

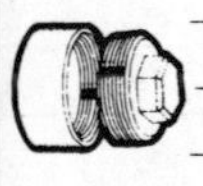

DWV Fittings Solder-type for copper drainage systems

Fig. No. / Nominal Size	Pack Quantity	Pack Weight	Master Pack	Master Weight
818 A.S.A. PLUG — Threaded				
1½	25	5.5	—	—
3	10	8.2	—	—
951 CLOSET FLANGE — Copper				
3	5	4.2	50	42.0
4	5	3.2	—	—
851-C CLOSET FLANGE				
4 x 3	1	1.2	—	—
960 ELBOW — 60° — Copper x Copper				
1¼	10	1.5	100	15.0
1½	10	2.5	100	25.0
2	5	2.1	50	21.0
3	2	1.9	30	29.0
876 RETURN BEND W/CLEANOUT — Copper x Slip Joint				
1½	1	1.2	50	57.0
878 RETURN BEND W/CLEANOUT — Copper x Copper x Cleanout				
1½	—	—	50	47.0
2	—	—	25	32.0

Fig. No. / Nominal Size	Pack Quantity	Pack Weight	Master Pack	Master Weight
880 P-TRAP W/CLEANOUT — Copper x Slip Joint x Cleanout W/Plug				
1¼	1	1.4	—	—
1½	—	—	50	92.0
2	—	—	25	61.0
884 P-TRAP W/CLEANOUT — Copper x Copper x Cleanout W/Plug				
1¼	—	—	50	61.5
1½	—	—	50	72.5
2	—	—	25	54.0
3	—	—	10	57.4
891 DRUM TRAP 3″ x 6″ SWIVEL DRUM — Copper x Copper				
1½	—	—	25	69.0
2				
892 P-TRAP W/UNION JOINT — Copper x Slip Joint				
1½	—	—	25	45.5
892-3 P-TRAP W/UNION JOINT — Female x Slip Joint				
1½	—	—	25	51.2

Accessories

Fig. No. / Nominal Size	Pack Quantity	Pack Weight	Master Pack	Master Weight
518 FLARING TOOL				
3/8	6	1.2	—	—
1/2	6	1.8	—	—
3/4	6	3.0	—	—
1	6	5.4	—	—
755 FITTING CLEANING BRUSH				
3/8	12	1.2	—	—
1/2	12	1.2	—	—
5/8	12	1.2	—	—
3/4	12	1.2	—	—
757 TUBE END CLEANING BRUSH				
3/8	6	.6	—	—
1/2	6	.6	—	—
3/4	6	.6	—	—

Fig. No. / Nominal Size	Pack Quantity	Pack Weight	Master Pack	Master Weight
765 SAND CLOTH — 1½″ Wide				
10-Yard Roll	1	.5	—	—
25-Yard Roll	1	1.3	—	—
50-Yard Roll	1	2.5	—	—
766 NOKORODE — Paste Flux				
1.7 oz. can	12	1.5	144	16.0
1 lb. can	1	1.0	12	12.0

Valves Plumbing & Heating (for use with copper and iron pipe)

Fig. No.	Nominal Size	Class	Pack Quantity	Pack Weight	Master Pack	Master Weight

Gate Valves

T-22 U-VALVE — Female Thread x Female Thread
S-22   Copper x Copper
"Full flow like a Gate — Throttles like a Globe"

	Nominal Size	Class	Pack Quantity	Pack Weight	Master Pack	Master Weight
	½	A	5	3.0	100	50.0
	¾	A	5	4.3	100	80.0
	1	A	1	1.8	40	68.0
	1¼	B	1	2.7	25	65.0
	1½	C	1	3.9	20	74.0
	2	C	1	6.9	10	65.0

T-22K U-VALVE W/CROSS HANDLE — Female Thread x Female Thread
"Full flow like a Gate — Throttles like a Globe"
For use with sprinkling systems

	Nominal Size	Class	Pack Quantity	Pack Weight	Master Pack	Master Weight
	½	B	5	3.2	—	—
	¾	B	5	4.4	—	—
	1	C	1	1.9	—	—

T-29 FLAT TOP GATE VALVE — Female Thread x Female Thread
S-29   Copper x Copper
Low profile, full flow gate valve

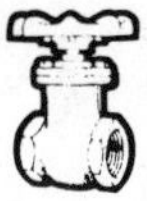

	Nominal Size	Class	Pack Quantity	Pack Weight	Master Pack	Master Weight
	½	Ⓐ	5	2.5	100	55.0
	¾	Ⓐ	5	3.5	100	83.0
	1	B	1	1.3	40	58.0
	1¼	B	1	1.6	25	45.0
	1½	C	1	2.1	20	47.0
	2	C	1	3.0	10	33.0

T-180 RING GATE VALVE — Female Thread x Female Thread
S-180   Copper x Copper
Teflon seal for easier operation

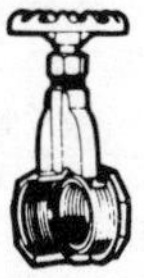

	Nominal Size	Class	Pack Quantity	Pack Weight	Master Pack	Master Weight
	⅜	B	5	2.0	100	40.0
	½	A	5	2.5	100	50.0
	¾	A	5	3.0	100	60.0
	1	A	1	1.1	50	55.0
	1¼	B	1	1.5	25	37.5
	1½	C	1	2.2	20	44.0
	2	C	1	3.3	20	66.0

Ball Valves

T-580 RING BALL VALVE — Female Thread x Female Thread
S-580   Copper x Copper

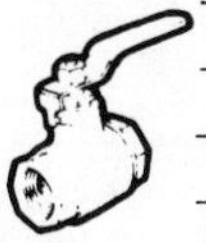

	Nominal Size	Class	Pack Quantity	Pack Weight	Master Pack	Master Weight
	½	B	2	1.1	100	55.0
	¾	B	2	1.4	100	70.0
	1	B	2	2.6	50	65.0
	1¼	C	1	1.9	25	47.5
	1½	C	1	2.9	10	30.0
	2	C	1	4.5	10	45.0

Check Valves

T-480 IN-LINE CHECK VALVE — Female Thread x Female Thread
S-480   Copper x Copper
Can be used in horizontal or vertical position.

	Nominal Size	Class	Pack Quantity	Pack Weight	Master Pack	Master Weight
	½	A	5	1.5	50	15.0
	¾	B	5	2.5	50	25.0
	1	C	5	4.0	40	32.0
	1¼	C	1	1.2	25	30.0

Hex Shoulder Hose Bibbs

56 HOSE BIBB — S.S.S. — Male Thread x Hose

	Nominal Size	Class	Pack Quantity	Pack Weight	Master Pack	Master Weight
	½	A	12	7.5	72	45.0
	¾	B	12	8.3	72	50.0

57 HOSE BIBB — S.O.T. — Male Thread x Hose

	Nominal Size	Class	Pack Quantity	Pack Weight	Master Pack	Master Weight
	½	C	12	9.0	72	54.0
	¾	C	12	10.3	72	62.0

Angle Lawn Faucets

63 ANGLE SILL FAUCET W/FLANGE — Female Thread x Hose
Quality Angle Sill Faucet

	Nominal Size	Class	Pack Quantity	Pack Weight	Master Pack	Master Weight
	½	Ⓐ	10	5.0	100	50.0
	¾	B	10	6.0	100	60.0

763 ANGLE SILL FAUCET W/FLANGE — Copper x Hose
Quality Angle Sill Faucet

	Nominal Size	Class	Pack Quantity	Pack Weight	Master Pack	Master Weight
	½	B	10	5.8	100	58.0
	¾	C	10	6.2	100	62.0

T-53 ANGLE SILL FAUCET — Female Thread x Hose
S-53   Copper x Hose
Economy Angle Sill Faucet

	Nominal Size	Class	Pack Quantity	Pack Weight	Master Pack	Master Weight
	½	Ⓐ	10	4.0	100	40.0
	¾ (T-53 only)	B	10	4.0	100	40.0

Garden Hose Valve

61 GARDEN HOSE VALVE — Female Thread x Hose

	Nominal Size	Class	Pack Quantity	Pack Weight	Master Pack	Master Weight
	½	B	10	7	100	70.0
	¾	Ⓐ	10	6	100	60.0

Valves Plumbing & Heating (for use with copper and iron pipe)

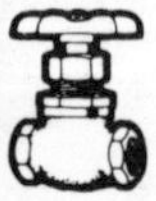

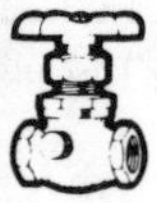

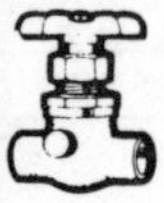

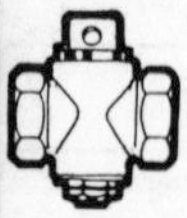

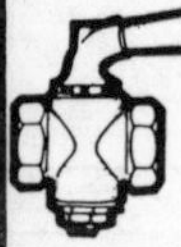

Fig. No.	Nominal Size	Class	Pack Quantity	Pack Weight	Master Pack	Master Weight

Frost Proof Lawn Faucets

52 FROST PROOF LAWN FAUCET — Threaded for Garden Hose
½ (½ Copper x ½ Male Thread)
¾ (¾ Male Thread x ½ Female Thread)

8" length		C	1	1.2	20	31.0
10" length		Ⓐ	1	1.3	20	32.0
12" length		A	1	1.4	20	36.0

Drain Valves
(Also for use with washing machine hook-ups.)

72 DRAIN VALVE — Copper x Hose

	½	B	10	5.0	100	53.0
	¾	B	10	5.0	100	54.0

73 DRAIN VALVE — Female Thread x Hose

	½	B	10	5.0	100	52.0
	¾	B	10	5.0	100	57.0

74 DRAIN VALVE — Male Thread x Hose

	½*	Ⓐ	10	5.0	100	51.0
	¾	B	10	5.0	100	56.0

*These drains have ½" Male IPS threads and are also machined for ½" nominal copper tubing.

4464 DRAIN VALVE — Compression x Hose

	½	C	10	7.0	100	70.0

For use with Copper or CPVC tube.
With CPVC, use 4732-2 transition fitting.

Non Kink Hose Faucets

54 NON KINK HOSE FAUCET — (Boiler or Drain Valve)
Male Thread x Hose

	½*	Ⓐ	10	3.1	100	30.0
	¾	B	10	3.5	100	31.0

*These drains have ½" Male IPS threads and are also machined for ½" nominal copper tubing.

55 NON KINK HOSE FAUCET (Boiler or Drain Valve)
Female Thread x Hose

	½	B	10	3.6	100	36.0
	¾	B	10	3.5	100	36.0

Stop Valves (Globe Pattern)

75 STOP VALVE — Female Thread x Female Thread

	⅜	B	10	4.0	100	39.0
	½	Ⓐ	10	5.0	100	56.0
	¾	Ⓑ	10	6.0	100	66.0
	1	C	5	8.5	50	88.0

725 STOP VALVE — Copper x Copper

	⅜	B	10	3.0	100	33.0
	½	Ⓐ	10	4.0	100	47.0
	¾	Ⓐ	10	5.0	100	55.0
	1	B	5	8.0	50	83.0

Stop & Waste Valves (Globe Pattern)

76 STOP & WASTE VALVE — Female Thread x Female Thread

	⅜	B	10	4.0	100	40.0
	½	Ⓐ	10	6.0	100	58.0
	¾	Ⓐ	10	7.0	100	69.0
	1	C	5	8.5	50	89.0

726 STOP & WASTE VALVE — Copper x Copper

	⅜	B	10	3.0	100	34.0
	½	Ⓐ	10	5.0	100	49.0
	¾	Ⓐ	10	5.0	100	58.0
	1	B	5	8.0	50	84.0

4476 STOP & WASTE VALVE — Compression x Compression

	½	Ⓑ	10	7.0	100	70.0
	¾	Ⓑ	10	8.0	100	80.0

For use with Copper or CPVC tube.
With CPVC, use 4732-2 transition fitting.

Gas Cocks

35 GAS COCK — FLAT HEAD — Female Thread x Female Thread

	⅜	B	12	4.4	288	110
	½	A	12	4.6	288	115
	¾	B	6	4.0	168	117
	1	C	4	3.1	144	115

36 GAS COCK — LEVER HANDLE — Female Thread x Female Thread

	⅜	B	12	4.6	288	115
	½	A	12	5.4	288	132
	¾	B	6	4.2	168	123
	1	C	4	3.1	144	115

Valves Plumbing & Heating (for use with copper and iron pipe)

Western States Valves

Fig. No.	Nominal Size	Pack Quantity	Pack Weight	Master Pack	Master Weight
H9054 WASHING MACHINE VALVE FILLER — By-Pass — Satin ½" Female Thread x ½" Male Thread w/¾" Hose Outlet on bottom					
	½	12	7.8	72	47.0
H9055 WASHING MACHINE VALVE FILLER — Reversible By-Pass — Satin ½" Female Thread x ½" Male Thread w/¾" Hose Outlet on side (Reversible either left or right.)					
	½	12	8.8	72	53.0

Fig. No.	Nominal Size	Pack Quantity	Pack Weight	Master Pack	Master Weight
H9058 WASHING MACHINE VALVE FILLER — Satin plated. Connect Automatic Washing Machine Hose to ½" Female Thread					
	½	24	9.5	144	57.0
H9059 WASHING MACHINE VALVE FILLER — Satin ½" Female Thread water inlet w/¾" hose outlet on bottom					
	½	12	6.0	72	36.0
H9073 EVAPORATIVE COOLER FAUCET — ¾" Female Hose Thread inlet x ¾" Male Hose outlet. Tapped ⅛" Female Thread thru side wall					
	¾	12	6.6	72	40.0

CPVC Hot & cold plastic fittings & pipe

Fig. No.	Nominal Size	Class	Pack Quantity	Pack Weight	Master Pack	Master Weight
4701 COUPLING — Plastic x Plastic						
	½	A	100	.3	1000	14.0
	¾	A	50	.8	500	20.0
	¾ x ½	B	10	.2	250	8.0
4704 ADAPTER — Plastic x Outside Thread						
	½	B	20	.2	1000	13.0
	¾	B	10	.4	250	11.0
4706 ELBOW — 45° — Plastic x Plastic						
	½	A	20	.4	500	10.0
	¾	B	20	.4	250	12.0
4707 ELBOW — 90° — Plastic x Plastic						
	½	A	100	.6	1000	24.0
	¾	B	20	1.1	500	29.0
	¾ x ½	B	10	.4	250	11.0
4707-2 ELBOW — 90° — Fitting x Plastic						
	½	A	10	.2	500	10.0
	¾	B	10	.5	250	13.0
4711 TEE — Plastic x Plastic x Plastic						
	½	A	100	.7	1000	31.0
	¾	A	20	.7	500	38.0
	¾ x ¾ x ½	B	25	1.9	250	19.0
	¾ x ½ x ¾	B	25	1.9	250	19.0
	¾ x ½ x ½	B	25	1.9	250	19.0

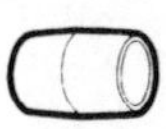

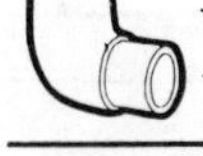

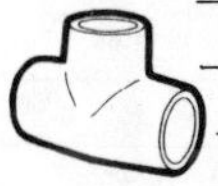

Fig. No.	Nominal Size	Class	Pack Quantity	Pack Weight	Master Pack	Master Weight
4717 CAP — Plastic						
	½	A	10	.1	1000	12.0
	¾	B	10	.2	500	14.0
4718 BUSHING — Fitting x Plastic						
	¾ x ½	B	10	.1	250	4.0
4724 STRAP — "SNAP-ON" — (Polypropylene)						
	½	A	100	.2	2000	7.0
	¾	A	100	.4	2000	8.0
4401 UNION COUPLING — Compression x Compression						
	½	A	10	2.2	—	—
	½ x ¼	C	10	1.6	—	—
	¾	C	5	2.5	—	—
4403 ADAPTER — Compression x Inside Thread						
	½	B	10	1.5	—	—
4404 ADAPTER — Compression x Outside Thread						
	½	B	10	1.9	—	—
	¾	C	5	1.4	—	—
4407-3-5 ELL — DROP EAR — Compression x Inside Thread						
	½	C	10	4.5	—	—

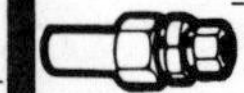

CPVC Hot & cold plastic fittings & pipe

Fig. No.	Nominal Size	Class	Pack Quantity	Pack Weight	Master Pack	Master Weight
4732-2 TRANSITION — Fitting x Compression						
	½	A	25	.5	250	6.0
	¾	A	10	.3	100	5.0
4733-4 UNION — Plastic x Outside Thread						
	½	A	25	3.5	—	—
	½ x ¾	Under Development				
	¾	Under Development				

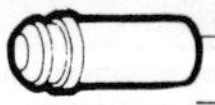

Permits Full Flow connection of tube to inside thread component. Must be used on hot water lines.

Fig. No.	Nominal Size	Class	Pack Quantity	Pack Weight	Master Pack	Master Weight
4464 WASHING MACHINE VALVE — Compression x Hose						
	½	C	10	7.0	100	70.0

For use with copper or CPVC tube.
With CPVC, use 4732-2 transition fitting.

Fig. No.	Nominal Size	Class	Pack Quantity	Pack Weight	Master Pack	Master Weight
4476 STOP & WASTE VALVE — Compression x Compression						
	½	B	10	7.0	100	70.0
	¾	B	10	8.0	100	80.0

For use with copper or CPVC tube.
With CPVC, use 4732-2 transition fitting.

Fig. No.	Nominal Size	Class	Pack Quantity	Pack Weight	Master Pack	Master Weight
4776 STOP & WASTE VALVE — CPVC to CPVC						
	½	B	10	6.8	100	68.0
	¾	B	Available March 31, 1975			

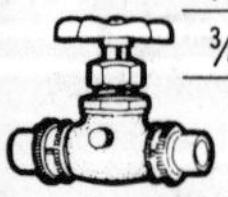

NEW!

Fig. No.	Nominal Size	Class	Pack Quantity	Pack Weight	Master Pack	Master Weight
47 CPVC PIPE						
	½ 10 ft.	A	50	44	1000 ft.	93.0
	¾ 10 ft.	B	50	73	500 ft.	84.0
4798 SOLVENT — CPVC With Applicator						
	¼ Pint	A	24	10.0	24	10.0
	½ Pint	A	24	17.0	24	17.0
	1 Pint	B	12	14.0	12	14.0
4899 PRIMER — PVC & CPVC With Applicator						
	½ Pint	A	24	17.0	24	17.0
	1 Pint	B	12	14.0	12	14.0

DWV Fittings for ABS and PVC drainage systems

Fig. No.	Nominal Size	Class	Master Pack	Master Pack Wt. (Lbs.)
4800 & 5800 ADAPTER — SOIL PIPE — Plastic x Hub				
	2	C	10	2.2
	3	B	10	5.5
	4	C	10	10.0
5800 CLAY — ADAPTER — SOIL PIPE — Plastic x Clay Pipe Hub				
	4	B	10	10.2
4800-SD & 5800-SD ADAPTER — Plastic x ASTMD 2852 Plastic				
	3 x 4	A	10	4.0

Fig. No.	Nominal Size	Class	Master Pack	Master Pack Wt. (Lbs.)
4801 & 5801 COUPLING — Plastic x Plastic				
	1¼	B	50	2.8
	1½	A	250	16.8
	1½ x 1¼	C	50	3.7
	2	A	100	9.0
	2 x 1½	B	100	10.3
	3	A	50	16.0
	3 x 1½	B	50	13.8
	3 x 2	C	50	14.0
	4	B	20	10.4
	4 x 1½	C	10	5.8
	4 x 2	C	10	4.8
	4 x 3	B	10	5.5

DWV Fittings for ABS and PVC drainage systems

Fig. No. / Nominal Size	Class	Master Pack	Master Pack Wt. (Lbs.)
4801-RP & 5801-RP COUPLING — REPAIR — Plastic x Plastic			
1½	B	100	6.0
2	C	100	8.2
3	C	50	16.0
4	C	10	4.9
4801-2-F & 5801-2-F BUSHING — Fitting x Plastic			
1½ x 1¼	C	100	3.1
2 x 1¼	C	25	1.9
2 x 1½	C	100	7.8
3 x 1½	B	50	14.0
3 x 2	C	50	15.0
4 x 2	C	10	5.1
4 x 3	C	10	4.0
4801-2-7 & 5801-2-7 ADAPTER — TRAP — Fitting x Slipjoint			
1¼	C	50	4.2
1½	A	200	18.0
1½ x 1¼	B	100	8.3
2	C	25	5.2
4801-7 & 5801-7 ADAPTER — TRAP — Fitting x Slipjoint			
1¼	B	50	6.1
1½	A	200	18.0
1½ x 1¼	A	200	18.0
2	C	100	18.4
4803 & 5803 ADAPTER — Plastic x Inside Thread			
1¼	C	50	3.3
1½	A	100	8.0
2	C	50	5.0
3	C	20	7.0
4	C	10	5.4
5803-TPA ADAPTER — TRAY PLUG — Plastic x NPSM			
1½	C	50	8.4
(NPSM — National Pipe Straight Mechanical)			
4803-2 & 5803-2 ADAPTER — Fitting x Inside Thread			
1¼	C	50	2.6
1½	A	100	7.0
2	C	50	4.3
3	C	20	6.6
4	C	10	5.3
4803-2-F & 5803-2-F ADAPTER — Fitting x Inside Thread			
1½ x 1¼	C	100	5.0
2 x 1½	C	50	5.0
3 x 1½	C	20	7.0
3 x 2	C	20	7.0
4 x 2	C	10	6.0
4 x 3	C	10	5.0

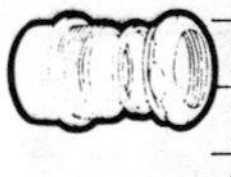

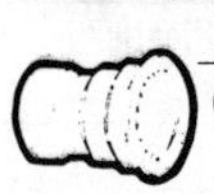

Fig. No. / Nominal Size	Class	Master Pack	Master Pack Wt. (Lbs.)
5803-2-TPA ADAPTER — TRAY PLUG — Fitting x NPSM			
1½	C	50	9.0
4804 & 5804 ADAPTER — Plastic x Outside Thread			
1¼	C	50	2.9
1½	A	200	12.4
1½ x 1¼	B	100	6.4
2	B	100	8.3
3	C	20	5.5
4	C	10	4.9
4804-2 & 5804-2 ADAPTER — Fitting x Outside Thread			
1¼	C	50	3.0
1½	B	100	6.7
1½ x 1¼	C	50	3.0
2	C	25	2.0
3	B	20	5.5
4	C	10	4.5
4805 & 5805 ADAPTER — SOIL PIPE — Plastic x Spigot			
1½ x 2	C	50	10.0
1½ x 3	B	50	20.0
2	C	50	11.0
2 x 3	C	20	8.0
3	C	10	5.3
3 x 4	B	10	7.2
4	C	10	8.5
4805-N & 5805-N ADAPTER — NO HUB SOIL PIPE — Plastic x No Hub			
1½ x 2	C	25	2.0
2	C	20	2.4
3	B	10	2.9
4	C	10	5.0
4806 & 5806 ELBOW — 45° — Plastic x Plastic			
1¼	B	50	5.5
1½	A	100	14.0
2	B	100	22.0
3	A	25	15.4
4	C	10	11.0
4806-2 & 5806-2 ELBOW — 45° — Fitting x Plastic			
1¼	C	50	5.1
1½	B	100	12.9
2	C	50	10.2
3	B	20	12.0
4	C	10	10.5

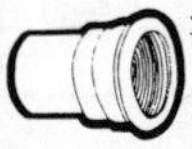

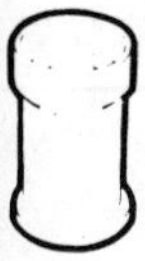

DWV Fittings for ABS and PVC drainage systems

Fig. No.	Nominal Size	Class	Master Pack	Master Pack Wt. (Lbs.)
4807 & 5807 ELBOW — 90° — Plastic x Plastic				
	1¼	C	50	6.6
	1½	A	100	16.0
	2	A	100	24.0
	3	A	20	15.0
	4	C	10	13.0
4807-CL & 5807-CL CLOSET ELBOW — 90° — Plastic x Plastic				
	4 x 3	C	10	10.3
4807-LT & 5807-LT ELBOW — 90° LONG TURN — Plastic x Plastic				
	1¼	C	20	3.1
	1½	C	100	21.1
	2	C	50	15.4
	3	C	20	18.0
	4	C	10	16.0
4807-V & 5807-V ELBOW — 90° VENT — Plastic x Plastic				
	1¼	C	100	10.7
	1½	B	100	13.3
	2	C	50	10.1
	3	B	25	14.5
	4	C	10	11.0
4807-2 & 5807-2 ELBOW — 90° — Fitting x Plastic				
	1¼	C	50	6.0
	1½	A	100	15.0
	2	C	50	12.0
	3	B	20	15.0
	4	C	10	13.0
4807-2-CL & 5807-2-CL CLOSET ELBOW — 90° — Fitting x Plastic				
	4 x 3	C	10	11.3
	4 x 3 W/Cap	C	10	11.7
4807-2-LT & 5807-2-LT ELBOW — 90° LONG TURN — Fitting x Plastic				
	1½	C	25	4.9
	2	C	20	5.9
	3	C	10	8.5
	4	C	10	14.8
4807-2-V & 5807-2-V ELBOW — 90° VENT — Fitting x Plastic				
	1½	B	100	12.0
	2	C	500	9.1
4807-7-LT & 5807-7-LT ELBOW — 90° LONG TURN — Plastic x Slipjoint				
	1½	C	25	6.9
	1½ x 1¼	C	25	5.1
4807-9 & 5807-9 ELBOW — 90° W/SIDE INLET — Plastic x Plastic x Plastic				
	3 x 3 x 1½	C	20	14.9
	3 x 3 x 2	C	20	15.9
4808 & 5808 ELBOW — 22½° — Plastic x Plastic				
	1½	B	100	10.1
	2	C	50	7.8
	3	C	20	9.4
	4	C	10	7.9
4808-2 & 5808-2 ELBOW — 22½° — Fitting x Plastic				
	3	C	20	8.6
	4	C	10	7.4
4810 & 5810 Y — 45° — Plastic x Plastic x Plastic				
	1¼	C	50	11.0
	1½	B	100	27.8
	2	B	50	22.0
	2 x 1½ x 1½	C	50	16.0
	2 x 1½ x 2	C	50	25.0
	2 x 2 x 1½	C	50	17.5
	3	A	20	24.6
	3 x 3 x 1¼	C	20	15.2
	3 x 3 x 1½	B	20	15.6
	3 x 3 x 2	C	20	17.5
	4	C	10	22.0
	4 x 4 x 1½	C	10	14.0
	4 x 4 x 2	C	10	13.3
	4 x 4 x 3	B	10	17.4
4810-13 & 5810-13 Y — 45° — Plastic x Female Thread x Plastic				
	1½	C	50	17.0
	2	C	20	16.0
	3	B	20	29.0
	4	C	10	22.0
	4 x 4 x 2	C	10	16.9
	4 x 4 x 3	C	10	18.0
4810-14 & 5810-14 Y — 45° — Plastic x Plastic x Inside Thread				
	1½	C	50	13.7
	3	C	20	25.5
4810-17 & 5810-17 Y — 45° — Inside Thread x Plastic x Plastic				
	3	C	10	12.0
	3 x 3 x 2	C	20	18.0

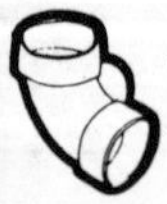

DWV Fittings for ABS and PVC drainage systems

Fig. No.	Nominal Size	Class	Master Pack	Master Pack Wt. (Lbs.)
4811 & 5811 TEE — Plastic x Plastic x Plastic				
	1¼	C	50	8.4
	1½	A	100	22.0
	1½ x 1¼ x 1¼	B	50	13.4
	1½ x 1¼ x 1½	B	50	12.0
	1½ x 1½ x 1¼	B	50	12.0
	2	B	50	15.6
	2 x 1½ x 1½	B	50	13.9
	2 x 1½ x 2	C	50	15.0
	2 x 2 x 1¼	C	50	15.4
	2 x 2 x 1½	B	50	14.6
	3	A	20	19.7
	3 x 3 x 1¼	C	20	12.6
	3 x 3 x 1½	A	20	11.7
	3 x 3 x 2	B	20	13.4
	4	C	10	17.5
	4 x 4 x 1½	C	10	11.0
	4 x 4 x 2	C	10	10.2
	4 x 4 x 3	C	10	14.0
4811-C & 5811-C TEE — 2 WAY CLEANOUT — Plastic x Plastic x Plastic				
	3	C	10	14.1
	4	C	5	12.3
4811-V & 5811-V TEE — VENT — Plastic x Plastic x Plastic				
	1¼	C	50	7.6
	1½	C	100	18.0
	2	C	50	11.4
	2 x 2 x 1½	C	50	11.3
	3	C	10	7.8
	3 x 3 x 1½	C	10	5.3
	3 x 3 x 2	C	10	5.9
	4	C	5	7.0
4812 & 5812 TY — LONG TURN — Plastic x Plastic x Plastic				
	1¼	C	25	5.9
	1½ x 1½ x 1¼	C	25	9.6
4812-LR & 5812-LR TY — LONG RADIUS — Plastic x Plastic x Plastic				
	1½	C	20	6.6
	2	C	20	10.9
	2 x 1½ x 1½	C	20	7.3
	2 x 1½ x 2	C	20	10.6
	2 x 2 x 1½	C	20	7.7

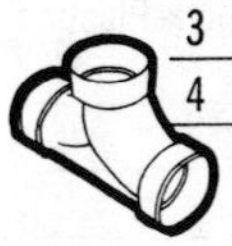

Fig. No.	Nominal Size	Class	Master Pack	Master Pack Wt. (Lbs.)
4812-LR & 5812-LR (continued)				
	3	C	10	16.1
	3 x 3 x 1½	C	10	7.6
	3 x 3 x 2	C	10	9.6
	4	C	5	15.6
	4 x 4 x 2	C	5	6.8
	4 x 4 x 3	C	5	10.6
4812-13 & 5812-13 TY — LONG TURN — Plastic x Inside Thread x Plastic				
	1¼	C	25	5.7
	1½	C	20	8.4
	3 x 3 x 1½	C	10	10.6
	4	C	5	20.0
	4 x 4 x 2	C	5	7.5
4812-13-LR & 5812-13-LR TY — LONG RADIUS — Plastic x Inside Thread x Plastic				
	2	C	20	10.7
	2 x 2 x 1½	C	20	8.5
	3	C	10	16.5
	3 x 3 x 2	C	10	11.7
	4 x 4 x 3	C	5	15.0
4812-17 & 5812-17 TY — LONG TURN — Inside Thread x Plastic x Plastic				
	3	C	10	16.4
	3 x 3 x 2	C	10	11.4
4814 & 5814 TEST TEE — Plastic x Plastic x Cleanout W/Plug				
	1½	C	50	11.4
	2	C	50	13.1
	3	C	10	8.7
	4	C	10	15.6
4816 & 5816 CLEANOUT — Fitting x Cleanout W/Threaded Plug				
	1¼	C	20	1.6
	1½	A	100	10.8
	2	C	50	7.3
	3	A	25	9.9
	4	C	10	8.0
4818 & 5818 PLUG — Outside Thread				
	1¼	B	100	2.9
	1½	A	100	4.7
	2	B	100	6.0
	3	A	50	7.2
	4	B	20	5.4

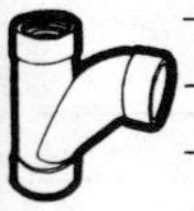

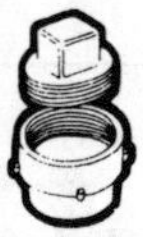

DWV Fittings for ABS and PVC drainage systems

4826 & 5826 PLUG — Fitting

Fig. No.	Nominal Size	Class	Master Pack	Master Pack Wt. (Lbs.)
	1½	A	100	4.3
	2	C	50	4.9

4827 & 5827 CAP — Inside Thread

Fig. No.	Nominal Size	Class	Master Pack	Master Pack Wt. (Lbs.)
	1¼	C	100	5.1
	1½	C	100	6.6
	2	C	100	8.6
	3	C	100	25.3

5829 NIPPLE — Outside Thread x Outside Thread

Fig. No.	Nominal Size	Class	Master Pack	Master Pack Wt. (Lbs.)
	3 x 3″ Length	C	50	13.6
	3 x 4″ Length	C	50	19.5
	3 x 5″ Length	C	50	25.5
	3 x 6″ Length	C	25	15.6
	3 x 7″ Length	C	25	18.8
	3 x 8″ Length	C	10	8.5
	3 x 9″ Length	C	10	9.8
	3 x 10″ Length	C	10	10.6
	3 x 12″ Length	C	10	13.3

4833-S & 5833-S UPTURN — SINGLE STACK — Plastic x Plastic x Plastic

Fig. No.	Nominal Size	Class	Master Pack	Master Pack Wt. (Lbs.)
	2	C	20	11.5
	2 x 2 x 1½	C	20	7.6
	3	C	10	16.2
	3 x 3 x 2	C	10	13.0

4834 & 5834 Y — 45° DOUBLE — Plastic x Plastic x Plastic x Plastic

Fig. No.	Nominal Size	Class	Master Pack	Master Pack Wt. (Lbs.)
	1½	C	25	8.5
	2	C	20	10.4
	2 x 2 x 1½ x 1½	C	25	10.7
	3	C	10	15.9
	3 x 3 x 1½ x 1½	C	10	8.1
	3 x 3 x 2 x 2	C	10	9.7
	4	C	5	14.3
	4 x 4 x 2 x 2	C	5	7.0
	4 x 4 x 3 x 3	C	5	10.3

4835 & 5835 TEE — DOUBLE — Plastic x Plastic x Plastic x Plastic

Fig. No.	Nominal Size	Class	Master Pack	Master Pack Wt. (Lbs.)
	1½	A	50	13.1
	2	C	20	8.4
	2 x 2 x 1½ x 1½	C	25	8.6
	3	A	10	12.6
	3 x 3 x 1½ x 1½	B	20	12.8
	3 x 3 x 2 x 1½	C	20	13.5
	3 x 3 x 2 x 2	C	20	14.8
	4	C	5	11.1
	4 x 4 x 2 x 2	C	5	5.4
	4 x 4 x 3 x 3	C	5	8.2

5835-A TEE — DOUBLE FIXTURE — Plastic x Plastic x Inside Thread x Inside Thread

Fig. No.	Nominal Size	Class	Master Pack	Master Pack Wt. (Lbs.)
	1½	C	25	11.0
	2 x 1½ x 1½ x 1½	C	25	12.0

4835-B & 5835-B TEE — DOUBLE FIXTURE — Inside Thread x Plastic x Plastic x Plastic

Fig. No.	Nominal Size	Class	Master Pack	Master Pack Wt. (Lbs.)
	1½	C	25	9.0
	2	C	25	10.8
	2 x 1½ x 1½ x 1½	C	25	9.6
	2 x 1½ x 2 x 2	C	25	14.6
	3	C	5	8.9
	3 x 2 x 3 x 3	C	5	8.4

4835-9 & 5835-9 TEE — DOUBLE W/ONE 90° SIDE INLET — Plastic x Plastic x Plastic x Plastic x Plastic

Fig. No.	Nominal Size	Class	Master Pack	Master Pack Wt. (Lbs.)
	3 x 3 x 3 x 3 x 1½	C	10	13.0
	3 x 3 x 3 x 3 x 2	C	10	13.9
	4 x 4 x 4 x 4 x 2	C	5	11.4

4835-9-9 & 5835-9-9 TEE — DOUBLE W/TWO 90° SIDE INLETS — Plastic x Plastic x Plastic x Plastic x Plastic

Fig. No.	Nominal Size	Class	Master Pack	Master Pack Wt. (Lbs.)
	3 x 3 x 3 x 3 x 1½ x 1½		10	17.0
	3 x 3 x 3 x 3 x 2 x 2		10	18.0
	4 x 4 x 4 x 4 x 1½ x 1½		5	18.0
	4 x 4 x 4 x 4 x 2 x 2		5	14.0

4836 & 5836 TY — DOUBLE LONG TURN — Plastic x Plastic x Plastic x Plastic

Fig. No.	Nominal Size	Class	Master Pack	Master Pack Wt. (Lbs.)
	3	B	5	11.5
	3 x 3 x 1½ x 1½	B	10	10.8
	3 x 3 x 2 x 2	C	10	13.9
	4	C	5	24.4

4837 & 5837 ELBOW — DOUBLE — Plastic x Plastic x Plastic

Fig. No.	Nominal Size	Class	Master Pack	Master Pack Wt. (Lbs.)
	1½	C	50	11.2
	2	C	25	12.4
	2 x 1½ x 1½	C	25	5.7
	3	B	10	10.4
	4	C	5	6.2

4851 & 5851 CLOSET FLANGE — Plastic

Fig. No.	Nominal Size	Class	Master Pack	Master Pack Wt. (Lbs.)
	4	B	50	23.4
	4 x 3	A	50	27.7

DWV Fittings for ABS and PVC drainage systems

Fig. No.	Nominal Size	Class	Master Pack	Master Pack Wt. (Lbs.)
4851-A & 5851-A CLOSET FLANGE — W/INSERT — Plastic				
	4	B	25	15.4
	4 x 3	A	25	19.8
4851-2-A & 5851-2-A CLOSET FLANGE — W/INSERT — Fitting				
	4	B	25	21.5
	4 x 3	B	25	20.0
4851-3 & 5851-3 CLOSET FLANGE — Inside Thread				
	4 x 3	C	50	23.3
4851-4 & 5851-4 CLOSET FLANGE — Outside Thread				
	4 x 3	C	50	21.5
4860 & 5860 ELBOW — 60° — Plastic x Plastic				
	1½	A	50	6.4
	2	C	20	3.8
	3	A	10	6.0
	4	C	10	10.1
4860-2 & 5860-2 ELBOW — 60° — Fitting x Plastic				
	4	C	5	4.7
4861 & 5861 ELBOW — 90° W/HIGH HEEL INLET — Plastic x Plastic x Plastic				
	3 x 3 x 1½	C	20	16.0
	3 x 3 x 2	C	20	15.2
4861-LH & 5861-LH ELBOW — 90° W/LOW INLET C Plastic x Plastic x Plastic				
	3 x 3 x 1½	A	20	15.7
	3 x 3 x 2	B	20	16.0
4870 & 5870 TEE — W/90° RIGHT & LEFT INLET — Plastic x Plastic x Plastic x Plastic x Plastic				
	3 x 3 x 3 x 1½ x 1½	B	10	10.9
	3 x 3 x 3 x 2 x 2	B	10	13.5
	4 x 4 x 4 x 1½ x 1½	C	5	9.0
	4 x 4 x 4 x 2 x 2	C	5	10.4

Fig. No.	Nominal Size	Class	Master Pack	Master Pack Wt. (Lbs.)
4871 & 5871 TEE — W/90° LEFT INLET — Plastic x Plastic x Plastic x Plastic				
	3 x 3 x 2 x 1½	C	10	13.0
	3 x 3 x 2 x 2	C	10	8.8
	3 x 3 x 3 x 1½	C	10	10.6
	3 x 3 x 3 x 2	C	10	10.6
	4 x 4 x 4 x 1½	C	5	9.7
	4 x 4 x 4 x 2	C	5	9.6
4872 & 5872 TEE — W/90° RIGHT INLET — Plastic x Plastic x Plastic x Plastic				
	3 x 3 x 2 x 1½	C	10	13.3
	3 x 3 x 2 x 2	C	10	13.4
	3 x 3 x 3 x 1½	C	10	10.6
	3 x 3 x 3 x 2	C	10	11.4
	4 x 4 x 4 x 1½	C	5	9.4
	4 x 4 x 4 x 2	C	5	9.4
4876 & 5876 RETURN BEND — TRAP W/CLEANOUT — Plastic x Slip Joint				
	1½	B	20	6.1
	2	C	10	6.2
4877 & 5877 RETURN BEND — TRAP — Plastic x Slip Joint				
	1½	B	20	5.8
	2	C	10	4.5
4878 & 5878 RETURN BEND — W/CLEANOUT — Plastic x Plastic x Cleanout W/Plug				
	1½	C	50	12.0
	2	C	10	4.4
4879 & 5879 RETURN BEND — Plastic x Plastic				
	1½	C	50	11.3
	2	C	10	5.0
	3	C	10	10.6
	4	C	5	11.4
4880 & 5880 P-TRAP — W/CLEANOUT — Plastic x Slipjoint x Cleanout W/Plug				
	1½	B	10	4.2
	2	C	10	8.1
4881 & 5881 P-TRAP — Plastic x Slipjoint				
	1½	B	10	4.9
	2	C	10	8.1
4884 & 5884 P-TRAP — W/CLEANOUT — Plastic x Plastic x Cleanout				
	1½	A	50	15.9
	2	C	10	6.1

DWV Fittings for ABS and PVC drainage systems

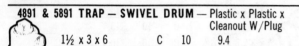

Fig. No.	Nominal Size	Class	Master Pack	Master Pack Wt. (Lbs.)
4885 & 5885 P-TRAP — Plastic x Plastic				
1½	A	50	18.0	
2	C	25	15.6	
3	C	10	17.5	
4	C	5	7.8	
5885-TPA P-TRAP — Plastic x NPSM				
1½	C	50	27.0	
(NPSM — National Pipe Straight Mechanical)				
4891 & 5891 TRAP — SWIVEL DRUM — Plastic x Plastic x Cleanout W/Plug				
1½ x 3 x 6	C	10	9.4	
4892 & 5892 P-TRAP — W/UNION JOINT — Plastic x Slip Joint				
1½	A	50	22.0	
1½ x 1¼	C	50	27.1	
(L. A Trap)				

Fig. No.	Nominal Size	Class	Master Pack	Master Pack Wt. (Lbs.)
4892-3 & 5892-3 P-TRAP — W/UNION JOINT — Inside Thread x Slipjoint				
1½	C	25	12.0	
1½ x 1¼	C	25	12.0	
4895 & 5895 P-TRAP — W/UNION JOINT — Plastic x Plastic				
1½	C	50	20.0	
2	C	25	16.8	
5895-TPA P-TRAP — W/UNION JOINT — Plastic x NPSM				
1½	C	50	33.0	
4895-3 & 5895-3 P-TRAP — W/UNION JOINT — Inside Thread x Plastic				
1½	C	25	10.8	

Fig. No.	Nominal Size	Class	Pack Quantity	Pack Weight
828 ROOF FLASHING — Neoprene				
1¼ x 1½	C	12	9.0	
2	B	12	10.0	
3	A	12	12.0	
4	C	12	13.0	

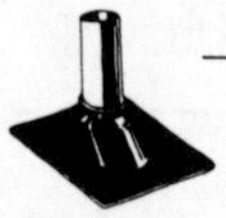

Fig. No.	Nominal Size	Class	Pack Quantity	Pack Weight
828-G ROOF FLASHING — Galvanized W/Neoprene Collar				
1¼ - 1½	C	12	11.0	
2	B	12	14.0	
3	A	12	16.0	
4	C	12	23.0	

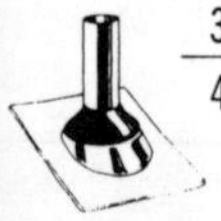

Fig. No.	Nominal Size	Class	Pack Quantity	Pack Weight
4898-2 SOLVENT — PVC — Clear With Applicator				
¼ Pint	A	24	10.0	
½ Pint	A	24	17.0	
1 Pint	B	12	14.0	
1 Quart	B	12	25.0	
4899 PRIMER — PVC & CPVC — With Applicator				
½ Pint	A	24	17.0	
1 Pint	A	12	14.0	
1 Quart	B	12	25.0	

Fig. No.	Nominal Size	Class	Pack Quantity	Pack Weight
5898 SOLVENT — ABS — With Applicator				
¼ Pint	A	24	10.0	
½ Pint	A	24	17.0	
1 Pint	A	12	14.0	
1 Quart	B	12	25.0	
5899 THINNER — ABS — Less Applicator				
1 Pint	A	12	14.0	
1 Quart	B	12	25.0	

DWV Fittings for ABS and PVC drainage systems

Fig. No.	Pack Quantity	Pack Weight		Fig. No.	Pack Quantity	Pack Weight
BATH KITS — Pipe and Fittings to connect a Three-Piece Bath				**BATH KITS** — Fittings only to connect a Three-Piece Bath		
48BKP (PVC)	1	57		**48BK** (PVC)	1	16
58BKP (ABS)	1	40		**58BK** (ABS)	1	11
***58BKP-V** (ABS)	1	48		***58BK-V**(ABS)	1	9.5

*For West of the Rocky Mountains. Should satisfy Uniform Plumbing Code requirements for dry vent.

*For West of the Rocky Mountains. Should satisfy Uniform Plumbing Code requirements for dry vent.

NEW PRODUCT LISTING

Fig. No.	Pack Quantity	Pack Weight		Fig. No.	Pack Quantity	Pack Weight
ELECTRIC SOLDERING TOOL						
LG-101						
Makes 1/8" thru 1" Solder Joints — 110/120 volts; 60 cycle; 7 amps.	1	20				
Replacement Electrodes: For Model LG-101						
CE-10 Carbon Electrodes	1 pair	—				

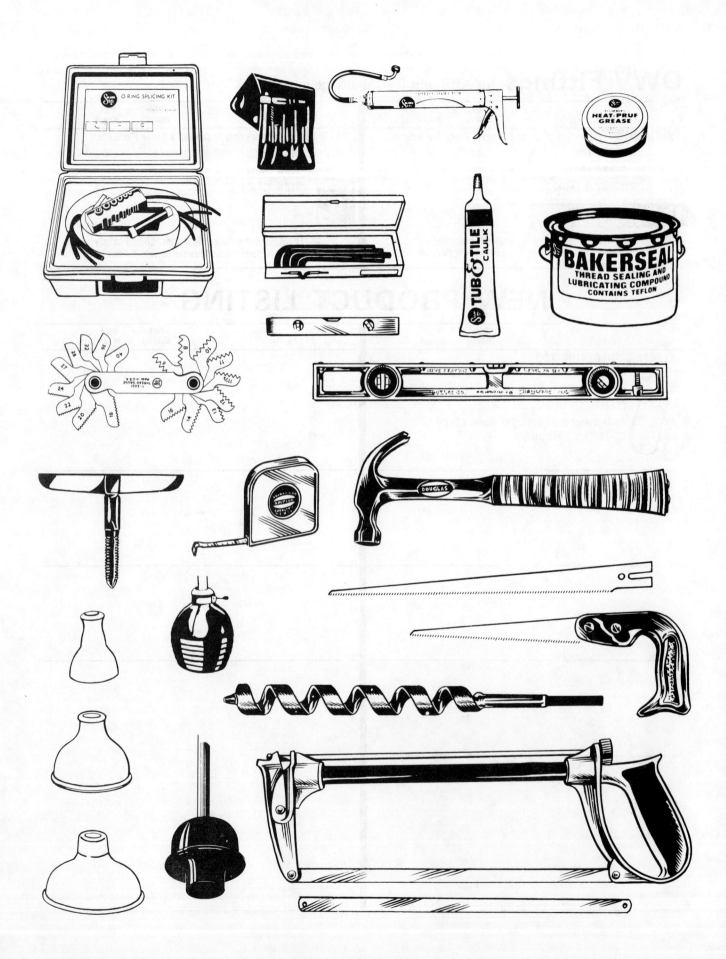

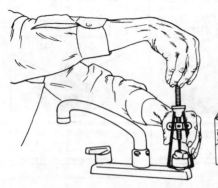

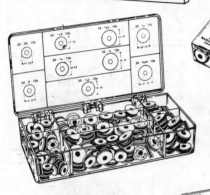